CYTOCHROME OXIDASE

DEVELOPMENTS IN BIOCHEMISTRY

_____

CYTOCHROME OXIDASE

Proceedings of the Japanese-American Seminar on Cytochrome Oxidase
held in Osaka, Japan on August 29 – September 2, 1978

Edited by
TSOO E. KING
State University of New York/Albany, Albany, NY, USA
YUTAKA ORII
Osaka University, Toyonaka, Osaka, Japan
BRITTON CHANCE
University of Pennsylvania Medical School, Philadelphia, PA, USA
KAZUO OKUNUKI
Osaka University, Toyonaka, Osaka, Japan

1979
ELSEVIER/NORTH-HOLLAND BIOMEDICAL PRESS
AMSTERDAM · NEW YORK · OXFORD

ISSN: 0-165-1714 (Series)
ISBN: 0-444-80100-6 (Volume 5)

Published by:
Elsevier/North-Holland Biomedical Press
335 Jan van Galenstraat, P.O. Box 211,
Amsterdam, The Netherlands

Sole distributors for the U.S.A. and Canada:
Elsevier North-Holland Inc.
52 Vanderbilt Avenue
New York, N.Y. 10017

Library of Congress Cataloging in Publication Data

Japanese-American Seminar on Cytochrome Oxidase,
   Osaka, 1978.
   Cytochrome oxidase.

   (Developments in biochemistry ; v. 5 ISSN
0165-1714)
   Includes indexes.
   1. Cytochrome oxidase--Congresses. I. King,
Tsoo E., 1916-   II. Title. III. Series.
QP603.C85J36 1978   574.1'9258   78-31986
ISBN 0-444-80100-6

PRINTED IN THE NETHERLANDS

Preface

The present proceedings are derived from a symposium formally known as the Japanese-American Seminar on Cytochrome Oxidase (JASCO), which was held from August 29 to September 2, 1978 at the summit of the serene Mt. Rokkosan, Kobe, Japan. It was officially sponsored by the United States-Japan Cooperative Science Program of the U. S. National Science Foundation and the Japan Society for the Promotion of Science.

This is the first symposium ever organized solely devoted to cytochrome oxidases. It is generally agreed that cytochrome oxidase plays a basic role in most oxygen-requiring phyla. If this role cannot be understood and the underlying mechanisms elucidated, it is literally meaningless to talk about life or life processes or even energy transduction in these organisms. Apart from these facts, the recent surge of new and renewed interest on various facets of this terminal oxidase makes the symposium even more timely. We had intended to gather *all* representatives of active workers in the most important areas of the subject in an attempt to clarify or even iron out disagreements, exchange research expertise and experience, and most importantly, ponder the directions of future developments and experimentations, if possible, in a not-too-big forum held under a relaxed and conducive atmosphere.

However, we soon discovered that the budget coupled with manifold other reasons made our original plans not practical. Nevertheless, on a limited scale, the Symposium highlights *inter alia* new developments in the structure, the reaction intermediates, and the mechanism of action of cytochrome oxidase. The problem has been approached by a multidisciplinary group of biochemists, biophysicists, chemists, and physicists who brought to bear upon the problems of cytochrome oxidase function the most modern of techniques particularly magneto-optical circular dichroism and the newest of all, X-ray photon absorption studies of fine structure and extended fine structure is now obtainable from synchrotron radiation. The manuscripts reflect the state of the art and science in studies of cytochrome oxidase by these new modalities, and in part, are impacted upon by the vigorous discussion that took place at the meeting. Regretfully the limited resources and time available to the editors compelled the omission of the valuable discussion in the Proceedings.

It may be pointed out that the term cytochrome oxidase is not used here in a restricted sense. Perhaps the plural, cytochrome oxidases, is more accurate. As a matter of fact, this semantics becomes more meaningful owing to the budding interest of workers to unravel the evolution of cytochrome oxidase; this topic naturally evolves from the very fruitful investigation of biogenesis in recent years.

The Symposium was made possible by financial assistance from the Japan Society for the Promotion of Science, Protein Research Foundation, Yamada Science Foundation, AMCO Inc., and the U. S. National Science Foundation. To them we acknowledge deep gratitude. To all staff members who arranged the meeting, checked the papers, and completed the publication in the shortest possible time we pay our tributes.

Tsoo E. King

Yutaka Orii

Britton Chance

Kazuo Okunuki

October, 1978

Participants

Dr. Shiro Akabori, Emeritus President of Osaka University, President's
Office, Osaka University, Toyonaka, Osaka 560, Japan.

Dr. Tadashi Asahi, Laboratory of Biochemistry, Faculty of Agriculture,
Nagoya University, Chigusa-ku, Nagoya 464, Japan.

Dr. G. T. Babcock, Department of Chemistry, Michigan State University, East
Lansing, Michigan 48823, U.S.A.

Dr. Helmut Beinert, Institute for Enzyme Research, University of Wisconsin,
1710 University Avenue, Madison, Wisconsin 53706, U.S.A.

Dr. William E. Blumberg, Bell Laboratories, 600 Mountain Avenue, Murray
Hill, New Jersey 07974, U.S.A.

Dr. Sunney I. Chan, The Chemical Laboratories, Division of Chemistry and
Chemical Engineering, California Institute of Technology, Pasadena,
California 91125, U.S.A.

Dr. Britton Chance, Johnson Research Foundation, University of Pennsylvania
A. N. Richards Building, Philadelphia, Pennsylvania 19174, U.S.A.

Dr. Marius Clore, Department of Biochemistry, University College London,
Gower Street, London WCIE 6BT, England.

Dr. Helen Conrad Davies, Department of Microbiology, University of
Pennsylvania, Medical School, Philadelphia, Pennsylvania 19119, U.S.A.

Dr. Maria Erecińska, Department of Biochemistry and Biophysics, University
of Pennsylvania, Philadelphia, Pennsylvania 19174, U.S.A.

Dr. Beverly Errede, Department of Radiation Biology and Biophysics,
University of Rochester, School of Medicine and Dentistry, Rochester,
New York 14642, U.S.A.

Dr. Shelagh Ferguson-Miller, Department of Biochemistry and Molecular Biology,
Northwestern University, Evanston, Illinois 60201, U.S.A.

Dr. Hans Frauenfelder, Department of Physics, Loomis Laboratory of Physics,
University of Illinois at Urbana-Champaign, Urbana, Illinois 61801, U.S.A.

Dr. O. Hayes Griffith, Institute of Molecular Biology, University of Oregon,
Eugene, Oregon 97403, U.S.A.

Dr. Bunji Hagihara, Department of Biochemistry, Osaka University School of
Medicine, Kita-ku, Osaka 530, Japan.

Dr. Masahiro Hatano, Chemical Research Institute of Nonaqueous Solutions,
Tohoku University, Sendai 980, Japan.

Dr. Tetsutaro Iizuka, Department of Biochemistry, Keio University School of
Medicine, Shinjuku-ku, Tokyo 160, Japan.

Dr. Akira Ikegami, Institute of Physical and Chemical Research, Wako-shi,
Saitama 351, Japan.

Dr. Yuzuru Ishimura, Department of Biochemistry, Keio University School of
Medicine, Shinjuku-ku, Tokyo 160, Japan.

Dr. Suguru Kawato, Institute of Physical and Chemical Research, Wako-shi,
Saitama 351, Japan.

Dr. Tsoo E. King, Department of Chemistry and Laboratory of Bioenergetics,
State University of New York at Albany, Albany, New York 12222, U.S.A.

Dr. Teizo Kitagawa, Institute for Protein Research, Osaka University, Suita,
Osaka 565, Japan.

Dr. Bernhard Kleffel, Institut für Biophysik und Electronenmikroskopie,
4000 Düsseldorf, Germany.

Dr. Jack Leigh, Department of Biophysics and Physical Biochemistry, University
of Pennsylvania, Philadelphia, Pennsylvania 19174, U.S.A.

Dr. E. Margoliash, Department of Biochemistry and Molecular Biology,
Northwestern University, Evanston, Illinois 60201, U.S.A.

Dr. Hiroshi Matsubara, Department of Biology Faculty of Science, Osaka
University, Toyonaka, Osaka 560, Japan.

Dr. Peter Mitchell, Glynn Research Laboratories, Bodmin, Cornwall, PL 30
4AU, England.

Dr. Tsunenori Nozawa, Chemical Research Institute of Nonaqueous Solutions,
Tohoku University, Sendai 980, Japan.

Dr. Kazuo Okunuki, Emeritus Professor at Osaka University, Toyonaka,
Osaka 560, Japan.

Dr. Yutaka Orii, Department of Biology, Faculty of Science, Osaka University,
Toyonaka, Osaka 560, Japan.

Dr. Takayuki Ozawa, Department of Biochemistry, Nagoya University School of
Medicine, Showa-ku, Nagoya 466, Japan.

Dr. Linda Powers, Bell Laboratories, 600 Mountain Avenue, Murray Hill, New
Jersey 07974, U.S.A.

Dr. Nobuhiro Sato, Department of Biochemistry and First Department of Medicine,
Osaka University School of Medicine, Kita-ku, Osaka 530, Japan.

Dr. Lucile Smith, Department of Biochemistry, Dartmouth Medical School,
Hanover, New Hampshire 03755, U.S.A.

Dr. G. Steffens, Abteilung Physiologische Chemie der Medizinischen Fakultat,
Med. Theoret. Inst. an der Rhein-Westf. Techn. Hochschule Aachen,
Melatenert Str. 211, D-5100 Aachen, Germany.

ix

Dr. Shigeki Takemori, Department of Environmental Sciences, Faculty of
    Integrated Arts and Sciences, Hiroshima University, Hiroshima 730, Japan.
Dr. Masaru Tanaka, Department of Biochemistry and Biophysics, John A. Burn
    School of Medicine, Snyder Hall, University of Hawaii at Manoa, 2538 The
    Mall, Honolulu, Hawaii 96822, U.S.A.
Dr. Akira Tsugita, European Molecular Biology Laboratory, Heidelberg, Germany.
Dr. Bob F. van Gelder, Universiteit van Amsterdam, Vakgroep Biochemie, B. C.
    P. Jansen Instituut, Plantage Muidergracht 12, 1018 TV Amsterdam-C, The
    Netherlands
Dr. Tore Vänngard, Department of Biochemistry and Biophysics, Chalmers
    Institute of Technology and University of Göteborg, Fack, S-402 20
    Göteborg, Sweden.
Dr. Keishiro Wada, Department of Biology, Faculty of Science, Osaka University,
    Toyonaka, Osaka 560, Japan.
Dr. G. C. Wagner, 417 Rogers Adams Labs, University of Illinois, Urbana,
    Illinois 61801, U.S.A.
Dr. Dale A. Webster, Department of Biology, Lewis College of Science and
    Letters, Illinois Institute of Technology, Chicago, Illinois 60616, U.S.A.
Dr. David F. Wilson, Department of Biochemistry and Biophysics, University of
    Pennsylvania, Philadelphia, Pennsylvania 19174, U.S.A.
Dr. Tateo Yamanaka, Department of Biology, Faculty of Science, Osaka University,
    Toyonaka, Osaka 560, Japan.
Dr. Kerry T. Yasunobu, Department of Biochemistry and Biophysics, University
    of Hawaii at Manoa, Honolulu, Hawaii 96822, U.S.A.
Dr. Takashi Yonetani, Department of Biochemistry and Biophysics, University
    of Pennsylvania, Philadelphia, Pennsylvania 19174, U.S.A.
Dr. Satoshi Yoshida, Department of Biophysics, Faculty of Engineering Science,
    Osaka University, Toyonaka, Osaka 560, Japan.
Dr. Shinya Yoshikawa, Department of Biology, Faculty of Science, Konan
    University, Higashinada-ku Kobe 658, Japan.

Contents

Cytochrome Oxidase, T.E. King et al. eds.
© *1979 Elsevier/North-Holland Biomedical Press*

A BRIEF RETROSPECT OF CYTOCHROME OXIDASE STUDIES

KAZUO OKUNUKI

Department of Biology, Faculty of Science, Osaka University, Toyonaka,
Osaka 560, Japan

It is almost half a century ago that Warburg, Keilin, and Shibata and
Tamiya had a triangular debate on the nature of a terminal oxidase of
cellular respiration. Since then, although the center of interest changed
from time to time there are still long-lasting controversies over the
identity of this enzyme.

Investigators in this field at that time agreed that experimentally the
terminal oxidase should satisfy requirements such as to be autoxidizable and to
form inactive complexes reacting with respiratory poisons like carbon monoxide
and cyanide. The enzyme was named cytochrome oxidase, leaving a problem
whether the enzyme was cytochrome itself or differed from ordinary cytochromes.
It is agreed generally that cytochrome oxidase of most aerobic organisms belongs
to an A-type cytochrome. Even so, is it cytochrome a_3 or cytochrome a of a
unitarian theory that constitutes the entity of cytochrome oxidase?

Cytochrome a, both purified and the intact one in mitochondrial inner membrane, exists as dimer if we define monomer as a unit having one heme a and
copper. Under certain conditions the dimer is dissociated into monomers accompanying a two to three fold increase in specific activity. This result can
be interpreted as suggesting that the monomer, which corresponds to cytochrome
a, exhibits autoxidizability in the presence of cytochrome c, although this
interpretation has not gained general acceptance.

On the contrary, characteristics of cytochrome a to form an oxygenated
compound is accepted generally in these twenty years but there is an
argument that this form does not participate in the catalytic turnover of
cytochrome oxidase as an intermediate. Formation of the oxygenated compound
which we reported in 1957 at the International Symposium on Enzyme Chemistry,
Tokyo-Kyoto, would have been confirmed much earlier if purified cytochrome a,
but not cytochrome c deficient mitochondria, was used in follow-up experiments.
The oxygenated compound can be formed easily if purified cytochrome a preparation is used in the complete absence of either cytochrome c or c_1. In the
presence of a trace amount of cytochrome c the oxygenated compound could not
be observed on a conventional spectrophotometer, because its decay was fast.
Since then several intermediates have been observed after reaction of reduced

cytochrome a with oxygen either in the stopped-flow measurements or at very low
temperatures. The oxygenated compound that we reported must be one of these
members and be involved in a catalytic turnover even though it is not a
compound with a bound oxygen molecule as was implied initially. If an
essential factor in an enzyme-catalyzed reaction is lost the reaction will be
stopped or slowed down and so on. Therefore, depletion of cytochrome c from the
reaction system or lowering of reaction temperature can be employed equally to
observe the oxygenated compound formation. In fact, by addition of cytochrome
c the oxygenated compound was decayed immediately. In this context the oxygenated intermediates observed in the stopped-flow measurements and at low temperatures will be correlated with each other in a near future.

Dr. Yakushiji and I found a small but distinct band at 660 nm for a very concentrated solution of purified and oxidized cytochrome a under a hand spectro-
scope. This band has now revived as a subject of the physico-chemical studies
such as reflectance spectrophotometry and magnetic circular dichroism (MCD).
From the MCD study this small band is assigned to a copper component which gives
an 830-nm band. This is just a simple example that the development of experimental techniques provides us a good deal of new information, with which we can
endeavor to understand the detailed molecular architecture and reaction
mechanism of cytochrome oxidase.

Protein chemistry also manifested its usefulness in elucidating the molecular structure of cytochrome a. It is now believed that the three subunits out
of usually seven are of mitochondrial origin and that the rest are of cytoplasmic origin. How is heme a or copper attached to which subunits? What are
the most suitable arrangement of these subunits for cytochrome oxidase to
reduce molecular oxygen to water? Is it true that protons are translocated
through mitochondrial membrane coupling with this function? In connection with
this, how the translocation of protons is coupled to the energy transduction
that is the ultimate function of respiration? I am confident that these
vitally important problems will be solved in a few years.

It is my great pleasure that at this most prosperous time in a history of
cytochrome oxidase studies I am given a chance to watch the distinguished
investigators from different fields but working on the same enzyme, cytochrome
oxidase, to come together and to discuss intensively on this enzyme in the
coming few days. Cytochrome oxidase fascinated and stirred up predecessor
investigators and still attracts our scientific curiosity much more than ever,
and this occasion will stir again an old-timer investigator, I am.

PROSTHETIC GROUPS

Cytochrome Oxidase, T.E. King et al. eds.
© *1979 Elsevier/North-Holland Biomedical Press*

STUDIES ON THE ORIENTATION OF THE HEMES OF CYTOCHROME C OXIDASE

MARIA ERECIŃSKA[1], DAVID F. WILSON[2], AND J. KENT BLASIE[2]

Department of Pharmacology[1] and Department of Biochemistry and Biophysics[2]
University of Pennsylvania, Medical School, Philadelphia, Pa. 19104.

Cytochrome c oxidase, a large protein tightly bound to the mitochondrial inner membrane[1-3] transfers reducing equivalents from ferro-cytochrome c to molecular oxygen with the formation of water and simultaneous synthesis of ATP. This reaction is one of the fastest thus far described in biological systems and occurs with the second order rate constant of about $10^8 M^{-1} sec^{-1}$[4,5]. Since the velocity of a reaction would be expected to depend on the distance between and orientation of the participants involved, information of such nature obtained on cytochrome c oxidase may greatly aid in understanding the mechanism of coupled electron flow at phosphorylation site 3.

"Membranous" cytochrome c oxidase[6] and mitochondrial membranes form upon slow partial dehydration oriented multilayers[7] in which the orientations of the redox centers with respect to the plane of the membrane layers can be investigated by a variety of physical techniques. This short paper presents a brief summary of the results accumulated during the past two years[7-11] on the orientation of the hemes of cytochrome c oxidase using three techniques: a) x-ray diffraction b) optical polarization spectroscopy and c) epr spectroscopy.

X-ray diffraction studies on the oriented multilayer of membranous cytochrome c oxidase. The electron density profile of membranous cytochrome oxidase has been determined directly by taking into account the various forms of disorder present in the hydrated oriented multilayers of these membranes to $\sim$20 Å resolution. Moreover, a most probable profile has been calculated to about 8 Å resolution. It was found that although the molecule of cytochrome oxidase extended over the entire membrane thickness, it was oriented asymmetrically in the membrane profile with a large portion of its scattering mass occurring within the extravesicular surface of the membrane[7]. Bundles of α-helical polypeptide chain fragments were also detected whose helical axes had an average orientation normal to the plane of the membrane.

Polarized optical spectroscopy. The electronic transitions responsible for the visible and Soret absorption in hemeproteins are predicted to be polarized in the plane of the molecule (i.e. x-y polarized) and approximately doubly

degenerate (the x and y axes being equivalent)[12,13]. This assumption was implicit in the determination of the orientation of the heme plane relative to the
crystal axes in myoglobin[14] hemoglobin[15,16] and cytochrome $\underline{c}$[17], using optical
absorption measurements with polarized light and was also used to interpret the
experimental results obtained on the oriented multilayers of cytochrome oxidase
and mitochondrial membranes.

Polarized absorption spectroscopy measurements[9] were carried out on oxidized
and reduced multilayers and in the presence of various ligands (CO, formate)
which allowed resolution of the spectral properties of the two hemes, $\underline{a}$ and $\underline{a}_3$.
Optical spectra recorded at an angle of 0° between the incident light beam and
the normal to the planes of the oriented multilayers, with the light polarized
vertically and horizontally (with respect to the laboratory frame) were within
experimental error the same (dichroic ratios of 1). On the other hand at any
other angle between 0° and 90°, the optical absorption with the light polarized
horizontally was higher than that with the light polarized vertically. At a
45° angle, the dichroic ratios were almost 2 in the visible region and 1.2-1.4
in the Soret region. Moreover, they were similar for both the oxidized ($a^{3+}a_3^{3+}$)
and the reduced ($a^{2+}a_3^{2+}$) state of the multilayers and in the presence of formate
($a^{2+}a_3^{3+}$ - HCOOH) and CO ($a^{2+}a_3^{2+}$ - CO). Thus it was suggested that the hemes of
cytochrome oxidase were oriented in such a way that the angle between the heme
normal and the membrane normal is approximately 90°.

EPR spectroscopy on the oriented multilayers of cytochrome c oxidase. The
epr absorption of the hemeproteins arises from the paramagnetic properties of
the iron in its ligand field. The heme(s) of cytochrome $\underline{c}$ oxidase in the fully
oxidized state exhibits a low spin resonance characterized by three g values
which correspond to the three principal components of the g tensor indicating
that the ligand field symmetry is lower than tetragonal[18,19]. The orientation
of the principal axes of the g tensor in relation to the plane of the heme can
be determined from the analysis of single crystals where the orientation of the
heme with respect to the crystal axes is known. It has been generalized[20] from
the results obtained on crystals of metmyoglobin-azide[21], metmyoglobin-cyanide[22],
metmyoglobinimidazole[22] and horse heart-cytochrome $\underline{c}$[23] that the component of the
g-tensor which gives rise to the largest g value, g_z, is approximately normal
to the plane of the heme (Fig. 1). However, the in-plane components of the g
tensor, g_x and g_y, were found to be oriented differently with respect to the
N-Fe-N direction in the various low spin derivatives and thus no $\underline{a}$ $\underline{priori}$ predictions could be made with respect to the molecular orientation of the g_y and
g_x axes.

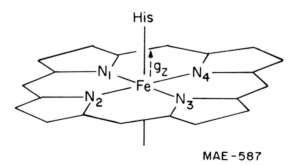

Fig. 1. The heme plane. The direction of z axis is shown.

EPR studies[9] on the oriented multilayers of cytochrome $\underline{c}$ oxidase in its oxidized state and in the half-reduced state in the presence of various ligands (azide, formate and sulfide) have allowed the determination of the orientation of the g_z, g_y and g_x axes with respect to the plane of the oriented multilayers. Moreover, since the g_z axis of the g tensor is normal to the plane of the heme, the results permitted the determination of the orientation of the heme itself with respect to the planes of the oriented multilayers. The results showed that normal to the heme plane of the cytochrome giving rise to the g_z=3.0 resonance in the fully oxidized oxidase, and generally assumed to arise from the absorption by cytochrome $\underline{a}$ (for discussion see 24), was approximately parallel to the plane of the oriented multilayer[7-11]. The same orientation was found for the g_z axis of the resonance observed in the aerobic, half-reduced oxidase in the presence of azide (g_z = 2.9) and sulfide (split resonances around g_z=2.5)[9,10]. Since these resonances are usually ascribed to ferric cytochrome $\underline{a}_3$-inhibitor complexes, a conclusion was drawn that cytochrome $\underline{a}_3$, like cytochrome $\underline{a}$, is oriented in such a way that the angle between the heme normal and the membrane normal is approximately 90°, in agreement with the optical studies. The resonances corresponding to the g_y and g_x axes of the g tensor were found to be oriented differently in the ferric low spin heme in the fully oxidized oxidase and in the half-reduced liganded oxidase[25]. The g_y axis of the ferric heme in the fully oxidized oxidase makes an angle of 30° with the membrane normal while the g_x axis is at 60° with respect to the membrane normal. The g tensor axes for the g=2.2 (g_y) and g < 2 (g_x) resonance lie at 0° and 90° respective to the membrane normal for the azide compound. It was suggested on the basis of this observation that the epr resonances seen in the two oxidase states (fully oxidized and half-reduced liganded) belong to two different hemes, those of cytochrome $\underline{a}$ and a_3.

ACKNOWLEDGEMENTS

This work was supported by USPHS HL 18708 and GM 12202. ME is an established

8

investigator of the American Heart Association.

REFERENCES

1. Lemberg, M.R. (1969) Physiol. Rev. 49, 48-121.
2. Nicholls, P. and Chance, B. (1974) in Molecular Mechanisms of Oxygen Activation (Hayaishi O. ed.) pp. 479-536, Acad. Press N.Y.
3. Caughey, W.S., Wallace, W.T., Volpe, T.A. and Yoshikawa, S. (1976) in The Enzymes (Boyer,P.D. ed) vol. 13, pp. 299-337, Acad. Press N.Y.
4. Gibson, Q.H. and Greenwood, C. (1963) Biochem. J. 86, 541-554
5. Chance, B. and Erecińska, M. (1971) Archiv. Biochem. Biophys. 143, 675-687.
6. Sun, F.F., Prezbindowski, K.S., Crane, F.L. and Jacobs, E.E, (1968) Biochim. Biophys. Acta. 153, 804-818.
7. Blasie, J.K., Erecińska, M., Samuels, S. and Leigh, J.S. Jr. (1978) Biochim. Biophys. Acta 501, 33-52.
8. Erecińska, M., Wilson, D.F. and Blasie, J.K. (1977) FEBS Lett. 76, 235-239,
9. Erecińska, M.,Wilson, D.F. and Blasie, J.K. (1978) Biochim. Biophys. Acta 501, 53-62.
10. Erecińska, M.,Wilson, D.F., and Blasie, J.K. (1978) Biochim. Biophys. Acta 501, 63-71.
11. Blum, H., Harmon, H.J., Leigh, J.S., Salerno, J.C. and Chance, B. (1978) Biochim. Biophys. Acta 502, 1-10.
12. Platt, J.R. (1956) in Radiation Biology (Hollaender,A. ed.) vol. 3, chap. 2, McGraw-Hill Book Co., N.Y.
13. Gouterman, M. (1961) J. Mol. Spectroscopy, 6, 138-163.
14. Kendrew, J.C. and Parrish, R.G. (1965) Proc. Roy. Soc. (London) Ser. A 238, 305-324.
15. Perutz, M.F. (1953) Acta. Cryst. 6, 859-864.
16. Makinen, M.W. and Eaton, W.A. (1973) Ann. New York, Acad. Sci. 206, 210-221.
17. Eaton, W.A. and Hochstrasser, R.M. (1967) J. Chem. Phys. 46, 2533-2539.
18. Weissbluth, M. (1966) Struct. Bonding 2, 1-125.
19. Peisach, J.,Blumberg, W.E., Adler, A. (1973) Ann. N.Y. Acad. Sci. 206, 310-327.
20. Taylor, C.P.S. (1977) Biochim. Biophys. Acta 491, 137-149.
21. Helcke, G.A., Ingram, D.J.E. and Slade, E.F. (1968) Proc. Roy. Soc. B. 169, 275-288.
22. Hori, H. (1971) Biochim. Biophys. Acta 251, 227-235.
23. Mailer, C. and Taylor, C.P.S. (1972) Canad. J. Biochem. 50, 1048-1055.
24. Erecińska,M. and Wilson, D.F. (1978) Archiv. Biochem. Biophys. 188, 1-14.
25. Erecińska, M.,Wilson, D.F. and Blasie, J.K. (1978) Biochim. Biophys. Acta in press.

Cytochrome Oxidase, T.E. King et al. eds.
© *1979 Elsevier/North-Holland Biomedical Press*

THE EFFECT OF SUBSTRATE BINDING ON THE ELECTRONIC STRUCTURE OF HEME AND IRON-LIGAND BOND IN CYTOCHROME P-450$_{cam}$

T. IIZUKA, H. SHIMADA, R. UENO AND Y. ISHIMURA
Department of Biochemistry, School of Medicine, Keio University, Shinjuku, Tokyo 160 (Japan)

ABSTRACT

The effect of substrate (d-camphor) binding on the electronic structure of ferrous carbon monoxide complex of cytochrome P-450$_{cam}$ was investigated by both flash photolysis and visible and infrared spectrophotometry. The results indicated that the substrate binding enhanced the overlap of iron dπ and ligand pπ orbitals. Significance of this finding in the catlytic reaction mechanisms of cytochrome P-450$_{cam}$ has been discussed.

INTRODUCTION

Cytochrome P-450 from <u>Pseudomonas putida</u> (P-450$_{cam}$) is a proto-heme protein, which catalyzes the d-camphor hydroxylation reaction[1,2]. As indicated in Table 1, the spin state of heme iron can be low or high spin state of ferric and ferrous ions. Starting from the ferric low spin state ①, the hemoprotein changes into ferric high spin form ② by the binding of the first substrate, d-camphor, then into ferrous high spin form ③ by accepting the first electron from putida redoxin, subsequently into ferrous low spin form ④ (oxygenated form) by the binding of the second substrate, molecular oxygen, and finally returns to the ferric low spin form ① after accepting the second electron from putida redoxin and yielding 5-exo-hydroxy camphor and H_2O (Fig. 1).

In the case of ferric enzyme, the effect of substrate is readily observed by EPR as a drastic spin change from $S=\frac{1}{2}$ to $\frac{5}{2}$.[3] On the other hand the oxygenated form in ferrous low spin state is silent against the direct magnetic measurements due to its diamagnetic

TABLE 1   SPIN STATE DIAGRAM

	Fe^{2+}	Fe^{3+}
HIGH SPIN	③ S = 2 P-450 (+cam)	② S = 5/2 P-450 (+cam)
LOW SPIN	④ S = 0 $\left(\begin{array}{l}\text{P-450 (+cam)}\\ O_2\end{array}\right.$	① S = 1/2 P-450 (−cam)

S, total spin value ; cam, camphor

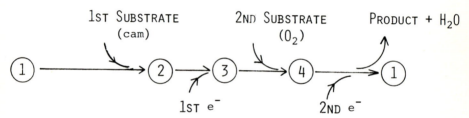

1ST SUBSTRATE (cam)   2ND SUBSTRATE (O_2)   PRODUCT + H_2O

1ST e^- 2ND e^-

Fig. 1. Reaction Sequence of P-450$_{cam}$. ① ~ ④ correspond to those in Table 1. e^- is an electron.

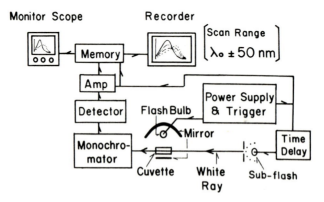

Fig. 2. Double flash technic for rapid detection

property (S=O). However, ferrous low spin complexes of hemoproteins with CO, O_2, NO and isocyanide have been reported to be
sensitive to light illumination[4] Among these hemoprotein derivatives, CO-complex is known to be most photosensitive. This property
has been utilized for studies on cytochrome c oxidase[5,6] and cytochrome P-450[7]

In the photodissociation process, light energy is absorbed by π
electron system of porphyrin ring and then utilized to break the
bond between ligand and heme iron. Therefore, analysis of this
process will give valuable information on electronic structures of
iron-ligand bond.

Though the oxygenated form of P-450$_{cam}$ is considered to be an
important intermediate in the hydroxylation reaction[2] it is not
so easy to apply physico-chemical methods for precise analysis due
to the unstable nature of the complex. Therefore, we choose the
carbon monoxide complex as a useful model of the oxygenated form
in order to understand the substrate-effect on the ferrous low spin
iron-ligand bond.

In this study we determined both quantum yields in the photodissociation process and CO stretching frequencies for the various
CO complexes of P-450$_{cam}$. Substrate-induced change of electronic
sturucture of ferric low spin heme was also detected in the cyanide
complex with 220 MHz proton NMR.

MATERIALS AND METHODS

P-450$_{cam}$ from Pseudomonas putida [ATCC 19453] was purified by
the method of Peterson[8] with a slight modification. The modified
method in our laboratory will be described elsewhere.
The absorbance ratio A_{392}/A_{280} of P-450$_{cam}$ was 1.26. The enzyme
freed of camphor was prepared by Sephadex G-25 gel filtration using
0.1 M Tris-HCl buffer (pH 7.4) without potassium chloride.

The photodissociation of the CO-complex of cytochrome P-450$_{cam}$
was observed by a pulse flash spectrophotometer (Union Giken RA
403) equipped with a double flash apparatus, a multichannel photodiode as a detector and a kinetic data processer (Union Giken

System 71). Fig. 2 illustrates the block diagram of this flash photolysis system. Xenon bulb (200-J) with a flash duration of about 30 μsec was used as a sourse of the exciting light, which intensity to irradiate the sample was varied with the aid of neutral density filters (Hoya Corporation). Here, the light intensity without filters was defined as the 100 %-intensity. As a light source for detection a small xenon bulb (0.2 J) with a duration time of about 3 μsec was adopted and fired just after the main flash for excitation expired. Sub-flash gave no disturbance to the CO complex of hemoprotein.

Absolute and difference spectra of CO forms of cytochrome P-450 under various conditions were recorded with a Union Giken SM-401 High Sensitivity Spectrophotometer. The CO stretching frequencies of P-450$_{cam}$·CO under various conditions were observed with JEOL 03F-FTIR spectrometer at 2 cm^{-1} resolution and at room temperature.

The 220-MHz proton NMR spectra were recorded at 22°C in a pulsed Fourier transform mode (PFT) by a Varian Associates HR-220 spectrometer equipped with a Nicolet TT-100 PFT instrument. The quadrature phase detection (QPD) method was used for the detection of the heme ring methyl signals in ferric enzyme with short pulse width (20 μs = 48° pulse). The spectra were obtained by 4K data points transformation of 40kHz spectral width after ca. 8192-16384 pulses with a repetition time of 0.05 s. Proton NMR chemical shifts were referenced externally with respect to the resonance of sodium 2,2-dimethyl-2-silapentane-5-sulfonate (DSS) and expressed in parts per million (ppm) assigning positive values to lower field resonance.

RESULTS AND DISCUSSION

Effect of substrate on the photosensitivity of CO complex of P-450$_{cam}$. In Fig. 3A is depicted a family of spectra of MbCO recorded with a double flash apparatus at 30 μsec after the start of the main flash for excitation. By the irradiation of main flash light stronger than 20 %-intensity, the spectrum of reduced myoglobin (Fe^{++}) was observed, which indicated the complete photodissociation of CO-complex. With the flash light weaker than 20 %-intensity, the CO complex remained appreciably, corresponding to the decrease in the intensity of excitation flash light. Fig. 3B illustrates that the degree of photodissociation increases hyper-

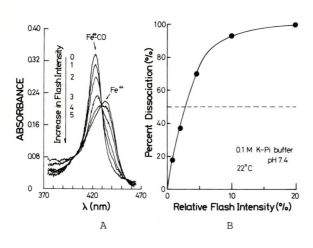

Fig. 3A, B. Photo-dissociation of Mb·CO. Fe^{++}-CO, carbon monoxide complex of myoglobin; Fe^{++}, reduced myoglobin.

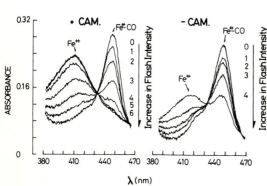

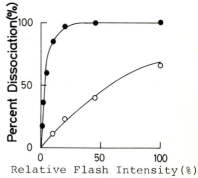

Fig. 4A. Photodissociation of P-450·CO in the presence (+cam) & absence (-cam) of camphor.

Fig. 4B. The effect of camphor on the photodissociation of P450·CO. ● and ○ in the presence and absence of camphor.

TABLE 2   QUANTUM YIELD AND CO STRETCHING FREQUENCY

CO-Complex	Quantum Yield Φ	ν_{CO} (cm^{-1})
Mb	1	1943
Hb	0.4	1951
P-450		
+ camphor	1	1940
- camphor	0.06	1959
+ α-picoline	0.06	1955
P-420	0.03	1965

bolically as the flash light intensity increases.

In order to characterize the photosensitivity of the hemoprotein CO complex, we define, here, $I_{1/2}$ as the relative flash light intensity required for photodissociating a half amount of total CO-complex, as if it were the dissociation constant in ligand titration experiment. The larger the value of $I_{1/2}$ is, the lower the photosensitivity is. Therfore, not $I_{1/2}$ but $I_{1/2}^{-1}$ is the criterion in the photosensitivity or the readiness of photodissociation. The photosensitivity $I_{1/2}^{-1}$ must be proportional to the extinction coefficient ε_{mM} of CO-complex, the ability to trap photons, and also to the quantum yield Φ in the photochemical reaction, the efficiency in breaking the iron-ligand bond by the use of the absorbed photon energy. Thus, the photosensitivity $I_{1/2}^{-1}$ is expressed by the equation (1);

$$I_{1/2}^{-1} = C \cdot \varepsilon_{mM} \cdot \Phi \qquad (1)$$

where C is a constant originated from the apparatus. In this study, we used the quantum yield of Mb·CO ($\Phi = 1.0$) as a wellknown standard for calculating those of other hemoproteins from the equation (1).

The photodissociation of P450$_{cam}$·CO was observed in the presence and absence of d-camphor, which were illustrated in Fig. 4A and B. In the presence of d-camphor, $I_{1/2}$ was 3.4 %, which was similar to that of Mb·CO, whereas $I_{1/2}$ was 56.5 % in the absence of d-camphor. In other words the photosensitivity ($I_{1/2}^{-1}$) of P-450·CO increased about tenfold with the binding of substrate. On the other hand the light absorption spectrum or ε_{mM} at Soret band of P-450·CO was not so much effected by substrate binding as examined below exactly. These results suggest that not ε_{mM} but quantum yield Φ in the equation (1) is greatly affected by the substrate binding. Values of Φ for several hemoproteins were caluculated and listed in Table 2.

The effect of substrate or its analogs on the Soret band of P-450·CO

Cytochrome P-450 was named after the Soret band peak appearing at about 450 nm in its reduced CO form.[9] This peak has been reported to be delicately different among various P450 species.[10] We found that the Soret band peak of P450$_{cam}$·CO was slightly but significantly changed by the binding of substrate or its analogs. In Table 3 were indicated absorption maxima (λ_{max}) and extinction

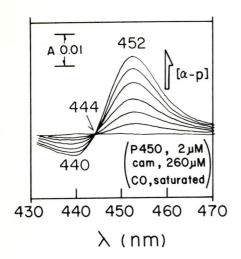

Fig. 5A. Titration of P-450$_{CO}^{cam}$ with α-picoline referenced with P-450$_{CO}^{cam}$. cam, camphor; α-p, α-picoline.

Fig. 5B. Double reciprocal plots, ΔA^{-1} vs [α-picoline]$^{-1}$, under the various concentrations of camphor.

REPLACEMENT REACTION

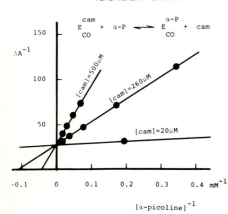

[α-picoline]$^{-1}$

TABLE 3. SORET ABSORPTION MAXMA OF 3 CO COMPLEXES

CO-COMPLEX	λ_{max} (nm)	ε mM
P-450 $_{CO}^{cam}$	446.4	105.0
P-450 $_{CO}^{free}$	447.5	110.0
P-450 $_{CO}^{α-p}$	448.0	110.3

cam, camphor; α-p, α-PICOLINE

coefficients (ε_{mM}) of Soret band for three kind of CO complexes. Here, we adopted α-picoline as one of the substrate analogs, but the other heterocyclic N-base such as imidazol, pyridine and their related compounds were able to effect the Soret band peak.

From this slight but definite change of Soret band, we confirmed the replacement reaction (2) in substrate site between camphor and α-picoline (α-p) for CO form;

$$P\text{-}450_{CO}^{cam} + \alpha\text{-}p \rightleftharpoons P\text{-}450_{CO}^{\alpha\text{-}p} + cam \qquad (2)$$

This experiment is illustrated in Fig. 5A, where the addition of α-picoline into the sample cuvette with $P450_{CO}^{cam}$ increased the difference spectrum referenced by $P450_{CO}^{cam}$. Fig. 5B shows the double reciprocal plot, ΔA^{-1} vs $[\alpha\text{-picoline}]^{-1}$, which means the competition between camphor and α-picoline. We comfirmed further the existence of $P\text{-}450_{CO}^{\alpha\text{-}p}$ by observing the quantum yield Φ by flash photolysis and the CO stretching frequency by FT-IR spectrometer (Table 2). The substrate analog such as α-picoline can effect the absorption spectrum (ε_{mM}), quantum yield Φ, and CO stretching frequency in the different manner from substrate.

The mechanism of photodissociation and the nature of the iron-ligand bond.

Using the values of quantum yield and CO stretching frequency (Table 2), a linear relationship between ν_{CO} and log Φ was obtained (Fig. 6). Fig. 6 indicates that ν_{CO} shifted to higher frequencies with the decrease in Φ value. Caughey et al. reported that the shift of ν_{CO} to lower frequencies is associated with the increase in π-bonding character of Fe-CO bond[11] and that the shift of ν_{CO} to higher values ($1966 cm^{-1}$) by the acidification (pH 3) of MbCO and HbCO was related to a transformation from a bent Fe-CO bonding (Fig. 7 (B)) of the native proteins to more nearly linear bonding (Fig. 7 (A)) of the unfolded ones at pH 3.[12] Bent structure of Fe-CO bond is also indentified in native Mb and Hb with X-ray structural analysis.[13] Therefore, it may be concluded here that the quantum yield Φ is correlated with the configuration of Fe-CO bond, more precisely with the degree of overlap between iron dπ and ligand pπ orbitals in Fig. 7 (B). The larger the π-bonding character is, the higher the quantum yield is. Through this π-type bonding, dπ electron of ferrous iron can delocalize on the ligand.

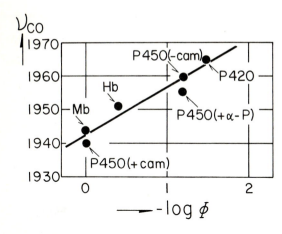

Fig. 6. Correlation between CO stretching frequency νco and quantum yield Φ.
α-p, α-picoline; cam, camphor.

Fig. 7. Two types of iron-ligand bond. White and shadowed orbitals are π- and σ- type ones, respectively.

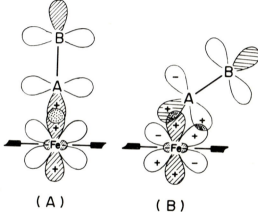

(A)   (B)

Fig. 8. 220 MHz proton NMR spectra of P-450 cyanide complex in the presence (A,C) and absence (B) of camphor. The bottom spectrum shows the NMR in protein region.

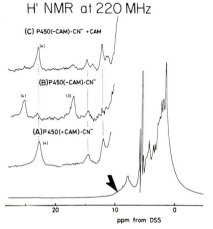

As seen in Table 2, the camphor binding to P450·CO drastically increases the quantum yield ϕ, favouring the bent structure and the increase in π-bonding character. This may be the reason why the molecular oxygen is stabilized as O_2^- in the presence of substrate compared with the case in the absence of substrate. It is also speculated that the 2nd electron in Fig. 1, goes on to 6th ligand O_2^- through the overlapped π type orbitals and finally splits the oxygen-oxygen bond.

NMR evidence of electron spin redistribution induced by substrate binding[14]

Though the ferrous low spin compound is silent against the direct magnetic measurement, ferric low spin one can be readily monitored by the paramagnetic shift of NMR signals. Fig. 8 illustrates the effect of substrate (camphor) on hyperfine shifted signals of ferric cyanide complex of cytochrome P-450$_{cam}$. In the absence of substrate, two methyl signals (4) and (3) appear in the lower field, whereas in the presence of substrate only one methyl signal (4) is observed in the higher field and other methyl signals are hidden in the protein region. This leads to following conclusion.

(i)   NMR can detect the difference between binary (enzyme-ligand) and ternary (enzyme-ligand-substrate) complexes.

(ii) Substrate (camphor) induces the change of electron distribution on heme-iron and porphyrin system, presumably through the conformational change of apo-protein.

Furthermore, it is suggested from the nature of hyperfine shift that the electron spin density in the center of heme is enhanced by the binding of substrate to the enzyme.

SUMMARY

(1) Flash photolysis experiment of the CO derivatives of P-450$_{cam}$ demonstrated that the photodissociation reaction was greatly affected by substrate binding. The quantum yields of the camphor -bound and -free forms were estimated to be 1.0 and 0.6, respectively. These values were correlated with the difference of CO stretching frequencies.

(2) The absorption spectrum of CO compound changed slightly but significantly upon the binding of d-camphor or its analogs (heterocyclic N-base) to the substrate site.

These results suggest that the binding of substrate or its
analogs significantly influences the electronic structure of heme.
By the binding of substrate the π-bonding character of iron-ligand
bond is enhanced.
(3) The electron spin redistribution of the ferric low spin compound (P-450·CN⁻) was clearly observed by 'H NMR as a definite
change of hyperfine-shifted methyl signals of heme periphery.

ACKNOWLEDGEMENTS

The authors would like to thank Prof. T. Kimura of Wayne State
Univ. for his critical reading of this paper. The results of IR
and NMR spectroscopy are a part of the cooperative studies with
Prof. T. Ohnishi of the Univ. of Tokyo and Prof. I. Morishima of
Kyoto Univ., respectively. The correlation between quantum yields
and CO stretching frequencies of P450·CO has been presented and
discussed in 6th International Biophysics Congress (1978, Kyoto).[15]

REFERENCES

1. Gunsalus, I.C., Meeks, J.R., Lipscomb, J.D., Debrunner, P. and
    Munk, E. (1974) in Molecular Mechanisms of Oxygen Activation.
    Hayaishi, O. ed., Academic Press, New York, pp. 561-613

2. Estabrook, R.W., Baron, J., Peterson, J. and Ishimura, Y.
    (1972) in Biological Hydroxylation Mechanisms, Boyd, G.S. and
    Smellie, R.M.S. eds., Academic Press, New York, pp. 159-185.

3. Tsai, R., Yu, C.A., Gunsalus, I.C., Peisach, J., Blumberg, W.,
    Orme-Johnson, W.H. and Beinert, H. (1970) Proc. Natl. Acad.
    Sci. USA, 66, 1157-1163.

4. Antonini, E. and Brunori, M. (1971) Hemoglobin andMyoglobin
    in Their Reaction with Ligands, North-Holland, Amsterdam pp 1-
    436.

5. Warburg, O. (1926) Biochem. Z., 177, 471-486.

6. Warburg, O. (1932) Z. Angew, Chem., 45, 1-6.

7. Estabrook, R.W., Cooper, D.Y. and Resenthal, O. (1963) Biochem.
    Z., 338, 741-755.

8. Peterson, J.A. (1971) Arch. Biochem. Biophys., 144, 678-693.

9. Omura, T. and Sato, R. (1962) J. Biol. Chem. 237, 1375-1376.

10. Alvares, A.P., Schilling, G., Leyin, W. and Kunzman, R. (1967)
    Biochem. Biophys. Res. Commun., 29, 521-526.

11. Caughey, W.S., Eberspaecher, H., Fuchsman, W.H., McCoy, S. and
    Alben, J.O. (1969) Ann. N. Y. Acad. Sci. 153, 722-737.

12. Caughey, W.S., Barlow, C.H., O'Keeffe, D.H. and O'Toole, M.C.
    (1973) Ann. Acad. Sci. 206, 296-309.

13. Heidner, E.J., Ladner, R.C., and Perutz, M.F. (1976) J. Mol.
    Biol. 104, 707-722.

14. Ishimura, Y., Iizuka, T., Morishima, I. and Hayaishi, O. (1978)
    in Polycyclic Hydrocarbons and Cancer Vol. 1, Gelboin and Ts'o
    eds., Academic Press, New York, pp 317-329.

15. Shimada, H., Ueno, R., Iizuka, T. and Ishimura, Y. (1978) Sixth
    International Biophysics Congress Abstracts, p. 403.

Cytochrome Oxidase, T.E. King et al. eds.
© 1979 Elsevier/North-Holland Biomedical Press

INTERMEDIATES IN THE REACTION OF CYTOCHROME O (VITREOSCILLA) WITH OXYGEN.

BERNADETTE TYREE and DALE A. WEBSTER

Dept. of Biology, Illinois Institute of Technology, Chicago Ill. 60616 USA

ABSTRACT

Cytochrome o from Vitreoscilla, MW 27,000 contains two identical 13,000 MW subunits and two moles of heme b. Reduced cytochrome o shows cooperative CO binding, but oxidized cytochrome o has two identical non-interacting cyanide binding sites, suggesting heme-heme interaction only in the reduced protein. When oxidized cytochrome o was titrated anaerobically with dithionite, two types of spectral transitions occurred, and each required approximately one electron; first, a decrease in the absorption bands at 540 and 405 nm; second, a shift in the absorption maxima to 555 and 425 nm, respectively. The midpoint potentials of the hemes, estimated using equilibrium photochemical titrations with EDTA, FMN, and either toluylene blue or indigo carmine, were +118 and -122 mV, respectively. Titration of reduced cytochrome o with oxygen at 23° resulted in recovery of the oxidized form and consumed 0.5 moles oxygen per mole of cytochrome. Two intermediates in this reaction were trapped by titrating at 0°. The first intermediate, compound D, absorption maxima 576, 548, and 418 nm, appeared on addition of the first 0.5 mole of oxygen per mole of cytochrome. Further addition of oxygen resulted in a sharp transition to the oxygenated compound, absorption maxima, 576, 543, and 414 nm. Compound D could be formed from the oxygenated compound by chemical or enzymatic reduction. Compound D, observed during the purification of cytochrome o chromatographed on Sephadex G-75 with a MW of 50,000. Addition of CO showed that the hemes were still in the ferrous state in compound D but were half oxidized in solutions of the oxygenated compound. The mixed valence cytochrome did not react with oxygen so it is unlikely that the oxygenated compound is a mixed valence species, but rather that these solutions of the oxygenated compound are heterogeneous. It is proposed that the first product of the reaction of oxygen with reduced cytochrome o is the dimeric compound D, $(b'^2 b^2)_2 O_2$. Oxygen then catalyzes the conversion of compound D to the oxygenated compound, $b'^2 b^2 - O_2$, oxidized cytochrome $b^3 b^3$, and peroxide anion.

Cytochrome o, a terminal oxidase widely distributed in many microorganisms, is functionally the same as cytochrome oxidase, so it is appropriate for the properties of this enzyme to be discussed at a symposium on cytochrome oxidase. The cytochrome o from Vitreoscilla, a filamentous myxobacterium that contains only protoheme IX pigments, is the only o-type cytochrome that has been studied in detail. Actually, this cytochrome o may not be a "cytochrome oxidase" in the sense originally meant by Keilin because there is at the present time no evidence that another cytochrome precedes it in the respiratory chain of Vitreoscilla. Nevertheless, it is the terminal member of the respiratory chain and the one that reacts with oxygen, and its relatively simple molecular structure makes it a model terminal oxidase for mitochondrial cytochrome oxidase. A review of its properties before the discussion of its reaction with oxygen in order.

Cytochrome o can be readily isolated from Vitreoscilla by blending the cells with alumina to solubilize the enzyme from the cell membrane. The purified protein, molecular weight 27,000, contains two heme b prosthetic groups but does not contain copper, flavin, nonheme iron, or cysteine. The two hemes in the reduced protein bind CO in a cooperative manner (Fig. 1). A Hill plot of the data in Fig. 1 has a slope of 2. The oxidized cytochrome has two identical but non-interacting binding sites for cyanide (Fig. 2). These data suggest that subunit or heme-heme interaction occurs in the reduced but not in the oxidized protein.

The reduction of cytochrome o by anaerobic dithionite titration occurred in two phases (Fig. 3); first, a decrease in the absorption bands of the oxidized cytochrome at 540 and 405 nm; second, a shift in the absorption maxima to 555 and 425 nm, respectively. Since each phase required approximately one electron this is evidence that the two hemes are reduced independently and sequentially. This is supported by the finding that the midpoint potentials of the two hemes differ by more than 200 mV. Two redox dyes, indigo carmine ($E_o' = -125$ mV) and toluylene blue ($E_o' = +115$ mV), with midpoint potentials close to those of the high and low potential hemes, respectively, were selected for the equilibrium photochemical redox titration. This method used EDTA, FMN in catalytic amounts, one of the two dyes, and cytochrome o in anaerobic solution (Figs. 4 and 5). Nernst plots for each titration were linear and midpoint potentials of +118 and -122 mV for the high and low potential hemes, respectively, were estimated from them.

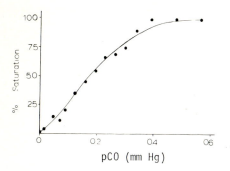

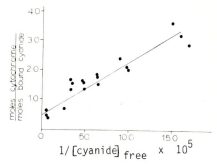

Fig. 1. Binding curve of reduced
cytochrome o to CO. An anerobic
solution of 4 µM cytochrome o in 0.1M
sodium phosphate, pH 7.0, reduced
with dithionite, was titrated with
CO saturated buffer at 23°. The
amount of CO compound was estimated
by the increase in A_{420}.

Fig. 2. Scatchard plot for the
binding of cyanide to cytochrome o
determined by equilibrium dialysis.
Dialysis tubing (12 x 0.9 cm) containing 7.5 µM cytochrome o and
50 µM potassium ferricyanide in
3.0 ml 0.2 M sodium phosphate,
pH 7.0, was suspended in 3.0 ml of
the phosphate buffer containing 200
to 4000 DPM $K^{14}CN$ at 0° for 48 hr.
50 µl from the outside of the bag
were counted in 5 ml Aquasol (New
England Nuclear) with a Beckman
liquid scintillation counter and
the bound cyanide was estimated by
difference.

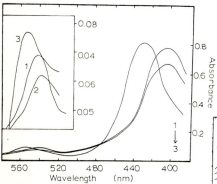

Fig. 3. Anerobic titration of
cytochrome o with dithionite. A
solution of 3.8 µM cytochrome o
in 0.1 M sodium phosphate, pH 7.0
in an argon atmosphere was titrated
by microliter additions of an anaerobic
solution of dithionite. Spectrum 1,
oxidized cytochrome; Spectrum 2,
half-reduced cytochrome, after
addition of approximately one
electron equivalent of dithionite;
Spectrum 3, fully reduced cytochrome, after additon of approximately two electron equivalents
of dithionite. The reduction of
the first heme can be monitored by
the decrease in the absorption
maxima at 540 and 405 nm, that of
the second heme by the increase at
555 and 425 nm. These spectra and
in all subsequent Figs. were recorded
with a Zeiss DMR-21 spectrophotometer.

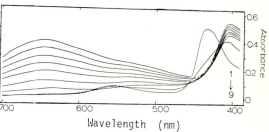

Fig. 4. Equilibrium photochemical
redox titration of high potential heme
of cytochrome o. The reaction mixture
contained 5 nmol cytochrome o, 40 nmol
toluylene blue, 1 nmol FMN, 2 units
glucose oxidase, 2 units catalase,
and 3.5 units superoxide dismutase in
1.6 ml of 0.1 M sodium phosphate, pH
7.0 in an argon atmosphere. A solution
(0.4 ml) containing 1.0 M glucose and
0.05 M EDTA in the phosphate buffer was
added in the dark. Spectra were recorded
after each 45 s of illumination.

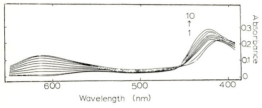

Fig. 5. Equilibrium photochemical redox titration of cytochrome o. Same conditions as Fig. 4 except reaction mixture contained 3 nmol cytochrome o and 20 nmol indigo carmine in place of the toluylene blue.

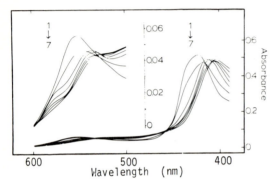

Fig. 6. Titration of reduced cytochrome o with oxygen at 23°. Cytochrome o (2.35 µM) in 0.1 M sodium phosphate, pH 7.0, in an argon atmosphere was reduced stoichiometrically by the addition of an equivalent amount of dithionite. Spectrum 1, fully reduced; Spectra 2-7, after successive microliter additions of air saturated buffer Temperature was maintained with a Haake circulating pump.

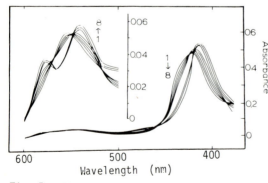

Fig. 7. Titration of reduced cytochrome o with oxygen at 0°. Same as Fig. 6 except at 0°.

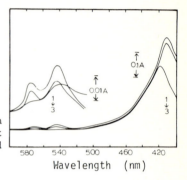

Fig. 8. Appearance of compound D during the NADH-cytochrome o oxidase reaction at 23°. A small amount of solid NADH was added to 2 nmol cytochrome o in 1 ml 0.02 M sodium phosphate, pH 7.4, containing 50 µg NADH-cytochrome o reductase. Spectra 1,2 and 3 are 5,30 and 60 min after NADH addition, respectively. Spectrum 3 is similar to the spectrum of compound D (Figs. 7 and 11).

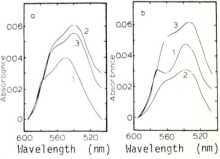

Fig. 9. Reaction of oxygenated intermediates of cytochrome o with CO. Solutions containing compound D (a) or oxygenated compound (b) were formed stoichiometrically at 0° and bubbled with CO for 3 min. Spectrum 1, Compound D (a) or oxygenated compound (b); Spectrum 2, after saturation with CO; Spectrum 3, after addition of dithionite.

Let us now turn to the reaction of the cytochrome with oxygen. An intermediate, oxygenated form of the cytochrome was first observed as the steady state intermediate during the aerobic oxidation of NADH by partially purified preparations of cytochrome o[1]. This compound was later shown to be the predominant form of the cytochrome in respiring whole cells of Vitreoscilla[2]. Both visible and infrared spectra of oxygenated cytochrome o suggest the similarity of the heme iron-oxygen bond to that present in oxymyoglobin and oxyhemoglobin[3,4]. When the reduced cytochrome was titrated with oxygen at 23°, the oxidized form was recovered after the addition of 0.5 moles of oxygen per mole of cytochrome (Fig. 6). No spectrally distinct intermediate was detectable during this titration, rather the spectra appear to be mixtures of more than one species of the cytochrome. When this titration was performed at 0° two intermediates were observed (Fig. 7). The first intermediate, compound D, appeared on the first addition of oxygen, was maximally formed after the addition of 0.47 mole of oxygen per mole of cytochrome, and had absorption maxima at 548, 418, and a shoulder at 576 nm. The further addition of oxygen resulted in formation of the oxygenated compound, absorption maxima 576, 543, and 414 nm, which is stable at 0°. The conversion of reduced cytochrome to compound D showed a linear dependence on the amount of oxygen added with isosbestic points at 569, 552, 453, and 388 nm. The abrupt conversion of compound D to the oxygenated cytochrome on the further addition of small amounts of oxygen suggested that oxygen was acting catalytically. Compound D could be formed from the oxygenated compound by back titrating with dithionite at 0°, and it has been detected during the oxidation of NADH by cytochrome o preparations at room temperature. For this experiment, excess NADH was added to an aerobic solution of cyto-chrome o containing NADH-cytochrome o reductase, and the reaction monitored spectrally as the oxygen was consumed. A sharp transition from the oxygenated compound, which is the steady state form, to compound D was observed just before the solution became anaerobic (Fig. 8).

The valence of the heme in solutions of these two oxygen intermediates of cytochrome o was determined with carbon monoxide. When a solution containing compound D formed stoichiometrically at 0° was bubbled with CO the extent of the reduced CO complex of cytochrome o formed showed that compound D contains only ferrous heme (Fig. 9a). In contrast, half of the heme in the solution of the oxygenated compound was oxidized (Fig. 9b). Since the mixed valence form of cytochrome o formed by anaerobic titration with dithionite

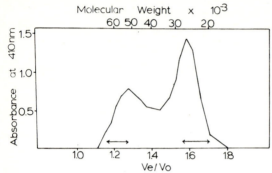

Fig. 10. Molecular weight of compound D estimated on Sephadex G-75. 100 mg of a DEAE-cellulose purified extract was prepared as described previously[1], applied to a 2 x 110 cm column of Sephadex G-75, and eluted with 0.05 M sodium phosphate-0.1 M NaCl, pH 7.5, at 4°. Five ml fractions were collected and pooled as indicated by the arrows. When the low molecular weight component was rechromatographed it eluted as a single peak of molecular weight 27,000, but when the high molecular weight component was rechromatographed it split into two fractions, similar to Fig. 10.

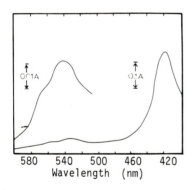

Fig. 11. Spectrum of high molecular component from Sephadex chromatography of Fig. 10.

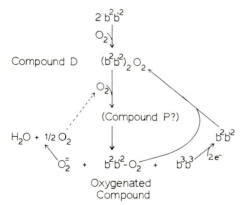

Fig. 12. Proposed model for the reaction of reduced cytochrome o with oxygen and possible catalytic cycle.

does not react with oxygen it is unlikely that the oxygenated compound is oxygen bound to half-reduced cytochrome o.

The stoichiometry of formation of compound D suggests that it consists of one molecule of oxygen bound to two molecules of cytochrome o. There is supporting evidence for this. A fraction of cytochrome o, observed frequently as a side band during the chromatographic steps of the purification, elutes from Sephadex G-75 with a molecular weight of 50,000 and was spectrally identical to compound D (Figs. 10 and 11). The oxygenated compound chromatographed with a molecular weight of 27,000 on Sephadex G-75.

Our proposed mechanism for the reaction of reduced cytochrome o with molecular oxygen in solution postulates that two molecules of cytochrome o act in concert to reduce molecular oxygen to hydrogen peroxide; the latter has been detected as an intermediate and possibly end-product of the NADH-cytochrome o oxidase system[5]. The direct reaction product of the oxygen binding step is the dimeric compound D (Fig. 12). Additional oxygen then acts semi-catalytically to convert this to the oxygenated compound, the oxidized cytochrome, and peroxide anion; oxygen could be regenerated from the latter. Reduction of the oxidized cytochrome and reaction of the reduced cytochrome with the oxygenated compound would enable reformation of compound D. Other intermediates in this catalytic cycle are likely, especially an unstable peroxidic compound of the cytochrome (compound p?).

ACKNOWLEDGEMENTS

This work was supported by National Science Foundation Grant BMS 74-12227 and National Institutes of Health Grant GM 20006.

REFERENCES
1. Webster, D. A., and Liu, C. Y. (1974) J. Biol. Chem. 249, 4257-4260.
2. Webster, D. A., and Orii, Y. (1977) J. Biol. Chem. 250, 1834-1836.
3. Liu, C. Y., and Webster, D. A. (1974) J. Biol. Chem. 249, 4261-4266.
4. Choc, M. G., Caughey, W. S., and Webster, D. A. (1978) Fed. Proc. 37, 1325.
5. Webster, D. A. (1975) J. Biol. Chem. 250, 4955-4958.

Cytochrome Oxidase, T.E. King et al. eds.
© *1979 Elsevier/North-Holland Biomedical Press*   29

HEME a AS THE PROSTHETIC GROUP OF AN ARTIFICIAL L-TRYPTOPHAN 2,3-
DIOXYGENASE (PYRROLASE) OF RAT LIVER*

KAYOKO SUDO[†], RYU MAKINO, TETSUTARO IIZUKA AND YUZURU ISHIMURA
Department of Biochemistry, School of Medicine, Keio University,
Shinjuku-ku, Tokyo 160 Japan

ABSTRACT
    Artificial L-tryptophan 2,3-dioxygenases containing various
natural and synthetic metalloporphyrins have been reconstituted
from highly purified apoenzyme preparations of rat liver. Among
several metalloporphyrins tested, heme a, an essential component
of cytochrome c oxidase was found to be the best in substituting
protoheme IX, the natural prosthetic group of this dioxygenase;
the combination of heme a with apoprotein resulted in a full
restoration of the catalytic activity, i.e. to the same extent to
that by protoheme IX. Reconstitution with mesohemin restored
about one-third of the full activity, whereas deuterohemin, 2,4-
diacetyldeuterohemin and cobaltptoporphyrin IX were found to be
innert for the reconstitution of an active holoenzyme.

INTRODUCTION
    L-tryptophan 2,3-dioxygenase, commonly known as tryptophan
pyrrolase, is a protohemoprotein which catalyzes the insertion of
molecular oxygen into the pyrrol ring of L-tryptophan giving L-
formylkynurenine as the product (1,2). It has been shown by us
that the protoheme IX is the sole prosthetitic group of this enzyme
(3,4) and also that the ferrous heme in the enzyme reacts with oxygen to form an oxygenated form of the heme in the presence of L-

---

* This investigation has been supported in part by the Scientific
  Research Fund of the Ministry of Education of Japan and grants
  from the Naito Research Foundation and Chiyoda Mutual Life
  Research Foundation.
† Present adress: The Department of Laboratory Medicine, Hamamatsu
  University, School of Medicine, Hamamatsu 431-31 Japan

tryptophan (5,6). The oxygenated form was shown to be an obligatory intermediate of the overall reaction and, thus, the role of
heme in the catalysis has been unambiguously established (7,8).

In order to obtain further information on the mechanism of
oxygen activation in this dioxygenase reaction, we have substituted
the protoheme IX in the enzyme with various metalloporphyrin compounds and some preliminary results are herein reported. Method
for the preparation of apoenzyme is also described.

MATERIALS AND METHODS

Livers of femal Wister rats, which received combined administ-
rations of cortisone acetate and L-tryptophan at 4 and 8 hours
before killing (9) were used as the source of L-tryptophan 2,3-
dioxygenase. The enzyme was purified from homogenates of the
livers approximetely 500-fold as holoenzyme by heat treatment,
EDAE-cellulose chromatography, calcium phosphate gel treatment
and finally by Sephadex G-200 chromatography (4,9). All of the
buffer used during the purification contained L-tryptophan in a
final concentration of either 10 or 20 mM to stabilize the enzyme
unless otherwise noted.

L-Tryptophan 2,3-dioxygenase activity was determined by measuring the increase in optical density at 321 nm due to the formation
of L-formylkynurenine (10). The standard assay medium contained
50 mM potassium phosphate buffer, pH 6.8, 13 mM of L-tryptophan,
2 mM of L-ascorbate and an appropriate amount of enzyme in a
total volume of 1.0 ml under normal atmospheric conditions. One
unit of enzyme was defined as the amount which catalyzes the conversion of 1 μmol of L-tryptophan to L-formylkynurenine per min at
25° under the standard assay conditions. Specific activity of the
enzyme was expressed as units/mg of protein and the final enzyme
preparations obtained by our purification procedure, exhibited the
specific activities of 0.8 to 1.3 units/mg of protein. Protein was
determined by the Biuret method (11) using bovine serum albumin as
a standard. Heme content of enzyme preparations was determined as
pyridine ferrohemochrome according to the method of Paul et al (12).
The turnover number (μmol of product formed/min/mol of heme) of our
enzyme preparations was 175 to 233 min^{-1} which were comparable to

or slightly better than that of a homogeneous enzyme preparation
described by Schutz and Feigelson (13).

Formylkynurenine formamidase which was free from hemoproteins
was obtained from rat liver according to the method of Knox (14).
Apomyoglobin was prepared by the methyl ethyl ketone-HCl method
reported by Yonetani (15). Protohemin was purchased from Sigma
and used without further purification. Deuterohemin was prepared
according to a slightly modified method of Chu and Chu (16) and
mesohemin was according to Fischer and Pützer (17). 2,4-Diacetyl-
deuterohemin was prepared from deutrohemin according to Fischer
and Orth (18). Heme a and cobaltprotoporphyrin IX were the kind
gifts of Dr. Y. Orii (Osaka University) and Dr. T. Yonetani (Univ.
of Penn. School of Medicine), respectively. The concentrations of
these heme derivatives were determined by the use of following
millimolar extinction coefficients of α-band of their reduced
pyridine hemochromes; 32 at 557 nm for protohemin (19), 32 at 547
nm for mesohemin (19), 25 at 551 nm for hematohemin (19), 26 at
544 nm for deuterohemin (19), 14 at 575 nm for 2,4-diacetyldeutero-
hemin (20), 28.2 at 587 nm for heme a (21), and 17 at 569 nm for
cobltprotoporphyrin (15), respectively. Other chemicals used
were of highest grade commercially available. All the spectrophotometric measurements were done with a Union Giken automatic
recording spectrophotometer model SM-401 equipped with a RA-450
digital data processor.

RESULTS AND DISCUSSION

Preparations of apo L-tryptophan 2,3-dioxygenase have previously been described by Greengard and Feigelson (22) and also by
Schimke (23). However, neither of these was found to be satisfactory with respect to the yield and purity of the reconstitutable
apoenzyme and, therefore, the method of Greengard and Feigelson
was modified in the following way.

Holoenzyme (1-2 mg) in a buffer with 10 mM L-tryptophan was
diluted with 50 mM Tris-HCl, pH 7.2, to give a protein concentration less than 0.2 mg/ml and L-tryptophan concentration below 1
mM, and then ca. 4 g (dry weight) of DEAE cellulose (DE 52,
Whatmen) equilibrated with the same buffer was added to the enzyme

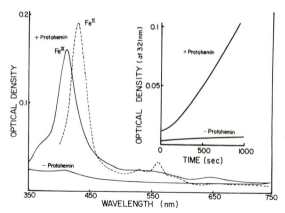

Fig. 1. Effects of protohemin on the spectra and activity of apo L-tryptophan 2,3-dioxygenase. Spectra- the sample solution contained 0.25 mg of apoprotein in 50 mM Tris-HCl, pH 7.8. To the apoenzyme solution, were added successively protohemin (17 μM) and sodium dithionite. Inserted figure- 2 μg of apoenzyme was assayed in the presence and absence of 0.6 μM protohemin under the standard assay conditions at pH 6.8.

solution. The mixture was kept at 0° for 3 hrs with occaisional stirring and centrifuged at 2,000 x g for 10 min. The pellets of DE 52 obtained were resuspended into 10-20 ml of 50 mM Tris-HCl, pH 7.2, containing 10 mM L-tryptophan and placed on the top of a DE 52 column (5x20 cm) which had been equilibrated with the same buffer described above. The protein was then eluted with 0.3 M KCl in 50 mM Tris-HCl, pH 7.2, containing 10 mM L-tryptophan and concentrated by ultrafiltration (Centriflo, Amicon).

The apoenzyme thus obtained was found to be essentially free from bound protohemin and also from the catalytic activity. As seen in Fig. 1, the apoenzyme showed almost no absorption in the entire visible region and the ratio of absorbancy at 407 nm to that at 280 nm was less than 0.1 which had been close to unity before the treatments. When the activity of such a preparation was examined, the reaction catalyzed was negligible in the absence of protohemin but was fully restored by the addition of protohemin into the reaction mixture (Inserted Fig. of Fig. 1). The reaction was carried out at pH 6.8 to prevent the increase in absorbancy at 321 nm due to nonenzymatic oxidation of ascorbate in the presence of hemins at a higher pH. A lag phase observable in the Fig. was the time required for the conversion of an inactive ferric form as reconstituted to an active ferrous form by the action of reductants in the assay medium. The spectrum of a hemoprotein could also be restored by the addition of protohemin to the enzyme solution as shown also in Fig. 1; the spectrum of the ferric form was reminiscent of that of the original holoenzyme but the spectrum of the dithionite reduced form differed greatly from that of the holoenzyme. The latter difference might probably be due to the protohemin nonspecifically bound to the protein, since we added slightly an excess amount of protohemin to the solution.

In the next series of experiments, we examined the effects of various other metalloporphyrins on the activation of apo L-tryptophan 2,3-dioxygenase. Those metalloporphyrins employed were deuterohemin, mesohemin, 2,4-diacetyldeuterohemin, cobalt protoporphyrin IX and heme a from beef heart cytochrome c oxidase. Unexpectedly, we found that the addition of heme a to apoenzyme in the reaction mixture resulted in a significant activation of the enzyme. Mesohemin was also found to activate apo L-tryptophan 2,3-dioxygenase, while the other three were totally ineffective under comparable experimental conditions.

Fig. 2 shows the effect of these hemin concentrations on the activation of apo L-tryptophan 2,3-dioxygenase. As can be seen, the concentration of heme a required for the half maximal activation was over 10 times greater than that concentration of protohemin indicating that the affinity of apoenzyme for heme a is considerably less than that for protohemin. However, the maximal

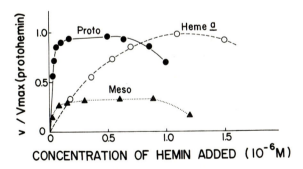

Fig. 2. Activation of apo L-tryptophan 2,3-dioxygenase as a function of hemin concentrations. Experimental conditions were similar to those described in Fig. 1 (inserted Figure) exept that hemin concentrations were varied.

catalytic activity obtained by heme a was almost equal to that by protohemin which is the natural prosthetic group of L-tryptophan 2,3-dioxygenase. Mesohemin showed an affinity comparable to that of protohemin but the Vmax was about one-third of that by proto-hemin.

The product of the reaction catalyzed by these artificial en-zymes was proved to be really formylkynurenine by confirming its conversion to kynurenine by the use of formylkynurenine form-amidase. It was also confirmed that the activation by "heme a" was not due to the contamination of protohemin in the heme a preparation; the effect of heme a on the activity was not antago-nized by the addition of an excess amount of apomyoglobin which was sufficient enough to remove 2 nmol/ml of protohemin. Thus, heme a which has a long isoprenoid sidechain is shown to serve as an artificial prosthetic group of L-tryptophan 2,3-dioxygenase. Heme a alone in the absence of apoprotein or an albumin complex of heme a showed no L-tryptophan 2,3-dioxygenase activity. In Fig. 3 are shown the absorption spectra of the artificial enzyme recon-stituted with heme a, although the apoenzyme preparation employed was not completely free from holoenzyme. Nevertheless the Soret absorption maximum of the ferric form at 424 nm and that of the ferrous form at 444 nm are noteworthy in relation to the spectra of cytochrome c oxidase. It should be also mentioned that

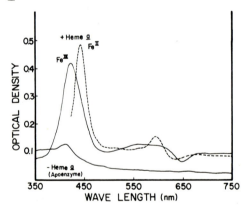

Fig. 3 Spectra of the enzyme reconstituted with heme a at pH 7.8. The solution contained 1 mg/ml of the DE 52-treated enzyme (apo/apo + holo = 0.7). Heme a (17 µM) was added to both sample and reference cuvettes to record the difference spectrum. Fe^{III} ____, oxidized form; Fe^{II} ---, dithionite- reduced form.

Km value for L-tryptophan of the enzyme reconstituted with heme $\underline{a}$
was greater than that of the natural enzyme by almost two order of
magnitude as listed in the Table 1. The enzyme reconstituted with
mesohemin showed the Soret absorption maximum at 398 nm in its
ferric form and Km for L-tryptophan between those of the enzymes
with protohemin and heme $\underline{a}$ (see also Table 1).

Table 1   Summary of the properties of artificial enzymes

Hemins	Kd (or Ki) ($\underline{M}$)	Km for L-tryptophan ($\underline{M}$)	Relative activity (%)	Soret band (Ferric) (nm)
Deutro	1.1×10^{-8}	——	n.d.*	396
Meso	2.7×10^{-8}	3.4×10^{-3}	ca. 30	398
Proto	1.3×10^{-8}	7.1×10^{-4}	ca. 90	409
Diacetyl deutro	8.5×10^{-8}	——	n.d.*	413
Heme $\underline{a}$	4.5×10^{-7}	1.2×10^{-2}	ca. 90	424
Holo-TPO	——	3.4×10^{-4}	100	407

* not detectable.

As already mentioned, none of deuterohemin, 2,4-diacetyldeutero-
hemin and cobalt protoporphyrin activated apo L-tryptophan 2,3-
dioxygenase under our experimental conditions. However, all of
these were found to inhibit the binding of other metalloporphyrins
in a competitive manner. Thus, they are likely to bind with the
apoprotein at the same site with that of protohemin, namely at
the proper site, but appear unable to activate molecular oxygen.
Their binding constants as measured as the Ki values against
protohemin binding were also described in Table 1. Further studies
on the properties of artificial L-tryptophan 2,3-dioxygenases
reconstituted with various metalloporphyrins are in progress.

Acknowledgements-- We thank Mr. Y. Tanizaki and Miss. K. Watanabe
for their excellent technical assistance.

36

REFERENCES

1) Hayaishi, O., Rothberg, S., Mehler, A.H., and Saito, Y. (1957) J. Biol. Chem. 229, 889-896

2) Tanaka, T. and Knox, W.E. (1959) J. Biol. Chem. 234, 1162-1170

3) Ishimura, Y. and Hayaishi, O. (1973) J. Biol. Chem. 248, 8610-8612

4) Makino, R. and Ishimura, Y. (1976) J. Biol. Chem. 251, 7722-7725

5) Ishimura, Y., Nozaki, M., Hayaishi, O., Tamura, M and Yamazaki, I. (1967) J. Biol. Chem. 242, 2574-2576

6) Ishimura, Y., Nozaki, M., Hayaishi, O., Tamura, M and Yamazaki, I. (1968) Advan. Chem. Ser. 77, 235-241

7) Ishimura, Y., Nozaki, M., Hayaishi, O., Nakamura, T., Tamura, M. and Yamazaki, I. (1970) J. Biol. Chem. 245, 3593-3602

8) Hayaishi, O., Ishimura, Y., Fujisawa, H., and Nozaki, M. (1971) In T. E. King, H. S. Mason and M. Morrison (Editors), Oxidases and Related Redox Systems, University Park Press, Baltimore Vol. I, p. 125

9) Schimke, R.T., Sweeney, E.W., and Berlin, C.M. (1965) J. Biol. Chem. 240, 322-331

10) Ishimura, Y. (1970) Methods Enzymol. 17A, 429-434

11) Layne, E. (1957) Methods Enzymol. 3, 450-451

12) Paul, K.G., Theorell, H., and Akeson, A. (1959) Acta Chem. Scand. 7, 1284-1287

13) Schutz, G., and Feigelson, P. (1972) J. Biol. Chem. 247, 5327-5332

14) Knox, W.E. (1955) Methods Enzymol. 2., 246-249

15) Yonetani, T., Yamamoto, H., and Woodrow, G. (1974) J. Biol. Chem. 249, 682-690

16) Chu, T.C., and Chu, E.J.H. (1952) J. Amer. Chem. Soc. 74, 6276-6277

17) Fischer, H., and Pützer, B. (1926) Hoppe-Seyler's Z. Physiol. Chem. 154, 39-63

18) Fischer, H., and Orth, H. (1937) In Die Chemie des Pyrrols, Akademische Verlagsgesellschafts M.B.H., Leipzig, P 304.

19) Antonini, E., Brunori, M., Caputo, A., Chiancone, E., Rossi-Fanelli, A. and Wyman, J. (1964) Biochim. Biophys. Acta, 79, 284-292

20) Makino, R. and Yamazaki, I. (1972) J. Biochem. 72, 655-664

21) Lemberg, R., Morell, D.B., Newton, N., and O'Hagen, J.E. (1962) Proc. Royal Soc. 155, 339-355

22) Greengard, O., and Feigelson, P. (1962) J. Biol. Chem., 237 1903-1907

23) Schimke, R.T. (1970) Method. Enzymol. 17A, 421-428

PRIMARY STRUCTURES

Cytochrome Oxidase, T.E. King et al. eds.
© *1979 Elsevier/North-Holland Biomedical Press*

CHARACTERIZATION OF SUBUNITS OF MITOCHONDRIAL CYTOCHROME OXIDASE

TAKAYUKI OZAWA, MASAKO TADA and HIROSHI SUZUKI

Department of Biomedical Chemistry, Faculty of Medicine, University of Nagoya,

Nagoya 466 (Japan)

ABSTRACT

Cytochrome oxidase purified from beef heart mitochondria using immobilized cytochrome c or phenyl-Sepharose CL-4B, a hydrophobic affinity ligand, contained 12 - 14 nmoles heme a per mg protein. Polyacrylamide gel electrophoresis in the presence of 0.1% sodium dodecyl sulfate (SDS) and 4M urea showed six subunits in the oxidase: I = 40,000 daltons, II = 25,000, III = 19,000, IV = 12,000, V = 9,000, IV = 5,000. The purified oxidase was split into two catalytic components, component A and B, in the presence of 5% SDS and 4M urea, and they were separated by chromatography using Sephadex G-75 superfine, in the presence of 0.1% SDS and 4M urea. Component A consisted of subunits I and II, and component B, of subunits III - VI. Further treatment of component B with 5% SDS and 4M urea and then rechromatography with Sephadex G-75 superfine yielded a non-catalytic component C which contained subunits V and VI. Component A was reduced by dithionite giving distinguishable α-peak at 597 nm. Both CO-bound components A and B were reduced by dithionite giving α-peaks at 600 nm and 602 nm, respectively. Component C showed similar spectral changes to component B. Analyses of heme a and copper demonstrated that components A and B contained one mole each of heme a and copper. In component B, one mole of heme a located in component C (*viz.* either in subunit V or VI) and one mole of copper in either subunit III or IV. Both components A and B oxidized the reduced cytochrome c with Km = 19 μM and 34 μM, respectively, but component C did not. Their activities were also demonstrated by oxygen-uptake. Addition of phospholipids stimulated their activities by about two fold, component A was more susceptible than component B. KCN and NaN$_3$ inhibited the activities of components A and B, however, the concentrations of inhibitors which give 50% inhibition were markedly different: 5 μM of KCN for component A but 5 mM for component B; 50 μM of NaN$_3$ for component A but 50 mM for component B. To CO, component B was more susceptible than A.

These data suggest that cytochrome oxidase is a complex of two catalytic units which contain equally heme a and copper, but having different enzymatic properties, such as Km, phospholipids requirement and susceptibility to inhibitors, KCN, NaN$_3$ and CO. Component A may correspond to the component a_3, and B to a.

INTRODUCTION

Cytochrome oxidase (ferrocytochrome c: oxygen oxidoreductase, EC 1.9.3.1), the terminal enzyme of mitochondrial electron transfer chain, which contains two heme a and two copper irons, and is composed of number of different subunits per functional unit, has been the subject of a large number of recent studies.

Our own study has been concentrated on the role of the subunits in the function of the enzyme by means of a controlled and progressive resolution of the oxidase into simpler forms lacking some of the subunits of the original enzyme but retaining the enzymic activity, and on the reconstitution of the oxidase together with a comparison of the properties of the simpler forms of the oxidase.

Although the different subunit components of mammalian cytochrome oxidase have been separated and partially characterized, the localization of copper and heme a among the subunit is still uncertain and very little information has been available concerning the functional group associating subunit. MacLennan and Tzagoloff[1] separated a polypeptide, molecular weight of about 25,000, containing about two atoms of copper per molecule, by gel filtration of succinated cytochrome oxidase, and then Tanaka et al.[2] purified a similar polypeptide, molecular weight of about 22,000, by dissociation with 5% SDS. Schatz et al.[3] have taken advantage of the Schiff-base reaction to link the heme a to protein covalently and have found that it is associated with a subunit of molecular weight 11,500. Recently, Yu et al.[4] isolated a heme a binding subunit by pyridine treatment, which has a molecular weight of 11,600. As for separation of simpler components from cytochrome oxidase, which retained heme a, Komai and Capaldi[5] obtained simpler component composed of only two polypeptides of 14,000 and 11,500 daltons which had the cytochrome oxidase activity. Yamamoto and Okunuki[6] succeeded in enriching the heme a content of a bovine cytochrome oxidase preparation, without damaging its enzyme activity and spectral properties, by removal of extraneous proteins from an original preparation by chymotrypsin digestion. The minimum molecular weight of this proteinase-treated cytochrome oxidase was calculated to be 70,000 from the amino acid composition of about 600 residues per heme a. Yamamoto and Orii[7] reported that the proteinase-treated oxidase gave predominantly two components of 14,000 and 11,000 daltons and retained the same enzyme activity as the original enzyme. Phan and Mahler[8] obtained a four-subunit enzyme from yeast, capable of catalyzing the oxidation of chtochrome c, which showed similar enzymic and spectral properties to those of the original enzyme.

In this communication, resolution of the cytochrome oxidase into three sim-

pler components which retain heme α and/or copper and their characterization
are reported.

MATERIALS AND METHODS

Purification of cytochrome oxidase and its resolution into simpler components

Cytochrome oxidase was isolated from bovine heart mitochondria according to
the method of Fowler *et al.*[9] and then purified with a hydrophobic affinity
chromatography. The isolated ammonium sulfate-fractionated enzyme (S_1), 100 mg,
was applied to phenyl-Sepharose CL-4B (Pharmacia Fine Chemicals) column (1.6 X
12.5 cm) previously equilibrated with 0.1 M bicarbonate buffer, pH 8.0, containing 0.05% phytic acid, an antioxidant. The column was washed with 100 ml of
0.1 M bicarbonate buffer, pH 8.0, containing 0.05% phytic acid and 1.0% deoxycholate (DOC). The cytochrome oxidase was then eluted with 100 ml of the
0.1 M bicarbonate buffer, pH 8.0, containing 0.05% phytic acid in which Triton
X-100 concentration was increased to 1.0% in a linear gradient. Alternatively,
the oxidase was purified by an affinity chromatography using immobilized cytochrome c, as reported previously[10]. The purified oxidase (S_2) contains 12-14
nmoles heme α/mg protein. To resolve S_2 into simpler components which retain
heme α, urea and SDS were added to 1.5 ml of S_2 (10 mg/ml) to a final concentration of 4 M and 5%, respectively, and then dissolved completely at 30°C. The
enzyme solutions were frozen at −20°C for 1 hr and then thawed at 30°C for 30
min. This freeze-thaw treatment was repeated 3 times. The urea-SDS treated
enzyme solutions thus obtained were applied to Sephadex G-75 superfine column
(1.6 X 80 cm) previously equilibrated with 0.1 M bicarbonate buffer, pH 8.0,
containing 4 M urea, 0.1% SDS and 0.05% phytic acid and then eluted with the
same buffer. The column chromatography was conducted at approximately 4°C.

Other procedures

Heme α content was determined from the difference in absorbance at 603 and
630 nm of dithionite-reduced cytochrome oxidase dissolved in 0.1 M bicarbonate
buffer, pH 8.0 containing 0.5% DOC, using the millimolar extinction coefficient
of 16.5[11]. Heme α concentration was also determined by the pyridine hemo-
chromogen method[12]. Ion and copper contents in S_2 and in the simpler components
were measured with a Hitachi 308 two wavelength atomic absorption spectrophotometer.

Digestion and SDS-polyacrylamide gel electrophoresis in SDS and urea were
carried out essentially according to the procedures of Swank and Munkres[13].

Cytochrome oxidase activity was measured by monitoring the rate of decrease
in absorbance at 550 nm[14] and also by oxygen-uptake using oxygen electrode.

Phospholipids were determined by assaying for phosphorus according to the method of Chen[15]. As for calculating the phospholipid content, it was assumed that 1 µg of lipid phosphorus was equivalent to 25 µg of phospholipid.

Protein concentration was determined according to a biuret method[16] and in some case by the method of Lowry et al.[17] with some modifications[18] using bovine serum albumin as the standard.

RESULTS

Purification of cytochrome oxidase and its separation into simpler components

Cytochrome oxidase was isolated from beef heart mitochondria according to the method of Fowler et al.[9] and then purified with a hydrophobic affinity chromatography (Fig. 1) as described in Materials and Methods. As shown in Table I, heme a content of the original enzyme (S₂) obtained after hydrophobic

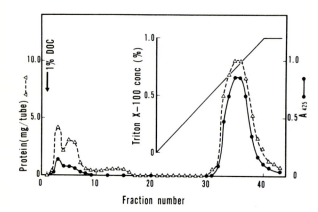

Fig. 1. Chromatography of cytochrome oxidase (S₁) on phenyl-Sepharose CL-4B. S₁, 94.5 mg, was applied to a phenyl-Sepharose 4B column. Fractions (33–38 in the number) were pooled as phenyl-Sepharose 4B-purified oxidase (S₂).

TABLE I

PURIFICATION OF CYTOCHROME OXIDASE

Fractions	Total protein (mg)	Total activity (sec^{-1}/ml)	Total heme a (nmoles)	Specific activity (sec^{-1}/mg/ml)	Heme a/ protein (nmole/mg)
S₁*	819	4341	6400	5.3[a]	7.8
S₂**	497	5467	6110	11.0[a]	12.3

*S₁: (NH₄)₂SO₄-fraction; **S₂: Phenyl-Sepharose CL-4B eluate.
[a] The activities of cytochrome oxidase were assayed in the presence of mitochondrial phospholipids (0.5 mg/mg protein).

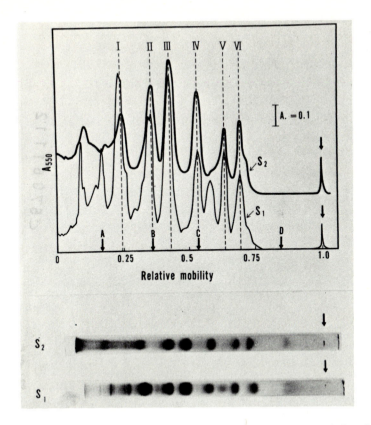

Fig. 2. Electrophoresis patterns of cytochrome oxidase in a SDS-polyacrylamide gel. The arrows indicate the position of the tracking dye. The Roman numerals mark peaks with the following molecular weight: I, 40,000; II, 25,000; III, 19,000; IV, 12,000; V, 9,000; VI, 5,000. S_1 = ammonium-sulfate-fractionated oxidase; S_2 = phenyl-Sepharose 4B-purified oxidase. A = bovine serum albumin (67,000), B = chymotrypsinogen A (25,000), C = cytochrome c (12,500) and D = insulin A chain (2,300) as standard proteins were run in parallel with cytochrome oxidase.

affinity chromatography was 12.3 nmoles per mg protein and increased 1.6 fold as compared with that of ammonium-sulfate-fractionated enzyme (S_1). The specific activity of S_2 increased in parallel with the increase in heme a content. Fig. 2 shows the typical scan of SDS-gel electropherogram of S_1 and S_2. S_2 contained six different subunits with the following molecular weights: 40,000, 25,000, 19,000, 12,000, 9,000 and 5,000. For convenience, these subunits were numbered from I to VI according to the descending molecular weights. S_1 and S_2 contained about 4.6% and 0.1% phospholipid, respectively.

S_2 was treated with 4 M urea and 5% SDS and then chromatographed on a Sephadex G-75 superfine column. The urea-SDS treated oxidase was resolved into

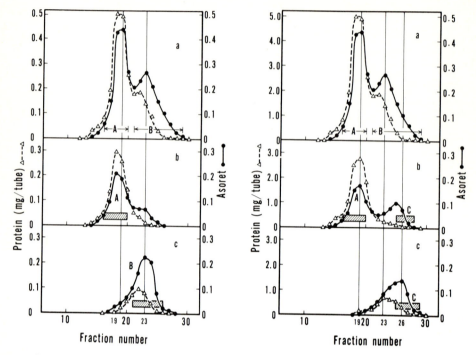

Fig. 3. Separation of cytochrome oxidase (S₂) into components A and B on Sephadex G-75 superfine column in the presence of 4 M urea and 0.1% SDS. a: A typical elution pattern of S₂ treated with 4 M urea and 5% SDS. b and c: Typical rechromatographed elution patterns of A (fraction number 16-20 in a) and B (fraction number 21-29 in a), respectively. ▨: Fractions in this regions were collected as components A and B.

Fig. 4. Isolation of component C from component B retreated with 4 M urea and 5% SDS. a: A typical elution pattern of S₂ treated with 4 M urea and 5% SDS. b and c: Typical rechromatographed elution patterns of A (fraction number 16-20 in a) and B (fraction number 21-29 in a) after re-treatment with 4 M urea and 5% SDS, respectively. ▨: Fraction in this region were collected as components A and C.

simpler components as described in Materials and Methods. Fig. 3 and 4 show typical elution patterns of Sephadex G-75 superfine chromatography. The recovery of heme a was found to be about 60% of original S₂. The fraction number 16 to 20 and 21 to 29 in Fig. 3-a were collected and rechromatographed on the same column as shown in Fig 3-b and -c. Component A and B were collected, and then component C was obtained from the component B retreated with 4 M urea and 5% SDS followed by rechromatography on Sephadex G-75 superfine in the presence of 4 M urea and 0.1% SDS (Fig. 4-c). These components were dialyzed three times to remove SDS and urea against 2,000 ml of 0.1 M bicarbonate buffer, pH 8.0, containing 0.05% DOC at 4°C for overnight.

Characterization of the simpler components

   Molecular weight and subunit composition. In order to indentify the actual
subunit composition of these components, SDS-polyacrylamide gel electrophoresis
were performed. Analyses of stained gels revealed two major bands (I and II)
and four minor bands (III-VI) for component A, four bands (III-VI) for component
B and only two bands (V and VI) for component C (Fig. 5).

   The molecular weight of components A, B and C were calculated from their
copper contents and subunits composition to be 61,000 - 65,000, 45,000 - 47,000
and 14,000, respectively. An assumption is made that the molar ratio of subunits I - VI is 1, as shown in Table II.

   Spectral properties. Characteristics of absorption spectra for three components A, B and C were distinguishable in the oxidized, reduced forms and reduced
CO-complex (the oxidized form was bubbled with CO and then reduced with sodium
dithionite) (Fig. 6), and are summarized in Table III. In the case of cytochrome oxidase (S_2), reduction by dithionite shifted the α-band of the oxidized
form from 600 to 603 nm and the Soret band from 423 to 444 nm. Upon reaction
with carbon monoxide, the spectrum of the reduced form exhibited the characteristic shoulder at 592 nm and the absorbance at 603 nm decreased slightly, whereas the Soret band shifted to 433 nm and a distinct shoulder remained at 440 nm.
For the component A, in the oxidized state, the Soret band was at 422 nm. The
reduced form of the component A showed α-band at 597 nm and the Soret band was
shifted to 440 nm accompanied by a lowering of the absorbance by about 10%.
Upon bubbling the oxidized form of the component A with CO then reducing with
dithionite, the spectrum showed a 10-20% increase in the intensity of the Soret
band with no appearance of the shoulder. The Soret band was simultaneously
shifted to 433 nm while α-band appeared at 600 nm. For the component B, in the
oxidized state, the Soret band was at 417 nm and a broad band appeared at 635
nm. The reduced form of the component B showed the Soret band at 440 nm accompanied by a lowering of the absorbance by about 35% but showed no α-band. Upon
bubbling the oxidized form of the component B with CO then reducing with dithionite, the spectrum showed a similar intensity of the Soret band with no appearance of a shoulder. The Soret band was simultaneously shifted to 432 nm and
an α-band appeared at 602 nm. For component C, characteristics of the absorption spectra for the various forms were essentially similar to those of component B.

   Localization of copper and heme a. Contents of copper and iron were measured
by atomic absorption spectrophotometer (Table IV). It was concluded that the

46

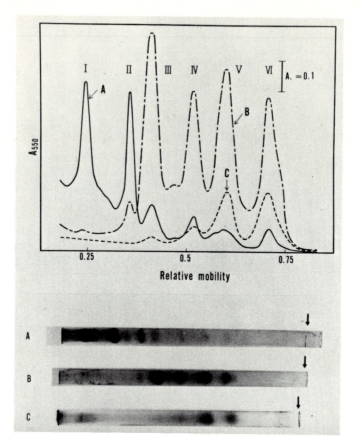

Fig. 5. Electrophoresis patterns of three simpler components of the oxidase in a SDS polyacrylamide gel. The arrows indicate the position of the tracking dye. The Roman numerals are the same as described in the legend of Figure 2. A = Component A; B = Component B; C = Component C.

TABLE II

MOLECULAR WEIGHTS OF COMPONENTS A,B AND C

	from copper contents[a]	from subunit composition[b]
Component A	61,000	65,000
Component B	47,000	45,000
Component C	---	14,000

[a] Calculated from copper contents in Table IV.
[b] Calculated from assumption that molar ratio of subunits I-VI is 1.

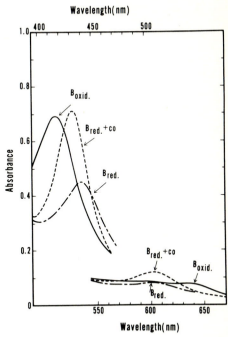

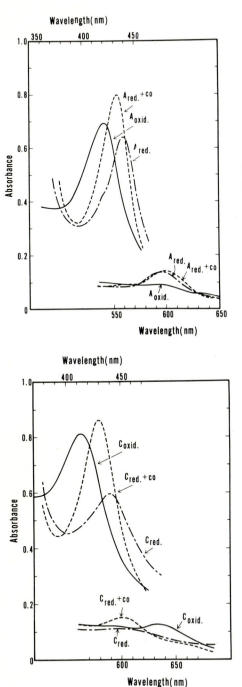

Fig. 6. Absorption spectra of the oxidized, reduced and the CO-reduced components. The components were dissolved in 0.1 M bicarbonate buffer, pH 8.0, containing 0.5% DOC. Oxid, red and red+CO represent the oxidized, reduced and CO-reduced forms, respectively. A, B and C represent components A, B and C, respectively.

TABLE III

CHARACTERISTICS OF ABSORPTION SPECTRA OF S_2 AND COMPONENTS A, B AND C

| | wavelength (nm) of peak at | | | | | | | | |
| | Soret band | | | α-band | | | other band | | |
	oxid	red	red-CO	oxid	red	red-CO	oxid	red	red-CO
S_2	423	444	433 (440)	–	603	603 (592)	–	–	–
Component A	422	440	433	–	597	600	–	–	–
Component B	417	440	432	–	–	602	635	–	–
Component C	413	440	430	–	–	602	635	–	–

"–" represents "not detectable".
Figures in parentheses represents wavelength (nm) at a shoulder.

copper was bound to both components A and B but not to component C, and that three components contained irons as heme a, because the specific copper content in component C is low suggesting copper contmination from subunits III or IV, whereas iron contents in components A, B and C are in parallel with heme a. The specific copper contents of components A and B are 16.3 and 21.2 nmoles per mg protein, respectively, and the specific iron contents of components A, B and C are 9.7, 11.8 and 24.0 nmoles per mg protein, respectively. Since S_2 contains two heme a and two copper ions per functional unit, we could conclude that subunits I and/or II (major subunits of components A) contain both copper and

TABLE IV

CONTENTS OF HEMES, Fe^{++} AND Cu^{++} IN S_2 AND COMPONENTS A, B AND C

Samples	Heme a (nmole/mg)	Fe^{++} (nmole/mg)	Cu^{++} (nmole/mg)
S_2	$12.1^a(5)^b$	11.5 (2)	12.3 (3)
Component A	5.1 (5)	9.7 (2)	16.3 (3)
Component B	9.6 (5)	11.8 (2)	21.2 (3)
Component C	17.4 (5)	24.0 (2)	8.8 (3)

[a] Mean values.
[b] Number of estimation in parentheses.

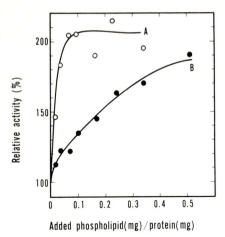

Fig. 7. Effect of phospholipids concentration on activities of Component A and B at 30°C. The activity was measured by following the rate of oxidation of reduced cytochrome c^{14} in 50 mM phosphate buffer, pH 7.0.

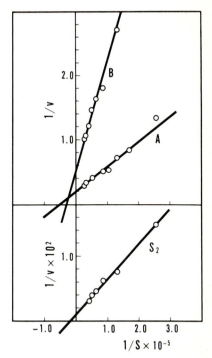

Fig. 8. Reciprocal plots of specific activities of S_2, Component A and B *versus* concentration of cytochrome c in the presence of 0.5 mg phospholipids/mg protein.

heme a, either subunit III or IV is copper-containing and either subunit V or VI is heme a-containing.

   Enzymatic properties. Both components A and B oxidized the reduced cytochrome c, but component C could not. Their activities were also demonstrated by oxygen-uptake. The optimum pH for S_2 was found to be 6.0 - 6.5, whereas that for component A or B, 4.0 - 4.5. Cytochrome c oxidase activities of components A and B in the presence of phospholipids are presented in Figure 7. The activation pattern of component A was different from that of component B. About 1.8% phospholipids was retained in component B, but was absent in component A. The Km values of S_2, components A and B are 50, 19 and 34 μM, respectively, as shown in Figure 8. KCN and NaN_3 inhibited the activities of S_2, components A and B. However, the concentrations of these inhibitors which gave 50% inhibition were markedly different: about 5μM of KCN for S_2 and component A, but about 5mM for component B; about 50 μM of NaN_3 for S_2 and component A, but 50 mM for component B, as shown in Figure 9. On the other hand, component B was more suscetible to CO than component A, as shown in Table V.

DISCUSSION

   We purified cytochrome oxidase by a hydrophobic ligand (Fig. 1,2 and Table I).

50

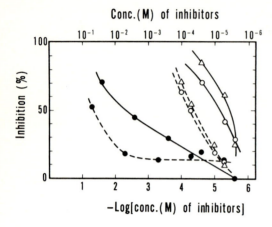

Conc.(M) of inhibitors

Fig. 9. Inhibition of KCN and NaN₃ on the activities of S₂, Component A and B. The reaction was started by the addition of 39 nmoles of reduced cytochrome c in the reaction mixture containing the enzyme sample, phospholipids and inhibitor at pH 7.0, 30°C. $\triangle$ = S₂, $\bigcirc$ = Component A, $\bullet$ = Component B, ——— = KCN ·········· = NaN₃.

TABLE V

INHIBITION (%) OF ACTIVITIES OF COMPONENT A AND B BY CARBON MONOOXIDE

	Component A	Component B
Air	0	0
CO	26	75

The activity of each component was measured as described in Fig. 7, except the sample solutions bubbled by air or CO at 25°C for 30 min. before assey.

We obtained three simpler components A, B and C from the purified cytochrome oxidase, S₂, upon treatment with 4 M urea and 5% SDS followed by gel filtration on Sephadex G-75 superfine (Figs. 3 and 4). All three components retained heme a. The possibility that these components retaining heme a as a result of denaturation of the native subunits and then reassociation of copper and heme a with another subunit which does not contain them originally could be eliminated, since, in Figures 3-b and -c, rechromatography of components A and B without retreatment did not cause a shift of the peak of heme a elution from the column. In those samples treated with urea and SDS (Fig. 4-b, -c), rechromatography of component A also did not cause the shift. Moreover, retention of activities of components A and B after removal of urea and SDS by dialysis indicated that heme a and copper are functionally bound to the native subunits. Treatment of S₂ with 4 M urea and 5% SDS led to the dissociation of cytochrome oxidase into component A containing larger subunits I and II, component B containing smaller subunits III – VI, and component C containing the smallest subunits V and VI (Fig. 5). Their molecular weights were calculated to be in thousand 61 – 65, 45 – 47 and 14, respectively (Table II). We are interested in the possible

reconstitution of component A with component B, since in the case of yeast
cytochrome oxidase, the three larger subunits, I - III, are quite hydrophobic
and are synthesized on mitochondrial ribosomes[19].

Judging from absorption spectral properties (Fig. 6 and Table III), it is
reasonable to conclude that heme a in component A is spectroscopically distinguishable from that in component B or C.   The difference of two hemes in the
oxidase has been established by the data of EPR[20], MCD spectra[21] and magnetic
susceptibility[22], whereas chemical extraction of the porphyrin yields a single
type of heme a[23].   Analyses of iron and copper contents in the three components
(Table IV) demonstrated that either subunit I or II is a heme and/or copper
containing subunit.   Either subunit III or IV is a copper-containing subunit,
and either subunit V or VI is heme containing.   Location of copper on subunit
II has been reported by MacLennan and Tzagoloff[1], Tanaka et $al.$[2] and Yu and Yu[24].
Bisson et $al.$[25] reported a photoaffinity labelling of cytochrome c with the II.

Both components A and B could oxidize reduced cytochrome c and then transfer
electrons to oxygen.   However, it was found that there were definite differences between Component A and B in enzymic properties such as effect of mitochondrial phospholipids, Km values for cytochrome c and inhibitory effects of
KCN, NaN$_3$ and CO (Fig. 7,8,9 and Table V), indicating that the oxidase consists
of two different catalytic components, A and B, having two different cytochrome
c binding sites.   Recently, Ferguson-Miller et $al.$[26] and Errede et $al.$[27] have
revealed two binding sites for cytochrome c by steady state kinetic studies.

We have for the first time isolated and characterized two different catalytic
entities as Component A and B from cytochrome oxidase.   However, up to now, we
cannot get any clear evidence for the existence of chemically different heme a.
Thus, our tentative conclusions are as follows: Keilin and Hartree's assumption[28] that the oxidase consists of two components, a_3 and a, and the component
a_3 differs from the component a by combining in the ferric state with cyanide,
hydrosulphide and azide is well documented by the isolation of Component A and
B, approximately corresponding to the component a_3 and a.   On the other hand,
Okunuki's contention[29] that there is no essential difference between two hemes,
but microenvironmental differences decide heme's susceptibility to the respiratory inhibitors is also supported by the demonstration that cytochrome oxidase
is a heterogeneous dimer with respect to the apo-enzyme protein.

REFERENCES

1. Maclennan, D. H. and Tzagoloff, A. (1965) Biochim. Biophys. Acta, 96, 166-
   168.
2. Tanaka, M., Haniu, M., Zeitlin, S., Yasunobu, K. T., Yu, C. A., Yu, L. and

  King, T. E. (1975) Biochem. Biophys. Res. Commun., 66, 357–367.

3. Schatz, G., Groot, G. S. P., Mason, T., Rouslin, W., Wharton, D. C. and
    Saltzgaber J., (1972) Fed. Proc., 31, 21–29.

4. Yu, C. A., Yu, L. and King, T. E. (1977) Biochem. Biophys. Res. Commun.,
    74, 670–676.

5. Komai, H. and Capaldi, R. A. (1973) FEBS Lett., 30, 273–276.

6. Yamamoto, T. and Okunuki, T. (1972) J. Biochem., 71, 435–445.

7. Yamamoto, T. and Orii, Y. (1974) J. Biochem., 75, 1081–1089.

8. Phan, S. H. and Mahler, H. R. (1976) J. Biol. Chem., 251, 270–276.

9. Fowler, L. R., Richerdson, S. H. and Hatefi, Y. (1962) Biochem. Biophys.
    Acta, 64, 170–173.

10. Ozawa, T., Okumura, M. and Yagi, K. (1975) Biochem. Biophys. Res. Commun.,
    65, 1102–1107.

11. Yonetani, T. (1961) J. Biol. Chem., 236, 1680–1688.

12. Morrison, M. and Horie, S. (1965) Anal. Biochem., 12, 77–82.

13. Swank, R. T. and Munkres, K. D. (1971) Anal. Biochem., 39, 462–477.

14. Smith, L., Glick, D. (1954) in Methods of Biochemical Analysis Vol. II,
    Glick, D. ed., Wiley (Interscience) New York, pp. 427–434.

15. Chen, P. S. Jr., Toribara, T. Y. and Warner, H. (1965) Anal. Chem., 28,
    1756–1758.

16. Gornall, A. G., Bardwill, C. J. and David, M. M. (1949) J. Biol. Chem., 177,
    751–766.

17. Lowry, O. M., Rosebrough, N. J., Farr, A. L. and Randall, R. J. (1951)
    J. Biol. Chem., 193, 265–275.

18. Smith, R. L. and Wang, C. S. (1975) Anal. Biochem., 63, 414–417.

19. Schatz, G. and Mason, T. L. (1974) Annu. Rev. Biochem., 43, 51–87.

20. Nicholls, P. and Chance, B. (1974) in Molecular mechanisms of Oxygen Activa-
    tion, Hayaishi, O. ed., Academic Press, New York, N. Y., pp. 479–534.

21. Babock, G. E., Vickery, L. E. and Palmer, G. (1976) J. Biol. Chem., 251,
    7904–7919.

22. Falk, K. E., Vanngård, T. and Angstrom, J. (1978) FEBS Letters, in press.

23. Caughey, W. S., Smythe, G. A., O'Keeffe, D. H., Maskasky, J. and Smith, M.
    L. (1975) J. Biol. Chem., 250, 7602–7622.

24. Yu, C. A. and Yu, L. (1977) Biochim. Biophys. Acta, 495, 248–259

25. Bisson, R., Azzi, A., Gutweninger, H., Colonna, R., Montecucco, C. and
    Zanotti, A. (1978) J. Biol. Chem., 253, 1874–1880.

26. Ferguson-Miller, S., Brautigan, D. L. and Margoliash, E. (1976) J. Biol.
    Chem., 251, 1104–1115.

27. Errede, B., Haight, G. P. Jr. and Kamen, M. D. (1976) Proc. Nat. Acad. Sci.,
    USA, 73, 113–117.

28. Keilin, D. and Hartree, E. F. (1939) Proc. Roy. Soc., B 127, 167–191.

29. Okunuki, K. (1966) in Comprehensive Biochemistry, Biological Oxidations,
    Florin, M. and Stotz, E. eds., Elsevier, Vol 14, pp. 232–308.

Cytochrome Oxidase, T.E. King et al. eds.
© *1979 Elsevier/North-Holland Biomedical Press*

SUBUNITS OF CYTOCHROME OXIDASE AND THEIR LARGE SCALE PREPARATIONS*

TSOO E. KING, LINDA YU, CHANG-AN YU, AND YAU-HUEI WEI
Laboratory of Bioenergetics, State University of New York, Albany, N.Y. 12222
U.S.A.

ABSTRACT

The preparation of "soluble" cytochrome oxidase dates back as early as 1940
by Okunuki *et al.* and Straub independently, but the studies of the subunit structure of this enzyme became vigorous only recently. A brief review is made about
the number and pitfalls using this technique to cause greatly disparate results.
Our laboratory has found that highly purified active oxidase contains seven subunits with a molar stoichiometry of unity as some others do. The molecular
weights of those subunits are estimated by us to be 40 (subunit I), 21 (II),
14.8 (III), 13.5 (IV), 11.6 (V), 9.5 (VI), and 7.6 (VII) thousands, using the
method of Osborn and Weber under our conditions. It must be noted that molecular weights determined by this method are difficult, if not impossible, to compare with those from other laboratories or by other methods. The final structural elucidation and the subunit molecular weights will not come before the
availability of amino acid sequence of each subunit. For that reason, we have
developed methods in isolation of quantities of subunits; more than enough for
conventional, accurate sequencing. Five subunits III, IV, V, VI, and VII
were purified without using SDS. Three subunits I, II, and III were obtained
by molecular sieving.

INTRODUCTION

Cytochrome oxidase was first reported by Keilin and Hartree in 1938 and solubilized by Okunuki *et al.*, and Straub independently[1,2]. From the techniques
available to Keilin and Hartree[1] at that time they called the component which
reacts with oxygen or carbon monoxide, a_3 and the non-reactive component, cytochrome a. In the last forty years investigators have failed to separate these
two hemoproteins. On the other hand, the subunit polypeptides have been cleaved
and separated mostly by analytical means. Whether this kind of separation of
subunits would minimize the polemics of the one *versus* two cytochrome theory depends upon the individual's view and would not affect the significance of the
original discovery nor the meaning of the difference of a and a_3.

*Abbreviations used in the text: SDS, sodium dodecyl sulfate

SUBUNITS OF CYTOCHROME OXIDASE

Since the wide use of polyacrylamide gel electrophoresis in the beginning of the nineteen seventies, the subunits of cytochrome oxidase have been the subject of intensive investigations. Consequently, numerous reports have appeared in the literature; some of the results are in good accord, whereas others greatly disparate. This is not surprising because gel electrophoresis was primarily developed[3,4] for non-membranous or soluble proteins. Essentially, this type of analytical method relies on two steps. The first process involves denaturation of the protein, its dissociation into subunits and cleavage of disulfide cross linkages. The originally recommended methods for dissociation and cleavage utilize aqueous SDS, β-mercaptoethanol and possibly another denaturating agent, at boiling water temperature for one to several minutes. Each of the dissociated subunits is supposedly bound with a constant amount of SDS per unit weight of polypeptide[5,6].

Secondly, these subunits are subjected to the so-called polyacrylamide electrophoresis and the separation is dependent upon the molecular weights. The latter is determined by comparing the mobilities on the gel to the mobilities of single polypeptide proteins of known molecular weights. It must be emphasized that the mobility (*i.e.*, the fraction of the distance the polypeptide travels on the gel over the distance travelled by a small molecular dye) is generally a linear function of the *logarithm* of molecular weight. For precaution the Ferguson plot[7] is employed to ascertain no abnormalities. Since the first publications[3,4], several modified methods have appeared, two of which are equally widely employed[8,9] (see also references cited in Table I). These methods differ in composition of dissociation media as well as in electrophoretic systems. Several critical papers have appeared both on methodological and theoretical aspects[10,11], discussing the advantages, precautions, and pitfalls of the methods in general. The polyacrylamide gel method has also been extended to a preparative scale (*cf*. System 16 A, B, Table I for the references, in this case actually a gel slab of about 1.0 x 9 x 20 cm was used with a few mg of cytochrome oxidase subunits) and to gel filtration without application of electric potentials.

We have collected some representative results on subunit studies of cytochrome oxidase in Table I. This was prepared not for the purpose of completeness but rather to see how various results by different investigators agree and deviate. Special attention is paid to the starting material and method of primary solubilization used for the purification of the oxidase. The conditions of dissociation of the enzyme and the actual electrophoresis are naturally important. This table purposely does not include the purity of the preparation

55

TABLE I

SUBUNITS OF CYTOCHROME OXIDASE REPORTED BY SOME INVESTIGATORS[a]

System (ref.)	Origin/ Methods	I / V	II / VI	III / VII	IV / VIII	Conditions — Dissociation	Conditions — PAGE or GF
1A (22)	BHM/DOC, cholate, Triton	38.0 / 6.0	19.0 / 8.6	25.0 / 4.3	13.8 / --	3% SDS, 10 mM ME, 37°, 2 hr	PAGE (10/0.27), O&W
1B (22)		33.0 / 12.5	21.3 / 8.5	19.0 / 4.9	14.0 / --		PAGE (10/0.38), FSW
1C (22)		35.3 / 12.1	25.2 / 6.7	21.0 / 3.4	16.2 / --		PAGE (10/0.34), S&M
2A (31)	BHM/DOC, cholate, Triton	36.0 / 12.5	22.5 / 11.2	-- / 8.0	17.2 / --	>2% SDS, 8M urea, 40 mM DTE, 10 mM Pi, pH 8.3	PAGE (12/0.3), 0.2 M TA, pH 6.4, 0.1% SDS
2B (31)		35.3 / 12.1	25.2 / 6.7	21.0 / 3.4	16.2 / --		PAGE (12.5/1.2), S&W
3A (32)	BHM/ Triton, cholate, DOC	36.0 / 9.7	22.5 / 5.3	17.1 / --	12.5 / --	Same as (2A)	Same as (2A)
3B (32)		35.0 / --	23.0 / --	-- / --	-- / --	Prot. 17 mg/ml in 10% SDS, 1 mM DTE	GF, Sephadex G-200, 10 mM SDS, 0.1 M Pi, pH 8.0
3C (32)		-- / 12.5	-- / 9.7	-- / 5.3	17.0 / --	IAA and 8 M Gd treated[b]	GF, Sephadex G-25, 6 M Gd, 0.5 mM DTE, 10 mM TC, pH 6.0
4A (33)	BHM/DOC + cholate	[36.0] / 14.0	-- / [12.8]	-- / 11.5	19.0 / [10.0]	3% SDS, 5 mM ME, 100°, 1 min	PAGE (10/?), 3% SDS, 5 mM ME, FSW
4B (33)	Purer 4A	[36.0] / 14.0	-- / [12.8]	-- / 11.5	[19.0] / [10.0]		Same as 4A
4C (34)	Purer 4A	-- / 14.0	-- / --	-- / 11.5	-- / --		PAGE (10/0.34) 0.1% SDS, 0.1 M Pi, pH 7.2
5 (27)	HMP/ cholate	40.0 / 11.6	21.0 / 9.5	14.8 / 7.6	13.5 / --	3 mg SDS/mg prot. 1% ME; 38°, 2 hr	PAGE (10/0.6), W&O
6 (12)	HMP/ cholate, emasol	39.6 / 11.7	20.7 / two diff. bands	15.2 /	13.4 /	Prot. 5 mg/ml, 1-5% SDS, 1% ME (4 M urea), 40°, 2 hr	PAGE (10/0.3), W&O
7A (35)	Various prep.	37.0[c] / 15.5	26.0 / 13.5	21.0 / 10.0	18.0 / 4.5	100 µg prot., 1% SDS, 1% ME, 8 M urea, 10 mM PO4-T pH 6.9, 70°, 15 min	PAGE (8/0.27) 1% SDS, S&W
7B (35)	Various prep.	42.0[c] / 13.0	24.0 / 13.0	18.0 / 15.5	17.5 / 14.5	Without urea	PAGE (10/0.27) W&O

8A (21)	HMP/ cholate	47.5/14.5	--/13.0	20.4/11.0	14.5/--	<0.1 mg prot. in 10 mM TC, pH 7.2, 3% SDS, 10% glycerol, 5% ME, 100°, 2 min	PAGE (8-16/0.27-0.54), W&O or FSW
8B (21)	YKHP/ cholate	42.4/14.6	34.1/12.3	24.7/10.6	14.6/--	Same as (8A)[d]	
9 (36, 37)	BH/ cholate	45.0/12.5	25.0/10.5	25.0/8.5[e]	16.0/--	Prot. in 0.068 M TC-2% SDS, pH 6.8 or IAA treated[f]	PAGE (15/0.4), 25 mM Tris-glycine, pH 8.8
10A (38)	BHM/DOC, in 0.05% Triton	34.5/7.3	19.0/--	13.5/--	9.8/--	1 mg prot. in 1% SDS, 1% ME, 10 mM Pi, pH 7.0, 10% glycerol, 0.004% bromphenol blue, 70°, 20 min	PAGE (7.5/0.2), W&O
		40.0/9.8	22.5/7.3	15.0/--	11.2/--		PAGE (10/0.19), W&O
10B (38)	YM/DOC	35.0/12.0	24.5/9.5	22.0/--	14.5/--		PAGE (7.5/0.2), W&O
		40.0/13.0	27.3/10.2	25.0/9.5	13.8/--		PAGE (10/0.19) W&O
11 (39)	BH/ cholate	37.0/10.0(4 bands)	19.0	14.0	14.0	Same as W&O	PAGE (10/0.27), W&O
12 (40)	YM/ cholate	43.0/12.0	34.0/12.0	24.0/4.5	14.0/--	Dialyzed in 6 M Gd at ~25°, 9 M Gd at 80°	PAGE 20% and 10% in lower and upper layer
13 (41)	YM/ cholate	(A) 37.5/13.8	33.6/12.7	24.2/5.2	14.6/--	Prot., 0.5 mg in 10 mM Pi, pH 7, 1% SDS 1% ME, 10% sucrose, 100°, 2 min[g]	PAGE (15/0.47)[h]
		(B) 41.0/13.4	32.7/12.4	23.9/4.6	14.1/--		PAGE (14/0.78)
		(C) 39.8/12.4	32.8/12.4	23.1/4.6	14.3/--		PAGE (12/0.375)
		(D) 41.6/13.1	28.3/13.1	21.7/4.5	14.7/--		GF, BioGel A 5M, 6 M Gd, 20 mM Pi, pH 6
		(E) 36.9/--	33.8/10.3	--/--	16.4/--		Amino acid analysis
14 (42)	Same as 13	42.0/12.5	34.5/9.5	23.0/--	14.0/--	2.25% SDS, as Ref.	PAGE (12/0.375), as Ref. 49
15 (43)	YM (Can- dida ut- lis)/ cholate	49.0/13.5	32.0/8.0	28.0/--	20.0/--	2% SDS at 100°, 2 min	PAGE (12/0.375), W&O
		46.0/12.5	35.0/7.8	23.0/--	19.0/--	2.5 mg prot., 5% SDS at 100°, 3 min	GF, Sephadex G-200, 50 mM TC, pH 7.5, 1 M NaCl, 0.5% SDS

57

No. (Ref)	Source/Prep					Conditions	Method
16A (44, 45)	N. crassa/ SMP, DOC	41.0 / 19.0	31.0 / 15.0	28.0 / 11.0	21.0 / --	5% SDS, 2% ME	PAGE (15/?), 1% SDS, 0.1 M TC, pH 8.0
16B (44, 45)	Mutant of (16A)[i]	-- / 14.0	-- / 12.0	-- / [10.0]	[16.0] / --		Same as (16A)
17 (20)	RLM/ Triton-114	38.0 / 12.4	24.5 / 10.3	20.5 / 8.5[j]	14.4 / --	1% SDS, 5% ME, 25°	PAGE (10/0.27), W&O
18 (46)	EBM/DOC	45.0 / 13.0	29.0 / 10.5	20.0 / --	17.0 / --		PAGE (12.5/0.31) Laemmli method Ref. 48
19 (47)	Pseudo-monas/ oxidase	63.0 / --	63.0 / --	-- / --	-- / --	Succinylation or pH < 4 or pH > 11	Hydrodynamic method (confirmed by SDS-PAGE

[a] We request the reader to correct, and forgive our inevitable errors due to the search of voluminous literature.

Abbreviations used: BH, beef heart; BHM, beef heart mitochondria; diff., diffused band; DOC, deoxycholate; DTE, dithioerythritol; FSW, S&M and W&O refer to methods listed in Refs. 8, 9 and 4, respectively; EBM, embryonic bovine trachea mitochondria; Gd, guanidine-HCl; GF, gel filtration method; HMP, the Keilin-Hartree heart muscle preparation, a kind of submitochondrial particle; IAA, iodoacetamide; ME, β-mercaptoethanol; NEM, N-ethyl maleimide; PAGE, polyacrylamide gel electrophoresis; (xx/yy), indicates % concentration of acrylamide and N,N'-methylenbisacrylamide in gel; PO_4-T, phosphate buffer, adjusted by Tris; prot., protein; RLM, rat liver mitochondria; SDS, sodium dodecyl sulfate; SMP, submitochondrial particles; TC, Tris-HCl; Triton, Triton X-100; YKHP, yeast Keilin-Hartree particle; YM, yeast mitochondria.

[b] [ ], in trace amount or minor bands. 50 mg oxidase dissolved in (2 mg/ml) 8 M Gd, 50 mM Tris-HCl, 50 mM DTE, and incubated at 50° for 4 hr. Then reacted with 0.2 M IAA, 20 min in dark, adjusted to pH 6.0, concentrated to 0.5-1.0 ml.

[c] Mol weights were calculated from re-electrophoresis of isolated subunits by system (7A) in the presence and absence of urea.

[d] Further purification in 8A removes 47.5 band, that in 8B removes 42.4 and 34.1 bands to give total mol wt (hydrodynamic) 124×10^3 and 170×10^3, respectively. "Recycling exclusion chromatography" on "leucine column" in the presence of 0.1% SDS, yields an [electron transport] active enzyme from 8B of mol wt 107×10^3 devoid of Bands I to III.

[e] Authors stated cytochrome oxidase "prepared in this lab consistently present... 10 [bands]. At pH 7.0 only 2 (Bands III and VIa) are labelled by NEM" (Ref. 37), Band V consists of 2 bands and VI consists of 3 bands.

[f] Ether-water extracted cytochrome oxidase dissolved in Tris-SDS, pH 6.8, reduced by 10 mM ME, alkylated with 40 mM IAA.

[g] For gel filtration, cytochrome oxidase was reacted with 6 M Gd at 25°, 4 hr or 8 M Gd at 50° for 4 hr and then treated with IAA.

[h] Double gel layers; low gel 7 cm, concentration as indicated; upper gel, 3 cm with concentration of (10/0.31). The electrophoresis was run with discontinued buffer as Ref. 11 system A.

[i] Cytochrome oxidase deficient mi-3 cytoplasma mutant of Neurospora crassa.

[j] Becomes 3 bands in the presence of 8 M urea during electrophoresis.

which is usually expressed in nmol of heme/mg of protein. It is sometimes almost meaningless to list this kind of data because various authors use different absorption coefficients to determine the purity as pointed out previously[12]; such as, commonly used absorbance index expressed in A^{red} (605) $-$ A^{ox} (605) varying from 7.6 to 12.5 cm^{-1} mM^{-1} (cɟ. Ref. 12).

Before the wide use of gel electrophoresis on the study of subunits, cytochrome oxidase has already been claimed to have been separated into smaller units[13-18]. Apparently these workers did not reach to the stage of single polypeptides. From Table I, one can see great variations obtained by different investigators. In addition to those important cautions pointed out by Weber and Osborn[10], the subunits and number of bands from electrophoresis of cytochrome oxidase are especially subject to errors, particularly in computation of their molecular weights. The earlier investigators in cytochrome oxidase obtained, by and large, a fewer number of bands. These facts are obviously due to the hydrophobic nature of some subunits and sensitivity to the composition of both dissociation and electrophoresis media as well as cross linkages of the gel.

For molecular weight determination, the subunits are compared with proteins of known molecular weights which are usually quite hydrophilic and the use of logarithm function is always not very sensitive. The polymerization of some subunits can introduce great errors. The original recommendations[3,4,8,9] for dissociation by boiling or treatment at high temperature even for a few minutes in SDS is enough to form aggregates of larger subunits of cytochrome oxidase (e.g. Refs. 19, 20). In our experience, relatively pure subunit I (about 40 x 10^3) cannot be subjected to further electrophoresis or gel filtration without aggregation. Most of the aggregates cannot even penetrate into the gel. Investigators who report smaller numbers of subunits might have overlooked the aggregates on the top of the column impenetrable to the gel. Whether this fact can also explain the 4 bands of the cytochrome oxidase purified by "reverse cycle column" from beef heart and yeast[21] remains to be seen. However, we suspect further purification of B_7 to B_5, or Y_7 to Y_5[21] may be due to similar causes.

Finally it should be mentioned that the so-called 2-dimensional gel electrophoresis[22] involves first conducting the usual analytical method of Weber and Osborn[4]. At the end of the experiment the gel columns are expelled and then laid on gel slabs prepared according to Swank and Munkres[8] or Fairbanks et al.[9]. Subsequently electrophoresis is performed for the second time. By this method, it has been found that the order of molecular weights of several subunits of cytochrome oxidase are reversed using Swank and Munkres' method as the basis of numbering (cɟ. Fig. 2 of Ref. 22). If any of these 3 methods is used alone,

this is not the case[19] as claimed[22]. That is the order of molecular weights
remains the same although inevitably small changes occur due to the different
systems involved. This phenomenon can be easily demonstrated by adding pure
subunits to a dissociated mixture of cytochrome oxidase then the measurement of
the enhanced intensities of the color of various bands is made[19].

Other aspects of the table are self explanatory. Workers in the field now
generally believe cytochrome oxidase prepared by currently available methods
from several species contains 6 or 7 subunits. It is our opinion that comparison of the molecular weights obtained from different laboratories is an unre-
warding task. One of the reliable means to obtain true molecular weights must
depend on the unraveling of the complete sequences of these subunits.

LARGE SCALE PREPARATION OF SUBUNITS

In order to conduct sequence studies, sufficient samples are a prerequisite.
Avoiding the use of sodium dodecyl sulfate is much preferred because SDS is,
indeed, a "nasty" compound in the sequence studies. We have succeeded in preparing five subunits in gram quantities suitable for sequence work without em-
ploying SDS at all. Unfortunately, at present, we cannot eliminate the use of
this detergent for the other two. The methods for large scale preparations of
subunits III, IV, V, VI, and VII are summarized in Chart I.

The separation of the 3 larger subunits is accomplished also from lipid-
depleted cytochrome oxidase. The enzyme is dissociated in a medium containing
3 mg SDS/mg protein, 4 M urea and 1% β-mercaptoethanol at a protein concentration of approximately 10 mg/ml, at 55° for 3 hours prior to applying to a Bio-
Gel P-150 column (6 x 70 cm). The column is eluted with 0.5% SDS, 50 mM phosphate buffer, pH 7.4 at room temperature. Caution must be exercised for subunit I since in a relatively pure form it easily aggregates; thus, we always
sacrifice the yield by collecting a very narrow range of fractions from column
eluate. So far we have obtained all 7 subunits which are pure (one band) on
analytical gel columns and their molecular weights are the same as those listed
in System 5 of Table I. It should be emphasized that we are not the only ones
to separate pure subunits from cytochrome oxidase, especially preparations from
yeast; several investigators did it before by other methods.

ARE ALL SEVEN SUBUNITS NEEDED FOR CYTOCHROME OXIDASE FUNCTIONS?

The literature has reported the number of subunits of cytochrome oxidase
ranging from 2 to more than 10; some results are listed in Table I. An obvious
question will be asked: Are all these subunits required to perform the cytochrome oxidase function *in vivo*? The answer cannot be unequivocally provided at

CHART I

ISOLATION OF PURE SUBUNITS III, IV, V, VI, AND VII

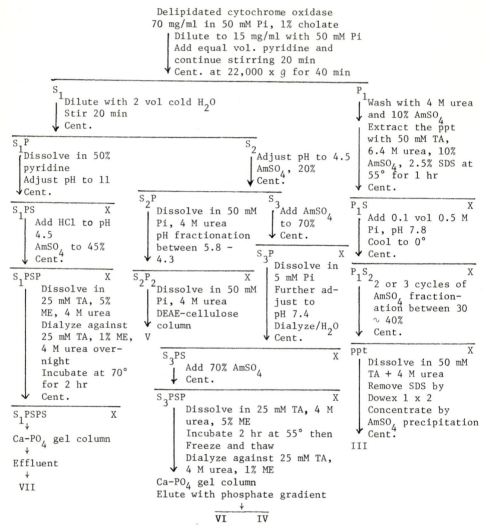

Delipidated cytochrome oxidase
70 mg/ml in 50 mM Pi, 1% cholate
| Dilute to 15 mg/ml with 50 mM Pi
| Add equal vol. pyridine and
| continue stirring 20 min
↓ Cent. at 22,000 x g for 40 min

S_1
  Dilute with 2 vol cold H_2O
  Stir 20 min
  Cent.

P_1
  Wash with 4 M urea
  and 10% AmSO$_4$
  Extract the ppt
  with 50 mM TA,
  6.4 M urea, 10%
  AmSO$_4$, 2.5% SDS at
  55° for 1 hr
  Cent.

S_1P | S_2
  Dissolve in 50%
  pyridine
  Adjust pH to 11
  Cent.

S_2
  Adjust pH to 4.5
  AmSO$_4$, 20%
  Cent.

P_1S X
  Add 0.1 vol 0.5 M
  Pi, pH 7.8
  Cool to 0°
  Cent.

S_1PS X
  Add HCl to pH
  4.5
  AmSO$_4$ to 45%
  Cent.

S_2P | S_3
  Dissolve in 50 mM
  Pi, 4 M urea
  pH fractionation
  between 5.8 –
  4.3

  Add AmSO$_4$
  to 70%
  Cent.

P_1S_2 X
  2 or 3 cycles of
  AmSO$_4$ fraction-
  ation between 30
  ∼ 40%
  Cent.

S_1PSP X
  Dissolve in
  25 mM TA, 5%
  ME, 4 M urea
  Dialyze against
  25 mM TA, 1% ME,
  4 M urea over-
  night
  Incubate at 70°
  for 2 hr
  Cent.

S_2P_2 X
  Dissolve in 50 mM
  Pi, 4 M urea
  DEAE-cellulose
  column
            V

S_3P X
  Dissolve in
  5 mM Pi
  Further ad-
  just to
  pH 7.4
  Dialyze/H_2O
  Cent.

ppt   X
  Dissolve in 50 mM
  TA + 4 M urea
  Remove SDS by
  Dowex 1 x 2
  Concentrate by
  AmSO$_4$ precipitation
  Cent.

            III

S_1PSPS X
  ↓
Ca-PO$_4$ gel column
  ↓
Effluent
  ↓
VII

S_3PS X
  Add 70% AmSO$_4$
  Cent.

S_3PSP X
  Dissolve in 25 mM TA, 4 M
  urea, 5% ME
  Incubate 2 hr at 55° then
  Freeze and thaw
  Dialyze against 25 mM TA,
  4 M urea, 1% ME
Ca-PO$_4$ gel column
Elute with phosphate gradient
                  ↓
            VI   IV

Abbreviations: AmSO$_4$, ammonium sulfate, the % refers to the saturation of
AmSO$_4$; Ca-PO$_4$, calcium phosphate gel; Cent., centrifuge at 22,000 x g for 20
min unless specified otherwise; ME, β-mercaptoethanol; Pi, phosphate buffer,
pH 7.4 unless specified otherwise; TA, Tris-acetate buffer, pH 7.8. Symbols
such as S_1, P_1S, PSP, etc. are our codes. All operations at 0-4° except as
noted. X symbolizes "discarded."

present. Most investigators have correlated these subunits generally only with electron transport activity and some even do not do this determination. We pointed out previously[23] that cytochrome oxidase functions more than as a respiratory carrier from cytochrome c to oxygen. However, no investigator, to our knowledge, has paid systematic attention to functions other than electron transfer while working on subunits. On the other hand, we believe, so far, that those who have reported less than 7 may be dealing with artifacts, whereas more than 7 could be due to proteolytic action as emphasized by Weber and Osborn, or due to low purity of the oxidase. *This interpretation, however, should not be construed as our belief that 7 is the number of subunits absolutely required.*

HEME a SUBUNITS AND COPPER SUBUNITS

The question concerning which subunits attach to heme a and which to copper, again cannot be answered definitively. However, the results, so far obtained, are not as gloomy as the essentiality of all subunits. We do not intend to review the literature but will present only our own argument briefly here. More detailed discourse and discussion will be published elsewhere.

When the oxidase is chromatographed in the presence of SDS without β-mercaptoethanol treatment, two fractions contain most of the copper. One is subunit I and the other is II. The latter fraction contains more than 80% subunit II and the impurity is mostly subunit III. The copper content of this fraction is 40 nmol per mg protein. If the copper was bound to subunit III, then the copper content of III would have been 200 nmol per mg protein, equivalent to a molecular weight of 5,000. This is a very unlikely event. Further purification in the *presence* of β-mercaptoethanol unfortunately results in the loss of copper. We believe that subunit II is a copper-containing polypeptide, although we do not wish at present to speculate whether this is a "visible" or "invisible" copper. Dr. B. Chance kindly informed us that his evidence would indicate this copper to be, indeed, "invisible" copper in order to fit his and co-workers' results from their recent experiments from X-ray absorption edge and EXAFS spectroscopy (see also the penultimate paragraph of the section).

When the lipid-free oxidase, after SDS treatment in the absence of β-mercaptoenthanol, is chromatographed, heme a attached to polypeptides is concentrated in two fractions. One is located at a position corresponding to subunit I and the other to V. Subunit I also possesses a high content of copper. All other fractions contain very little heme or copper. Even in the absence of β-mercaptoethanol but in the presence of SDS, some heme is released and slowly travels to the end of the gel column. The free heme a and the polypeptide associated heme a show different spectra; the free heme a eluted from the column exhibits the same spectra as that extracted by acid acetone from oxidase[24]. The free heme released is increased by treatment involving β-mercapto-ethanol. By a completely different procedure[25] without using SDS or the thiol, or other detergents except cholate, subunit V has been purified to a stage suitable for amino acid sequence work. Indeed, the sequence has been reported[26]. This subunit contains 40 nmol heme/mg tightly bound to the protein. The heme content, which is lower than the calculated value based on the molecular weight is apparently due to considerable loss of heme during purification, even though thiol or SDS is not employed. The amino acid composition of the fraction thus isolated is practically identical with that determined from Band V obtained directly by polyacrylamide gel electrophoresis[27].

Subunit I is most probably a polypeptide with heme-copper complex as a prosthetic group. However, subunit I is very hydrophobic and easily aggregates to such a size that it cannot be further gel-electrophoresized. The removal of copper by a thiol, such as β-mercaptoethanol, has been known for several decades. Apparently, the coordination or link of copper, similar to nonheme iron, by amino acid residues is not as strong as iron coordinated with prophyrin or chlorin. Moreover, evidence[19] seems to indicate that the disulfide linkages existing in the oxidase not only serve as cross links between subunits but may also play a role involving heme and copper. Moreover, there is no definitive evidence to rule out that the released heme or copper would not recombine with different polypeptides to different degrees. Nevertheless, no indications have been obtained that the other 5 subunits contain prosthetic groups.

In summary, the evidence from our experiments indicates that subunits I, II, and V contain prosthetic groups heme + copper, copper, and heme, respectively. Regardless of how plausible the above argument is, more direct proof must be searched and it may be a hard task.

The topology or arrangement of these subunits in the inner membrane of mitochondria has been studied[28]. If the claim of Frey et $al.$[29] can be applied to the inner membrane in intact mitochondria, then subunits II and III face the outer or C side. Does that not mean subunit III belongs to cytochrome a_3 and subunit II to invisible copper? It has been observed by several authors ($e.g.$ 30) that the oxidation states of a and visible copper on one hand and those of a_3 and invisible copper on the other show parallel behavior, respectively. On the other hand, according to the same authors[29] the solubilized oxidase in an artificial membrane face the interior, $i.e.$ the orientation is opposite to that found with intact mitochondria. It requires further evidence to correlate the well known phenomenon that the submitochondrial particles are "inside-out" ($i.e.$

reverse of the sideness after or in the cleavage of mitochondria to form sub-mitochondrial particles). This complexity is further compounded by the possible difference of the numbering system for the subunit used by various workers. Nonetheless, it seems to us that either subunit II representing visible or in-visible copper as well as the topology of the inner membrane needs much more experimentation. At any rate, very little chance exists that two heme mole-cules and two copper atoms are close enough to have direct electron transfer. This proposed mechanism becomes even more complex in the presence of those sub-units without prosthetic groups. All these findings may point out that there are some devices or mechanisms other than simple direct transfer of electrons through prosthetic groups. Naturally, at present only speculation may be made.

It should be also pointed out that very likely other possibilities exist. For example, the prosthetic groups (copper and heme a) may be caged or surrounded by the subunit polypeptides. Similarly, each, or some, of the prosthetic groups links more than one subunits. If either is true, then our thinking must be reorientated about the question concerning which subunits attach to heme and which to copper, visible or invisible.

ACKNOWLEDGEMENTS

This work is supported by NIH Grants GM-16767 and HL-12576.

REFERENCES

1. Keilin, D. (1966) The History of Cell Respiration and Cytochrome, Cambridge University Press, Cambridge.

2. Lemberg, R., and Barrett, J. (1973) Cytochromes, Academic Press, New York.

3. Shapiro, A. L., Vinela, E., and Maizel, J. V. (1967) Biochem. Biophys. Res. Commun., 28, 815-820.

4. Weber, K., and Osborn, M. (1969) J. Biol. Chem., 244, 4406-4412.

5. Reynolds, J. A., and Tanford, C. (1971) Proc. Nat. Acad. Sci., U.S., 66, 1002-1007.

6. Reynolds, J. A., and Tanford, C. (1970) J. Biol. Chem., 245, 5161-5165.

7. Ferguson, K. A. (1964) Metab. Clin. Exp., 13, 985-1002.

8. Swank, T. R., and Munkres, K. D. (1971) Anal. Biochem., 39, 462-477.

9. Fairbanks, G., Steck, L. T., and Wallach, D. F. H. (1971) Biochemistry, 10, 2606-2617.

10. Weber, K., and Osborn, M. (1975) in The Proteins, Neurath, H., and Hill, R., eds., Academic Press, New York, pp. 179-223.

11. Chrambach, A., and Rodbard, D. (1971) Science, 172, 440-451.

12. Kuboyama, M., Yong, F. C., and King, T. E. (1972) J. Biol. Chem., 247, 6375-6383.

13. MacLennan, D. H., and Tzagoloff, A. (1965) Biochim. Biophys. Acta, 96, 166-168.

14. Takamyama, K., MacLennan, D. H., Tzagoloff, A., and Stoner, C. D. (1965) Arch. Biochem. Biophys., 114, 223-232.

15. Davis, B. J. (1964) Ann. N. Y. Aca. Sci., 121, 404-427.

16. Tuppy, H., and Brikmayer, G. D. (1969) Europ. J. Biochem., 8, 237-243.

17. Goldberger, R., Smith, A. L., Tisdale, H., and Bostein, R. (1965) Biochim. Biophys. Acta, 96, 1-10.

18. Rutenbergs, S., and Wainio, W. W. (1972) Bioenergetics, 3, 403-409.

19. Unpublished results from this laboratory.

20. Hundt, E., and Kadenbach, B. (1977) Z. Physiol. Chem., 358, 1309-1314.

21. Phan, S. H., and Mahler, H. R. (1976) J. Biol. Chem., 251, 257-269; 270-276.

22. Capaldi, R. A., Bell, R. L., and Branchek, T. (1977) Biochem. Biophys. Res. Commun., 74, 425-433.

23. Yu, C. A., Yu, L., and King, T. E. (1975) J. Biol. Chem., 250, 1383-1392.

24. Takemori, S., and King, T. E. (1965) J. Biol. Chem., 240, 504-513.

25. Yu, C. A., Yu, L., and King, T. E. (1977) Biochem. Biophys. Res. Commun., 74, 670-676.

26. Tanaka, M., Haniu, M., Yasunobu, K. T., Yu, C. A., Yu, L., Wei, Y. H., and King, T. E. (1977) Biochem. Biophys. Res. Commun., 76, 1014-1019.

27. Yu, C. A., and Yu, L. (1977) Biochim. Biophys. Acta, 495, 248-259.

28. Racker, E. (1976) A New Look of Mechanisms in Bioenergetics, Academic Press, New York

29. Frey, T. G., Chan, S. H. P., and Schatz, G. (1978) J. Biol. Chem., 253, 4389-4395.

30. Yong, F. C., and King, T. E. (1972) J. Biol. Chem., 247, 6384-6388.

31. Downer, N. W., Robinson, N. C., and Capaldi, R. A. (1976) Biochemistry, 15, 2930-2936.

32. Briggs, M., Kamp, P.-F., Robinson, N. C., and Capaldi, R. A. (1975) Biochemistry, 14, 5123-5128.

33. Komai, H., and Capaldi, R. A. (1973) FEBS Lett., 30. 273-276.

34. Capaldi, R. A., and Hayashi, H. (1972) FEBS Lett., 26, 261-263.

35. Bucher, J. R., and Penniall, R. (1975) FEBS Lett., 60, 180-184.

36. Kornblatt, J. A., Baroff, G. A., and Williams, G. R. (1973) Can. J. Biochem., 51, 1417-1427.

37. McGeer, A., Lavers, B., and Williams, G. R. (1977) Can. J. Biochem., 55, 988-994.

38. Rubin, M. S., and Tzagoloff, A. (1973) J. Biol. Chem., 248, 4269-4274.

39. Keirns, J. J., Yang, C. S., and Gilmour, M. V. (1971) Biochem. Biophys. Res. Commun., 45, 835-841.

40. Birchmeier, W. (1977) Mol. Cell. Biochem., 14, 81-86.

41. Poyton, R. O., and Schatz, G. (1975) J. Biol. Chem., 250, 752-761

42. Mason, T. L., Poyton, R. O., Wharton, D. C., and Schatz, G. (1973) J. Biol. Chem., 248, 1346-1354.

43. Keyhani, J., and Keyhani, E. (1975) Arch. Biochem. Biophys., 167, 588-595.

44. Werner, S. (1977) Europ. J. Biochem., 79, 103-110.

45. Bertrand, H., and Werner, S. (1977) Europ. J. Biochem., 79, 599-606.

46. Yatscoff, R. W., Freeman, K. B., and Vail, W. J. (1977) FEBS Lett., 81, 7-9.

47. Kuronen, T., Saraste, M., and Ellfolk, N. (1975) Biochim. Biophys Acta, 393, 48-54.

48. Laemmli, U. K. (1970) Nature, 227, 680-685.

Cytochrome Oxidase, T.E. King et al. eds.
© *1979 Elsevier/North-Holland Biomedical Press*

SEQUENCE OF MITOCHONDRIAL MEMBRANE PROTEIN

AKIRA TSUGITA, IVAN GREGORE, ICHIRO KUBOTA & RUDOLF VAN DEN BROEK
European Molecular Biology Laboratory, Heidelberg, Germany &
Biozentrum, University of Basel, Basel, Switzerland

ABSTRACT

The amino acid sequences of cytochrome c oxidase subunits from baker's yeast and cytochrome c reductase subunits from Neurospora crassa were studied. New sequence techniques developed especially for use with membrane proteins have been used and are discussed.

INTRODUCTION

The mitochondrial inner membrane consists of a highly ordered network of enzyme complexes which together with lipid and other proteins are utilized in electron transport and oxidative phosphorylation.

Complex III (cytochrome c reductase) and Complex IV (cytochrome c oxidase) are examples of mitochondrial inner membrane enzyme complexes, and sharing cytochrome $\underline{c}$ as a substrate, function as reductase and oxidase activities, respectively. The cytochrome c reductase complex consists of at least seven subunits which include cytochrome b, cytochrome $\underline{c}_1$, an iron-sulfur protein and a number of subunits without known prosthetic groups[1]. Cytochrome c oxidase is also composed of about seven subunits, three high molecular weight subunits and four (or more) low molecular weight ones which include the cytochrome a and cytochrome a_3 heme associated poly-peptides[2]. The complexes were isolated from Neurospora crassa (Complex III) and yeast (Complex IV) mitochondria, thus there is the possibility of utilizing genetic approaches in future investigations.

The proteins of the mitochondrial inner membrane are either coded for by the nuclear genome and are synthesized on cytoplasmic ribosomes, or are specific by mitochondrial DNA and synthesized on mitochondrial ribosomes[3]. Cytochrome b in the cytochrome $\underline{c}$ reductase and the three large subunits I, II, III in the oxidase, are coded for by mitochondrial DNA. These protein subunits contain

a high proportion of hydrophobic amino acids and are thought to be
embedded in the interior of the mitochondrial inner membrane.
Cytochrome c_1 in cytochrome c reductase and the four low molecular
weight subunits of cytochrome oxidase are specified by nuclear DNA.
These proteins also contain a relatively high proportion of hydrophobic amino acids.

   Sequence studies on hydrophobic membrane proteins encounters
many difficulties. Several useful techniques have been developed
which help overcome some of the problems such as the use of solid
phase sequenators[4] for N-terminal sequence determinations and the
use of anhydrous hetptafluoroubutylic acid in the CNBr reaction[5].
In the present paper we present several new approaches which have
been developed recently in our laboratory, in order to aid in the
sequence determination of membrane proteins, and some results are
given.

MATERIAL AND METHODS
   Carboxy peptidase (C pase) A and B were obtained from
Worthington Biochemicals as diisopropylphosphate treated samples.
Cpase Y[6] was from Oriental Yeast and purified. Cpase P′ was from
Takara Shuzo. Phenylisothiocyanate (PITC), 4-sulfonyl-phenyliso-
thiocyanate (SPITC), isothiocyanate phenylthiocarbamyl aminopropyl
glass (DITC-glass), acetonitrile and trifluoroacetic acid (TFA)
were sequential grade, obtained from Pierce Chemical. Dimethyl-
allylamine (DMAA) was from Fluka finechemical. Quodrol buffer
(0.1M pH 9.0) was obtained from Beckman sequential grade and isopropanol and n-propanol were from Merck. Insulin and lysozyme
were purchased from Sigma Chemical and Myoglobin was from Serva.
Trypsin (TPCK treated) and chymotrypsin were purchased from
Worthington Biochemicals, Staphylococcal V_8 protease was donated
by Drapean[8], thermolysin was from Daiwa Kasei.

RESULTS AND DISCUSSION
Use of organic solvents
   Hydrophobic membrane proteins, such as subunits IV, VI and VII
of cytochrome oxidase and cytochrome b, can be theoretically, if
not always in practice, completely digested by trypsin, but not by
chymotrypsin under conventional conditions. These observations

indicate that the polar groups, such as the lysyl or arginyl residues, are exposed on the surface of the molecules whereas the apolar groups, required for chymotryptic digestion, are buried within the molecules. In some cases the addition of organic solvents can change the protein conformation such that the apolar groups, required for chymotrytic digestion, become surface exposed and thus susceptible to proteolytic attack. Organic solvents, such as propanol, when added to proteins presumably change the protein conformation by breaking intra-molecular hydrogen-bonding. (see Gutfreund and Knowles for discussion[9]).

The stability of various proteases to different concentrations of n-propanol was tested. Thermolysin, staph-protease and trypsin are stable in 30% or 60% n-propanol at $37^{O}C$ for 16 hrs at their pH optimum. Chymotrypsin retained 17% and 22% of its optimal activity in 30% and 60% propanol, respectively. After incubation under similar conditions Cpase B, Y and P appeared stable in both 30% and 60% propanol but Cpase A activity was reduced to 50% and 80% of its optimal activity in 30% and 60% propanol, respectively.

Subunit IV of the oxidase was found to be more susceptible to proteolytic cleavage in 50% propanol than in aqueous buffers, and the cleavage specifications were different in the two solvents. Cytochrome b in 30% isopropanol was digested by Cpase A and B 1.7 fold more than in aqueous buffer alone. These results further support the view that in organic solvents protein conformational changes can take place which increase the steric availability of hydrophobic residues.

The insolubility of membrane proteins in aqueous solution will of course limit the extent to which the proteins can be digested by proteases. However, even when the solubility of such proteins is increased by chemical modification, they are often still resistant to protease activity. For example, citraconilated subunit IV of cytochrome c oxidase is soluble in pH 8.0 aqueous buffer but it is less effectively digested by chymotrypsin than the untreated, water insoluble control. Addition of organic solvents to the modified subunit markedly increased the effectiveness of proteolytic digestion. This suggests that the surface of the modified protein is covered by polar groups and the apolar residues required for chymotrypsin activity are hidden inside. Denaturing detergents such as sodium dodecyl-sulfate (SDS) break intermolecular bonds and unfold the quaternary structures of proteins. Such detergents, however, do not always increase the effectiveness of proteolytic digestion because of reasons discussed below and because, in the case of SDS, bound detergent presumably masks those amino acids required for proteolytic activity. SDS inactivates some proteolytic enzymes, for example, almost all the Cpases are inactivated in 0.1% SDS. Some enzymes are more resistant to the denaturating effects of SDS and thus trypsin, chymotrypsin and staphylococcal protease can digest to limited extents with normal specificities in 0.1% SDS-solution[10].

Sequence of peptides

As mentioned above, only certain proteases can be used to digest lipophilic proteins, and thus obtaining overlapping peptides for sequence alignment is difficult. To overcome this difficulty we have developed a new method for the alignment of peptides. Firstly, a protein is labelled with a radioactive agent only at the N-terminus and then the labelled derivative is subjected to both complete and partial enzymatic digestion. Secondly, the digests are separated and their compositions and sequences determined by convential procedures. Comparing the composition of the radio-active peptides from the partial digestion with those from the complete digestion enables the linear alignment of peptides to be established. For example (see Fig 1), a protein consisting of 76 amino acids, including 7 lysyl and 3 arginyl residues, was modi-fied with s-PITC, which modifies all internal ϵ-lysyl residues and the N-terminal residue. The modified protein was then subjected to one step of the Edman degradation and the new exposed N-terminus labelled with C^{14} PITC. A sample of the C^{14}-derivative was digested completely with trypsin and the digest chromatographed on a Biogel P4 column using 0.1% NH_4HCO_3 as an eluent. Since all the lysyl residues had been blocked by reaction with SPITC the tryptic cleavage in this case was limited to arginyl residues, and 4 peptides A, B, C and D were obtained. (Peptides A & B were in poor yield and a peptide A-B was also found, which resulted from partial cleavage due to the inhibitory effects of a neighbourhood SPITC-lysyl residue). The peptides A and A-B

Fig 1

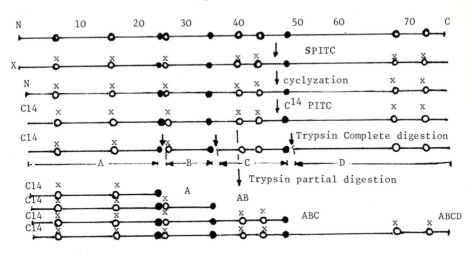

○ Lys , ● Arg, x SPITC, c^{14}; c^{14} PITC

contained the N-terminal residue since they were found to be radio-
active. The sequences of the 4-peptides were established including
the N-terminal peptides A and A-B. The remaining portion of the
c^{14} protein was partially digested with trypsin and the resulting
peptides separated on Biogel P10. The elution profile was established by plotting radioactivity vs. elution volume; peak fractions
were collected and amino acid compositions determined. Peptides
A-B and A-B-C were recovered in addition to peptide A and undigest-
ed protein. The composition and the molecular weight (see the
following section) of these peptides was compared with those of
the completely digested protein, thus enabling the linear alignment of the peptides to be established. This method has been
successfully applied to the precursor protein of T_4 phage internal
peptide II. Two other lipophilic proteins are under investigation
by this method. This method is not limited to the use of trypsin,
although tryptic digestion has certain advantages, but other
proteases and chemical cleavage procedures could in principle be
used as long as the reaction conditions do not require a low pH.
A limitation of this method is that the N-terminus of the protein
must be free. The details of the conditions for the chemical
modification will be described in the following section.

TABLE 1

Name of protein	Coupling condition for SPITC X40, 50°C, 30 min	Removal of SPITC (dialysis)	Coupling condition for PITC X400 50°C, 30 min	Removal of C^{14} PITC (dialysis)	MW* (No. of Ex)
Insulin (bovine pancreas)	0.4M DMAA pH 9.5	0.1N Col-Pyr-AcOH pH 8.0	0.4M DMAA pH 9.5 in 60% PrOH	benzen extraction X3 0.1N Col-Pyr-AcOH pH8	5,800 (4) (5,700)
Myoglobin (whale sperm)	0.4M DMAA pH 9.5	"	"	benzen extraction X3 0.1N Col-Pyr-AcOH pH8	15,700 (2) (17,200)
Lysozyme (chicken egg white)	4M Urea pH 9.5	"	0.1M Quadrol 0.6% SDS heat 100° 1 min.	benzen extraction X3 0.1N Col-Pyr-AcOH pH8 0.1% SDS	12,300 (2) (13,600)
Gp22 (T4 phage)	titrate to pH 9.5 with NaOH	"	0.4M DMAA pH 9.5 in 60% PrOH	Biogel P10 1% NH$_4$ HCO$_3$	27,900 (2) (27,500)
Cytochrome c$_1$ (Neurospora)	0.4M DMAA pH 9.5 0.1% SDS	0.1N Col-Pyr-AcOH 0.1% SDS pH 8.5	0.1M Quadrol 0.1% SDS	benzen extraction X3 0.1N Col-Pyr-AcOH pH8 0.1% SDS	30,200** (2) (32,000)
Asp-Ala	-	-	0.4M DMAA pH 9.5 in 60% PrOH	DEAE Sephadex	-

* ( ) lists the values obtained by other methods. ** Values without Cys and Trp

TABLE 2

	pH	Specificity
Cpase A	8.5 - 6.5	Stop at Arg, Pro-x. Practically stop at Gly, Ser, Asp, Glu.
Cpase B	8.5 - 6.5	Cleave Arg, Lys.
Cpase Y	7.5 - 6.0	Cleave Pro-x and Glu. Practically stop at Lys, Arg, Ser Gly. In the presence of PCMB (1mM) cleave Ser, Gly
Cpase P	6.5 - 2.5	Cleave Pro-x and Asp, Glu. Practically stop Ser, Gly.

Minimum molecular weight

The molecular weight of a protein can be determined by several methods which include SDS-gel electrophoresis, molecular sieve chromatography, in the presence or absence of SDS, sedimentation methods and rate of diffusion. We have developed a non-empirical method of quantitative N-terminal analysis. The principle involved in this determination is based on the original idea of Sanger and developed by other authors as well[11,12].

The NH_2 groups in a protein are firstly modified by the water soluble PITC reagent (SPITC) under the conditions described in Table 1. The modified protein is then freed of excess reagent by either dialysis or by gel filtration in the case of small molecules. [It should be noted that a ten-fold amount of PITC (400 moles per NH_2 residue) compared to SPITC (40 moles per NH_2 residues) is required for quantitative modification of NH_2 groups, as determined by the modification of the dipeptide Asp-Ala in 0.4M DMAA buffer pH 9.5 at 50° for 30 min.] The modified protein N-terminus is cyclized with TFA at 50° for 10 min. the protein lyophilized and then reacted with C^{14} PITC acetonitril solution (356 μ moles/ml; 1.1 μCi/μ mole) under the conditions described in Table 1. The excess C14 reagent is removed by dialysis or by gel filtration under mild alkaline conditions which prevent any peptide bond hydrolysis. Aliquots of the C^{14} derivatives are hydrolysed (6N-HCl at 105° for 24 hrs) and subjected to amino acid analysis and radioactivity determinations. The radioactivity (normalized experimentally using Asp-Ala, or based upon the specific radioactivity of the reagent) gives the molar concentrations of the protein or peptide and summation of the amino acids gives the minimum molecular weight. This method has been tested with known proteins and applied to new proteins and the results are listed in Table 1. The amount of protein or peptide required depends on the sensitivity of the amino acid analyser used, for example, $\approx$ 10 μg protein can be analysed satisfactorily. The method does of course require that the protein be soluble in the coupling reagent and during the dialysis or gel filtration stages.

This method although developed for general purposes was found to be especially suitable for use with hydrophobic proteins. The molecular weight estimation of membrane proteins by the popular

methods of SDS gel electrophoresis and gel filtration in the
presence of detergents can produce anomalous results for a variety
of reasons. The method described here is less likely to produce
anomalous values since it is a non-empirical method, however, one
limitation is that the protein must have a free N-terminus.

Edman degradation

Solid phase sequenator techniques have been found to be useful
for hydrophobic proteins. It has been possible to obtain N-
terminal sequences of Subunit VI of cytochrome c oxidase and cytochrome c_1 using a solid phase sequenator but not using a spinning
cup sequenator. The coupling yield for the solid phase method is
very dependent upon the solubility of the protein. For Subunit IV
coupling to glass was achieved in 12% pyridine, 2% SDS, pH adjusted
to 9.5 with NaOH (pre-heating for a few seconds at 100°) at 45^OC
for 1 hr. For cytochrome c_1. Coupling was carried out in 0.4M
DMAA buffer, pH 9.5, 60% propanol at 50^OC for 30 min. In each case
over 60% yields were achieved under these conditions. When the
conditions for the coupling of Subunit IV were applied to cytochrome c_1 only 10% of the latter was coupled.

The cytochrome c oxidase Subunit VI was sequenced by the spinning cup method as well. Cytochrome b is known to have a blocked
N-terminus[13] (see Table 3).

C-terminal Sequence

Since there is a need for a reliable C-terminal sequence method
we have spent a few years attempting to develop such a method.
This method, partially published elsewhere[14], is based upon the
combined, or sequential, use of a variety of specific carboxy-
peptidases, namely, CPase's A, B, Y and P. The specificities and
the pH's for optimal activities are listed in Table 2.
(Specificity of the Cpases is dependent on the neighbouring amino
acids which are not listed here in order to avoid unnecessary
complication). Firstly, the protein or peptide is attached to
DITC-glass through the N-terminus, attachment is limited to the
N-terminus by modifying the NH_2 groups with SPTTC, the derivatised
N-terminal residue is then cleaved to obtain a free N-terminal
residue, which is then the second residue. The modified protein

is reacted with DITC glass and then packed into a small column
(LKB 2mm X 15 cm) equipped with a heating jacket. The carboxy-
peptides, in combination or sequentially, in 0.1 M pyridin-colli-
dine-acetate (formate) buffer, at various pH's (see Table 2) are
used as the elution solvent. Enzyme concentrations are 10-fold
higher than those used in the batch-wise method. Secondly, the
column effluents are collected, fractious evaporated and directly
analysed by an amino acid analyser. The enzymes do not disturb
the analyser column when a ordinary wash programme is used. Preliminary experiments to determine choice of enzymes, suitable pH's,
temperatures and time are carried out using the batch-wise method.

The rate of reaction was found to be dependent on the concentration of enzyme(s), pH, temperature and the flow rate. The
amount of sample required for this method depends mainly on the
sensitivity of the amino acid analyser used and upon the yield
during the coupling to glass beads; we use 5-20 n moles of sample.
The digestion diagram plotted against fraction number shows
differential kinetics. Overlapping amino acids makes it difficult
to make accurate kinetic analysis, however, the method is still
under development including automated operation.

Sequences

We believe that the sequence determination of membrane proteins
is one of the major challenges facing protein chemists. The
sequence determinations we have achieved to date are summaried in
Table 3. These results are just a beginning and using the
approaches described in this paper and others we hope to rapidly
expand our programme.

SUMMARY

During sequential study of mitochondrial membrane proteins, we
developed several new methods specially for lipophilic protein
1) minimum molecular weight determination for protein and for
peptide 2) alignment of peptide fragments to linear sequence
3) C-terminal sequence method and 4) use of organic solvent for
enzymatic (and chemical) digestion.

TABLE 3

Cyt c Oxidase
  (yeast)

| | N | | | | 5 | | | | 10 | | | | 15 | |
VI (12K) Ser (Asp Ala His)Asp Glu Glu Thr Phe Glu Glu Phe Thr Ala Arg Tyr Glu-

                                                                            c
                              -Glu Glu Asp Glu Pro Phe Leu Ser Ser

          N   5   10
IV (14K) His Gln Asn Pro Val Val Leu Tyr Ala Leu-

Cyt c Reductase
  (Neurospors)
            N
  Cyt b   F-Met-Tyr-His-

                                                                    c
                    -(Val Glu Ala Asp Ser AAn Ser Gly) Leu Tyr Leu Lys

            N   5   10
  Cyt c$_1$ Met His Pro Ala Glu Glu - Leu Arg Ala Thr-

                              -(Arg Glu Lys Val Leu)-Ala Thr

        AAn: acidic amino acid amide

ACKNOWLEDGEMENTS

    This work is in a frame of collaboration with Prof. G. Schatz,
Biozentrum, for yeast cytochrome c oxidase and with Dr. H. Weiss,
EMBL, for Neurospora cytochrome c reductase. Support has been
partially made by a Swiss National Fond (3.189.73). We also thank
Dr. P. Wingfield, EMBL for discussion and editing this paper.

REFERENCES

1.   Weiss, H. (1976) Biochim. Biophys. Acta 456, 291-313.
2.   Mason, T.L., Shatz, G. (1973) J. Biol. Chem. 248, 1355-1360.
3.   Tzagoloff, A. et al. (1973) Biochimie 55, 779-792.
4.   Laursen, R.A. (1971) Eur. J. Biochem. 20, 89-92.
5.   Ozols, J. & Gerard, C. (1977) J. Biol. Chem. 252, 5986-5989.
6.   Hayashi, R. et al. (1973) J. Biol. Chem. 248, 2296-2302.
7.   Yokoyama, S. et al. (1975) Biochim. Biophys. Acta 397,
     443-448.
8.   Howard, J. & Drapeau, G.R. (1972) Proc. Natl. Acad. Sci. 69,
     3506-3509.
9.   Gutfreund, H., Knowles, J. (1967) Essays in Biochemistry 3,
     25-42.

10. Cleveland, D.W. et al. (1977) J. Biol. Chem. 252, 1102-1106.

11. Hohn, T. et al. (1976) J. Mol. Biol. 105, 337-342.

12. Banks, B.E.C. et al. (1976) Eur. J. Biochem. 71, 469-473.

13. Weiss, H. & Ziganke, B. (1976) Genetics and Biogenesis of chloroplasts and mitochondria. Th Büchel et al. eds. Elsevier/North Holland Biochem Press, pp. 256-266.

14. Isobe, T. et al. (1976) J. Mol. Biol. 102, 349-365.

Cytochrome Oxidase, T.E. King et al. eds.
© 1979 Elsevier/North-Holland Biomedical Press

CHEMICAL CONSTITUTION AND SUBUNIT FUNCTION OF POLYPEPTIDE II FROM CYTOCHROME-C-OXIDASE

G.J. STEFFENS and G. BUSE

Abteilung Physiologische Chemie, RWTH Aachen

Melatener Str. 211, D-5100 Aachen, Germany

INTRODUCTION

The cytochrome-c-oxidase enzyme complex is located in the inner membrane of mitochondria. The oxidase consists of a set of different components: cytoplasmically and mitochondrially synthesized polypeptides, heme a, copper and phospholipids. The knowledge of the primary structures of the protein components will help to identify the functional subunits and may elucidate the arrangement of the final cytochrome complex within the membrane layer.

We have focused our attention to the isolation and the chemical characterization of some polypeptides. The complete amino acid sequence of the smallest polypeptide (VIIIa) has been published[1,2]. Here the chemical constitution of polypeptide II is presented and its subunit function is discussed.

MATERIALS AND METHODS

The beef heart enzyme is prepared by a modification[3] of the ammoniumsulfate fractionation procedure described by Fowler[4] et al. and Yonetani[5]. The purified enzyme contains 5% phospholipids and 10-12 nMol heme a/mg protein. The heme a content corresponds to the copper content.

200 mg of the oxidase is dissociated and fractionated into the polypeptide components by gel filtration on Bio-Gel P-60 (columns 5 x 180 cm) in the presence of 2% SDS[3,6]. This procedure allows large scale preparation of most of the protein components, including polypeptide II.

To obtain a fractionation independent of the molecular weight, polypeptide II has been separated from the oxidase complex by Craig distribution[7] in the system butanol-(2):0,7 M acetic acid (1:1).

The isolation of the low molecular weight components was achieved by chromatography on Bio-Gel P-10 (columns 5 x 150 cm) running with 10% acetic acid[1]. Both separations are accomplished with a lyophilized enzyme of low lipid content. The preparations were checked for purity by SDS gel-electrophoresis[8] and N-terminal endgroup determination.

The N-terminal formylmethionine of polypeptide II was deblocked[9] in the reaction cup of the sequencer by adding 1 ml methanol 0,5 mol HCL at 20°C. After one hour the reaction was stopped by drying in high vacuum. The peptide was redissolved in 90% formic acid prior to the start of the Quadrol sequencing program[10].

Cleavage with cyanogen bromide[11]: Cleavage of 4 μmol polypeptide II was performed with cyanogen bromide (100-fold molar excess over methionine) at 20°C in 70% formic acid. The sample was held in the dark for 4 hours. The reagent and the solvent were removed with a stream of N_2 and the remaining light coloured product was suspended in 4 ml acetic acid (20%) and centrifuged from the insoluble material. The soluble cyanogen bromide peptides were separated by gel chromatography on Bio-Gel P-6 in 10% acetic acid. The residue could be redissolved and chromatographed on Bio-Gel P-6 in 60% formic acid.

Cleavage with trypsin: 3 μmol polypeptide II was incubated with 1 mg TPCK-trypsin (Serva) for 2 hours at 37°C in 0,1% ammonium-bicarbonate and 0,1% SDS, pH 8. The digestion was terminated by freeze-drying and suspending the resultant peptides in 4 ml 20% acetic acid. After centrifugation, the soluble tryptic peptides were purified by chromatography on Bio-Gel P-10 in 10% acetic acid. The residue could be redissolved and chromatographed on Bio-Gel P-6 in 50% acetic acid.

The combination of gel filtration on Bio-Gel P-6 or P-10 and a successive paper chromatography in the system n-butanol: acetic acid: water (4:1:5) was, in general, sufficient for purification of the tryptic and cyanogen bromide peptides.

Digestion with carboxypeptidase A: 0,1 μmol of polypeptide II was incubated with 0,1 mg C'ase A at 37°C in 0,1% ammonium bicarbonate and 0,1% SDS, pH 8. Aliquots of the solution were taken after 20,40 and 80 minutes. The digestion was stopped by adding the pH 2.2 starting buffer for amino acid analysis. The rates of release of the C-terminal amino acids were determined by amino acid analysis.

The amino acid compositions of the peptides and proteins were determined on acid hydrolysates of the samples according to Spackman et al.[9] using a Beckman amino acid analyzer (Multichrom).

Automated sequence analyses were performed on a sequencer Beckman model 890C. The N-terminal sequencer run of polypeptide II was performed with the fast Quadrol protein program[10]. For automated sequencing of the various peptide fragments obtained in the course of work, we used the peptide program of Crewther and Inglis[13], which was adapted to the 890C sequencer. This program allows us to sequence peptides of different length and hydrophobicity without prior chemical modification.

Conversion of the 2-anilino-5-thiazolinone amino acids and identification of the phenylthiohydantoins has been described earlier[1].

RESULTS AND DISCUSSION

Figure 1 shows the preparative separation of cytochrome oxidase on Bio-Gel P-60 into seven polypeptide fractions. It has been demonstrated[3,6], that beside fraction V (polypeptides V and VI) all fractions appeared to be homogeneous in the densitometric tracings of the SDS gel-electrophoresis.

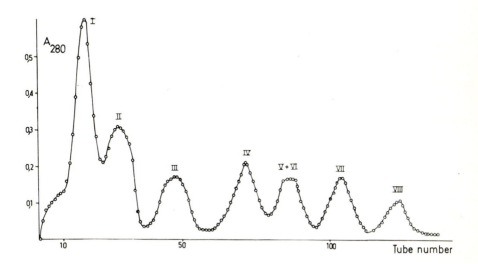

Fig. 1: Preparative gel filtration on Bio-Gel P-60 (-400 mesh) in the presence of 2% SDS. Roman numerals indicate the polypeptides as described in table 1.

The amino acid composition of these components have been des-cribed[3,6]. As observed in the yeast[14] and the neurospora[15] enzymes, we have isolated three hydrophobic polypeptides (I, II and III). These results were confirmed by Downer et al.[16] and Muijsers et al.[17]. Yu et al.[18], however, obtained only two polypeptides in this range of molecular weight. Fraction VIII is a mixture of three polypeptides (6000 Dalton) with the free N-terminal end-groups isoleucine, phenylalanine and serine[1,3,6].

A more straightforward chemical characterization is given by the N-terminal endgroup determination. The large polypeptides I and II have blocked N-termini. After treatment with methanolic HCL, the formyl-group is split off and both chains start with methionine on sequencing. This confirms the information on the site of biosyn-thesis given by Yatscoff and Freeman[19].

TABLE I. Polypeptide composition of beef heart
cytochrome oxidase

No in prep. separation	apparent mol.weight	synthesis	N-terminus
I	36000 D	mitochondrial	blocked, F-Met
II	24000 D	mitochondrial	blocked, F-Met
III	19000 D	(disputed)	Thr
IV	16000 D	cytoplasmic	Ala
V	11600 D	cytoplasmic	Ser
VI a,b,c,d	12500 –	probably	Ala, Pro
	10000 D	cytoplasmic	
	(substoichiometric extrinsic parts?)		
VII	9500 D	cytoplasmic	blocked
VIII a,b,c	6000 D	probably cytoplasmic	Ser, Ile, Phe

Table 1, as described earlier[1,2], comprises the polypeptide composition of beef heart cytochrome oxidase in the order of their molecular weights, including minor components (VI a,b,c,d). These data were obtained from preparative gel filtration in SDS or in 10% acetic acid and from SDS gel-electrophoresis in different systems[3,6,20]. The status of sequence work presently going on with all the components confirms these results. The N-terminal end-groups are given as a purity criterion.

Before starting sequence determination of polypeptide II, we wanted to isolate this component by a molecular weight independent procedure, to be sure working with a homogeneous product. Therefore we developed a distribution system, which allowed the isolation of polypeptide II from the entire enzyme.

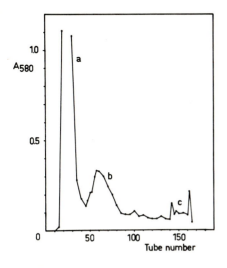

Fig.2:
Craig distribution of cytochrome oxidase in the system butanol-(2):0,7 M acetic acid (1:1)
a) bulk of the cytochrome oxidase polypeptides
b) polypeptide II
c) phospholipids and heme

As shown in figure 2, this chain is the only one extracted; the bulk of the enzyme remains at the start position, whereas the phospholipids and heme migrate with the front. The amino acid analyses of this chain, as obtained from different separation techniques, are shown in table 2.

The deformylation with methanolic HCl yields 70-80% of free N-terminal methionine. No other endgroup could be found. Unfortunately this medium esterifies side-chain carboxylgroups, which leads to an efficient blocking reaction for both aspartic- and glutamic acid/amide residues (probably by imide and pyroglutamic acid formation, respectively)[21]. In step 5(Gln), 10(Gln) and 11(Asp) repetitive yields are thus reduced drastically and we are not able to sequence more than 15-18 steps by the N-terminal sequencer run.

The time dependent release of C-terminal amino acids after treatment with C'ase A leads to the C-terminal sequence Met-Leu. This result is in agreement with the finding of free leucine after cyanogen bromide cleavage.

TABLE 2

Comparison of amino acid composition of polypeptide II with amino acid analyses (20 h hydrolysis) of this component, obtained from different isolation techniques.

Amino acid	Sequence data	Gelfiltration	Craig distribution
Asp	15	15.18	14.79
Thr	15	15.17	15.72
Ser	20	20.20	20.40
Glu	21	21.06	20.18
Pro	13	12.66	11.94
Gly	9	9.31	10.16
Ala	8	8.13	9.17
Cys	2	1.52 (a)	n.d.
Val	11	10.11	9.33
Met	15	14.32	14.94
Ile	11	10.07	9.82
Leu	28	28.69	27.75
Tyr	10	10.79	10.34
Phe	7	6.58	7.32
Lys	6	6.67	7.37
His	7	7.22	6.73
Arg	6	5.85	6.02
Trp	4	3.09 (b)	n.d.

a) after performic acid oxidation
b) after addition of 2% thioglycolic acid

The amino acid composition pointed out the further strategie of sequencing. Because of the unusually high methionine content, polypeptide II was cleaved with cyanogen bromide. Figure 3 shows the separation of the resulting fragments by chromatography on Bio-Gel P-6 in 10% acetic acid. Incomplete cleavage of some methionyl bonds leads to overlapping peptides in addition to the theoretically expected sixteen fragments. Further purification by paper chromatography was necessary in some cases. Fourteen cyanogen bromide peptides were isolated and sequenced. The last fraction (tubes 124-133) contains leucine (C-terminus), homoserinelacton and formylhomoserinelacton from the N-terminus. Formylhomoserinelacton was isolated by chromatographying this fraction on Dowex 50 x 2. Two cyanogen bromide peptides, insoluble in 20% acetic acid, were partially separated on Bio-Gel P-6 in 60% formic acid and sequenced.

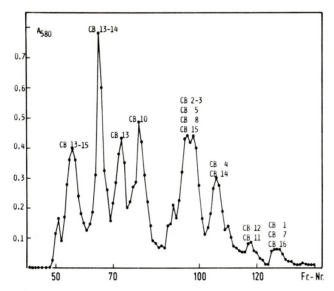

Fig. 3:
Gel filtration of the cyanogen bromide fragments of polypeptide II
on Bio-Gel P-6 in 10% acetic acid. CB: cyanogen bromide peptides
numbered as obtained from sequence data.

In order to obtain overlapping fragments and because of the relatively low contents of argenine and lysine tryptic digestion
appeared to be the most promising cleavage second. The tryptic
peptides were separated and purified by chromatography (figure 4)
in 10% acetic acid on Bio-Gel P-10 and successive paper chromatography. Nine tryptic peptides, out of thirteen expected, were
isolated and sequenced. Residual material, insoluble in 20% acetic
acid, contains, in general, two tryptic fragments, which could
partially be separated by gel filtration on Bio-Gel P-6 in 50%
acetic acid.

The yield of cleavage at the first lysine in the sequence is
very low, therefore we have isolated the N-terminal ditryptic peptide, consisting of 63 amino acids. Because of difficulties mentioned at deformylation, we were not able until now to sequence
this tryptic fragment. A gap of about fourty residues remains to
be solved. The other ditryptic peptide was isolated and sequenced,
it contains a Lys-Pro bond, resisting the tryptic attack. Table 3
presents the preliminary sequence data of polypeptide II.

Table 3: Preliminary sequence data of polypeptide II of cytochrome oxidase

10   20
F-Met-Ala-Tyr-Pro-Met-Gln-Leu-Gly-Phe-Gln-Asp-Ala-Thr-Ser-Pro-Ile-Met-Glu-Glu-Leu-

-Leu-His-Phe-

-Arg-Ile-Leu-Tyr-Met-Met-Asp-Glu-Ile-Asn-Asn-Pro-Ser-Leu-Thr-Val-Lys-Thr-Met-Gly-

-His-Gln-Trp-Tyr-Trp-Ser-Tyr-Glu-Tyr-Thr-Asp-Tyr-Glu-Asp-Leu-Ser-Phe-Asp-Ser-Tyr-

-Met-Ile-Pro-Thr-Ser-Glu-Leu-Lys-Pro-Gly-Glu-Leu-Arg-Leu-Leu-Glu-Val-Asp-Asn-Arg-

-Val-Val-Leu-Pro-Met-Glu-Met-Thr-Ile-Arg-Met-Leu-Val-Ser-Ser-Glu-Asp-Val-Leu-His-

-Ser-Trp-Ala-Val-Pro-Ser-Leu-Gly-Leu-Lys-Thr-Asp-Ala-Ile-Pro-Gly-Arg-Leu-Asn-Gln-

-Thr-Thr-Leu-Met-Ser-Ser-Arg-Pro-Gly-Leu-Tyr-Tyr-Gly-Gln-Cys-Ser-Glu-Ile-Cys-Gly-

-Ser-Asn-His-Ser-Phe-Met-Pro-Ile-Val-Leu-Glu-Leu-Val-Pro-Leu-Lys-Tyr-Phe-Glu-Lys-

-Trp-Ala-Ser-Ser-Met-Leu.

F = Formyl

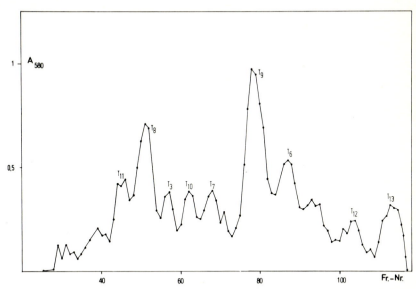

Fig. 4:
Gel filtration of the tryptic peptides of polypeptide II on Bio-Gel
P-10 in 10% acetic acid. T: tryptic peptides numbered as obtained
from sequence data.

Since Keirns et al.[22] have described the polypeptide composition
of cytochrome oxidase by SDS gel-electrophoresis, several laboratories[16,17,18,23] have actively tried to isolate and characterize
these polypeptides, and many attempts were done to identify the
functional subunits. It can be demonstrated from the literature[16,18]
that nearly all components have already been designated as heme
or copper subunits. In our opinion, however, there are only two convincing descriptions of a copper containing polypeptide, isolated
from the oxidase[24,25]. Knowledge of the primary structures of the
oxidase, however, provides the basis for detection of homologies
with structures of other proteins of wellknown function.

Table 4 shows a comparison of characteristic sequences of polypeptide II with those of azurin and plastocyanin. From sequences
around the invariant histidines clear homologies of azurin and
plastocyanin with component II can be infered. X-ray work[26] on
plastocyanin has established, that the copper is bond to two
histidines, one cysteine and one methionine. In the alignment
given in table 4 these functional amino acids are located at the
same position in polypeptide II, this leads to the conclusion of

88

Table 4: Comparison of characteristic sequences of polypeptide II with sequences of azurin and plastocyanin.

```
Pseudomonas fluorescens  -Thr-Phe-Thr-Val-Asn-Leu-Thr-His-Ser-Gly-Ser-Leu-Pro-Lys-Asn-Val-Met-Gly-His-Asn-Trp-
Polypeptide II           -Ile-Leu-Tyr-Met-Met-Asp-Glu-Ile-Asn-Asn-Pro-Ser-Leu-Thr-Val-Lys-Thr-Met-Gly-His-Gln-Trp-
French bean              -Ile-Val-Phe-Lys-Asn-Asn-Ala-Gly-Phe-Pro  -   -   -   -   -   - His-Asn-Val-

Pseudomonas fl.  -Val-Leu-Ser-Lys  -  Ser-Ala-Asp-Met-Ala-Gly-Ile-Ala-Thr-Asp-Gly-Met-Ala-Ala-Gly-Ile-Asp-Lys-Asp-
Polypeptide II   -Tyr-Trp-Ser-Tyr-Glu-Tyr-Thr-Asp-Tyr-Glu-Asp-Leu-Ser-Phe-Asp-Ser-Tyr-Met-Ile-Pro  -   - Thr-Ser-
French bean      -Val-Phe   -   -   -   -   -   - Asp-Glu-Asp-Glu-Ile-Pro-Ala-Gly-Val-Asp-Ala-Val-

Pseudomonas fl.  -Tyr-Leu-Lys-Pro-Gly-Asp-Ser-Arg-Val-Ile-
Polypeptide II   -Glu-Lys-Lys-Pro-Gly-Glu-Leu-Arg-Leu-Leu-
French bean      -Lys-Ile-Ser-Met-Pro-Glu-Glu-Leu-Leu-Leu-

Pseudomonas fl.  -Tyr-Glu-Phe-Phe-Cys-Ser  -  Phe-Pro-Gly  - His-Asn  - Ser-Met-Met-
Polypeptide II   -Tyr-Tyr-Gly-Gln-Cys-Ser-Glu-Ile-Cys-Gly-Ser-Asn-His  -   - Ser-Phe-Met-
French bean      -Tyr-Ser-Phe-Tyr-Cys-Ser  -  Pro   -   - His-Gln-Gly-Ala-Gly-Met-
```

+ One point mutations azurin ——→ polypeptide II

• One point mutations plastocyanin ←——→ polypeptide II

component II as one of the copper proteins of the oxidase, belonging to the azurin/plastocyanin protein family. It should be mentioned, that there is a second cysteine residue in component II close to the postulated copper binding site. The primary structure of stellacyanin also shows two cysteine residues and no methionine in this region[27]. So we cannot exclude, that two histidines and two cysteines bind to the copper as has been postulated for stellacyanines[27]. Other techniques, e.g. EPR and NMR measurements, may clear up the exact binding of copper in polypeptide II.

Another interesting feature is found in the sequences following the invariant histidine in position 83. The unusual repetitive alignment of aromatic residues (Tyr and Trp) suggest a highly ordered structure which may be involved in charge transfer within the inner mitochondrial membrane[6,28].

The identification of polypeptide II as a copper subunit and polypeptide VII as a heme subunit have been discussed elsewhere in view of a phylogenetic relationship to the electron transport chain of photophosphorylation[2,28].

ACKNOWLEDGEMENTS

Skilful technical assistance by Mrs. I. Dörner, J. Sporleder and J. Reumkens is acknowledged.
This work was supported by the Sonderforschungsbereich 160, "Biologische Membranen", of the Deutsche Forschungsgemeinschaft.

REFERENCES

1. Buse, G., Steffens, G.J. (1978) Hoppe-Seyler's Z. Physiol. Chem. 359, pp 1005-1009.

2. Buse, G., Steffens, G.J., Steffens, G.C.M. and Sacher, R. (1978) in Frontiers of Biological Energetics, in press Proceedings of a meeting held at the University of Pennsylvania. Academic Press.

3. Steffens, G.J. and Buse, G. (1976) Hoppe-Seyler's Z. Physiol. Chem. 357, pp 1125-1137.

4. Fowler, L.R., Richardson, S.H. and Hatefi, Y. (1962) Biochem. Biophys. Acta, 64, pp 170-173.

5. Yonetani, T. (1961) J. Biol. Chem. 236, pp 1680-1688.

6. Buse, G. and Steffens, G.J. (1976) in Genetics and Biogenesis of Chloroplasts and Mitochondria (Bücher, Th. et al., eds.) pp 189-194, North-Holland Publ. Comp., Amsterdam.

7. Hill, R.J., Königsberg, W., Guidotti, G.&Craig, L.C. (1963) Biochem. Prep. 10, pp 55-66.

8. Weber, K.& Osborn, M. (1969) J.Biol.Chem. 244, pp 4406-4412.

9. Sheehan, J.G. & Yang, D.H. (1958)J. Am. Chem. Soc. 80, pp 1154-1158.

10. Edman, P. (1975) in Molecular Biology, Biochemistry and Biophysics (Needleman, S.B., ed.) Vol. VIII pp 214-226, Springer-Verlag, Berlin.

11. Gross, E. and Witkop, B. (1969) J. Amer. Chem. Soc. 83, pp 1510-1514.

12. Spackman, D.H., Stein, W.H. & Moore, S. (1958) Anal. Chem. 30, pp 1190-1206.

13. Crewther, W.G. and Inglis, A.S. (1975) Anal. Biochem. 68, pp 572-585.

14. Sebald, W., Machleidt, W. & Otto, J. (1973) Eur. J. Biochem. 38, pp 311-324.

15. Schatz, G., Groot, G.S.P., Mason, I., Rouslin, W., Wharton D.C. & Saltzgraber, J. (1972) Fed. Proc. 31, pp 21-29.

16. Downer, N.W., Robinson, N.C. and Capaldi, R.A. (1976) Biochemistry 15, pp 2930-2936.

17. Muijsers, A.O., Boonman, J.C.P. and Biersma, R. (1977) 11th FEBS meeting, Copenhagen, abstract A 4-12 No. 660.

18. Yu, C.A. and Yu, L. (1977) Biochem. Biophys. Acta 495, pp 248-259.

19. Yatscoff, R.W. and Freeman, K.B. (1977) FEBS Lett. 81, pp 7-9.

20. Capaldi, R.A., Bell, R.L. and Branchek, T. (1977) Biochem. Biophys. Res. Commun. 74, pp 425-433.

21. Tarr, G.E. (1975) Anal. Biochem. 63, pp 361-364.

22. Keirns, J.J., Yang, C.S. & Gilmour, M.V. (1971) Biochem. Biophys. Res. Commun. 45, pp 835-841.

23. Tanaka, M., Haniu, M., Yasunobu, K.T., Yu, C.A., Yu, L., Wei, Y.H., and King, T.E. (1977). Biochem. Biophys. Res. Commun 76, pp 1014-1019.

24. Tzagoloff, A. and Mac Lennan, D.H. (1966) in The Biochemistry of Copper (Peisach, J., Aisen, P. and Blumbergs, W.E., eds.) pp 253, Academic Press, New York.

25. Tanaka, M., Hainu, M., Yasunobu, K.T., Yu, C.A., Yu, L. and King, T.E. (1976) Biochem. Biophys. Res. Commun. 66, pp 357-367.

26. Colman, P.M., Freeman, H.C., Guss, J.M., Murata, M., Noris, V.A., Ramshaw, J.A.M., and Venkatappa, M.P. (1978). Nature 272, pp 319-324.

27. Bergman, C., Gandvik, E.K., Nyman, P.O. and Strick, L. (1977) Biochem. Biophys. Res. Commun. 77, pp 1052-1059.

28. Buse, G., Steffens, G.J. and Steffens G.C.M., (1978) Hoppe Seyler's Z. Physiol. Chem. 359, 1011-1013.

Cytochrome Oxidase, T.E. King et al. eds.
© *1979 Elsevier/North-Holland Biomedical Press*

SEQUENCE STUDIES OF BOVINE HEART CYTOCHROME OXIDASE SUBUNITS

KERRY T. YASUNOBU, MASARU TANAKA, MITSURU HANIU, MUNEO SAMESHIMA, NEIL REIMER,
AND TATSUO ETO

Dept. of Biochemistry-Biophysics, University of Hawaii, Honolulu, Hi. 96822

TSOO E. KING, CHANG-AN YU, LINDA YU, AND YAU-HUEI WEI

Laboratory of Bioenergetics, State University of New York at Albany, Albany,
N.Y. 12222

ABSTRACT

The amino acid compositions, the end groups, the sequences of subunits
completed or nearly completed, the predicted conformations, the status of the
assignments of the prosthetic groups and possible active sites of the sequenced
subunits are presented or discussed.

INTRODUCTION

Our goal is to contribute towards the elucidation of the molecular
mechanism of oxidative phosphorylation. This is an enormous task which will
require the dedicated research efforts of physicists, chemists, biochemists,
and biologists. The expertise of our laboratory resides in the area of protein
chemistry and the results presented in this report concern our attempt to
determine the structure of the protein components involved in the third site
of oxidative phosphorylation in complex IV of the electron transport system
of the mitochondria. The protein components are cytochrome c, cytochrome c
oxidase, ATP synthetase and the structural proteins of the mitochondria. Thus
far, we have sequenced the bovine heart cytochrome c[1] and now the sequence
studies of bovine heart cytochrome oxidase are in progress. The studies are
still in progress and therefore, this is a status report rather than a final
report. Since Buse and Steffens[2] in Germany, and Boonman et al[3] in Amsterdam
are also sequencing bovine heart cytochrome oxidase and Tsugita is sequencing
yeast cytochrome oxidase, a comparison of the preliminary results obtained by
these groups will be useful.

EXPERIMENTAL PROCEDURES

Materials. Bovine heart cytochrome oxidase was isolated as reported
previously[4]. The procedures used for the isolation of the subunits of bovine
heart cytochrome oxidase have also been reported[5,6]. All of the reagents used
for the sequence determinations were of Sequanal quality and were purchased

from Pierce Chemical Co. Enzymes used were purchased from the Worthington
Biochemical Co. CNBr, mercaptoethanol and iodoacetate were purchased from
Nutritional Biochemical Co. Common reagents used were of reagent grade and
were obtained from standard chemical houses.

Methods. Methods used for protein cleavage, peptide purification, protein
sequencing have all been described in previous reports from our laboratory[7,8].
Amino acid analyses were determined on acid hydrolyzates in the Beckman Spinco
Model 120, 120C or 121MB automated amino acid analyzers as described by
Spackman et al[9]. Edman degradations were performed manually or in the Beckman
Spinco Model 890 Peptide/Protein Sequencer as described by Edman and Begg[10].
End groups were determined by published procedures[11-13].

RESULTS

The Number and Molecular Weights of the Subunits. Yu and Yu[6] have reported
that bovine heart cytochrome oxidase contains seven different subunits and have
characterized them with respect to molecular weight, amino acid composition and
prosthetic groups. The molecular weight data are summarized in Table I. Since
SDS-disc electrophoresis was used, the molecular weights are approximate and
the molecular weight which is determined from the amino acid or sequence data,
which are more reliable, are also shown in Table I. Numerous other investigators

TABLE I

NUMBER AND MOLECULAR WEIGHTS OF SUBUNITS OF BOVINE HEART CYTOCHROME OXIDASE

Subunit	Molecular Weights					
	I[a]	II[b]	IV[c]	V[d]	VI[e]	Sequence data
I	40,000	36,000		39,600	39,600	----
II	21,000	19,000	23,400	20,700	22,500	26,795
III	14,800	14,000	14,500	15,200	15,000	----
IV	13,500	12,800	13,300	13,400		----
V	11,600	11,500	10,700	11,700	11,200	12,436
VI	9,500	10,000	9,300	f	9,800	----
VII	7,600		6,600	f	7,300	----

[a]Yu and Yu[6]. [b]Capaldi and Hayashi[16]. [c]Rubin[15]. [d]Kuboyama et al[4].
[e]Rubin and Tzagoloff[17]. [f]Diffused.

have also reported the number and molecular weight of the subunits[4,6,14-17] of the oxidase and until they are all shown to be essential for activity, the actual number of subunits must be considered tentative. It is interesting to note that Yu and Yu[6] suggest that the molecular weight of cytochrome oxidase protomer is about 120,000 on the basis that there is one of each of the 7 subunits in the monomer while physicochemical methods have yielded diverse molecular weights (67,000-530,000) due to the ease with which the monomer aggregates[18-22].

    Amino Acid Composition of the Oxidase Subunits. There have been preliminary reports on the amino acid compositions of the bovine heart cytochrome oxidase[23] subunits[2,6]. The data differ significantly from our determined values which are shown in Table II. Part of this discrepancy may be a reflection of the purity of the preparations. Even the subunits which are isolated by rigorous purification procedures are not completely pure until they are converted to the carboxymethyl-cysteine derivative and further purified. Our results must also

TABLE II

AMINO ACID COMPOSITION OF SUBUNITS

Amino Acid	I[c]	II[d]	III[c]	IV[e]	V[d]	VI[c]	VII[e]	Totals
				Subunit No.				
Lysine	10	5	13	8	7	6	4	53
Histidine	12	6	3	4	3	1	2	31
Arginine	8	6	5	6	7	6	4	42
Aspartic Acid	24	16	12	11	13	12	7	95
Threonine	24	13	6	8	5	5	4	65
Serine	19	23	10	6	4	5	3	70
Glutamic Acid	16	22	15	15	12	10	8	98
Proline	18	13	7	8	7	4	4	61
Glycine	30	8	8	11	6	5	6	74
Alanine	30	9	11	9	8	7	5	79
Half Cystine	1^a	2^a	0.3^a	~3	1^a	3^a	0.24^a	~10
Valine	24	13	11	7	8	5	5	73
Methionine	15	15	2	2	1	1	1	37
Isoleucine	22	21	6	7	7	4	3	70
Leucine	47	36	12	9	11	3	7	125
Tyrosine	12	12	6	3	4	4	2	43
Phenylalanine	25	7	6	2	3	4	2	49
Tryptophan	5^b	6^b	5^b	3	$2^{b,d}$	2^b	1^b	24
Total Residues	342	233	138	122	109	87	68	1099 ~1110
MW (x 115)	39,330		15,870	13,524		10,005	7,800	~125,000
MW (sequence)		26,795			12,436			

^{a}CMC. ^{b}Spectrophotometric method. ^{c}24, 48, 72 hr hydrolyzates.
^{d}From sequence. ^{e}24 hr hydrolysis.

94

be considered tentative until the complete sequence of the seven subunits are
completed. Since the sequence of subunit V has been completed and subunit II
is nearly complete, the amino acid compositions are quite accurate. Since the
data for subunits I, III and VI have been determined from the 24, 48 and 72
hour hydrolyzates, they are better data than the ones reported for subunits IV
and VII, which are from the analysis of only 24 hour hydrolyzates. However,
from the appearance of nonstoichiometric ratios of some of the residues, it
is apparent that there are some impurities in many of the subunits. Despite
these limitations, we are getting closer to obtaining more accurate data
concerning the amino acid composition of cytochrome oxidase and its subunits.
As noted by others, subunits I and II are hydrophobic and contain high amounts
of leucine. In agreement with the acidic isoelectric points of yeast
cytochrome c oxidase subunits[24], the beef cytochrome c also contains a
preponderance of acidic over the basic amino acid residues. Methionine is
present in each of the subunits and its presence in the subunits is important
for cleaving with CNBr. The matter of the cysteine residues will be described
elsewhere in this report.

NH2- and COOH-Terminal Amino Acids. The end groups of some of the various
subunits of bovine heart cytochrome oxidase have been reported[2,3] and our
results are summarized in Table III. There are some disagreements concerning
the end-groups but this is understandable since it has been difficult to
isolate the subunits in pure forms and only recently, have the subunit
purification procedures been developed to stage where the subunits are at
least 90% pure. In our studies, the subunits are sufficiently purified to
establish clearly the NH2- and COOH-terminal amino acids. However, the COOH-
terminal amino acid of subunit VII must be confirmed due to the low yield of the
valine released (17%) by both hydrazinolysis and carboxypeptidase A.

TABLE III

END GROUPS OF THE BOVINE HEART CYTOCHROME OXIDASE SUBUNITS

Subunit No.	NH2-Terminal Amino Acid	COOH-Terminal Amino Acid
I	f-Met	Lys
II	f-Met	Leu
III	Ala	Lys
IV	Ala	His
V	Ser	Val
VI	Blocked	Ile
VII	Blocked	Val (?)

```
1                                    10                                    18   19
Ser-His-Gly-Ser-His-Glu-Thr-Asp-Glu-Glu-Phe-Asp-Ala-Arg-Trp-Val-Thr-Tyr-Phe-

                                     30                          36   37
Asn-Lys-Pro-Asp-Ile-Asp-Ala-Trp-Glu-Leu-Arg-Lys-Gly-Met-Asn-Thr-Leu-Val-Gly-

   40                                50                  54
Tyr-Asp-Leu-Val-Pro-Glu-Pro-Lys-Ile-Ile-Asp-Ala-Ala-Leu-Arg-Ala-Cys-Arg-Arg-

      60                                    70   72
Leu-Asn-Asp-Phe-Ala-Ser-Ala-Val-Arg-Ile-Leu-Glu-Val-Val-Lys-Asp-Lys-Ala-Gly-

      80                                    90   91
Pro-His-Lys-Glu-Ile-Tyr-Pro-Tyr-Val-Ile-Gln-Glu-Leu-Arg-Pro-Thr-Leu-Asn-Glu-

         100                             108 109
Leu-Gly-Ile-Ser-Thr-Pro-Glu-Glu-Leu-Gly-Leu-Asp-Lys-Val-COOH
```

Fig. 1. The amino acid sequence of bovine heart cytochrome oxidase subunit V [Tanaka et al[25]].

SEQUENCE STUDIES OF THE SUBUNITS

The amino acid sequence of subunit V, a single polypeptide chain consisting of 109 amino acids, has been sequenced by our laboratory[25]. For reference purposes, the sequence is shown in Fig. 1. In addition, the sequence of a 36 residue cysteine peptide from subunit II has been reported by our laboratory[26]. However, further studies have shown that the COOH-terminal region of the reported sequence was in error due to the fact that the peptide was not pure. The sequence of subunit II, as far as it has been sequenced, is shown in Fig. 2. The details of this sequence study will be published elsewhere. There is a 15 residue portion (residues 45-60) which has been very difficult to sequence. The cysteine containing peptide mentioned above occupies residues 195-223. Subunit II is a single polypeptide chain which consists of 233 amino acids with a f-Met NH_2-terminal residue which as others have pointed out[2] is an indication that it is synthesized in the mitochondria under the control of mitochondrial DNA. Further discussions of the possible structure-function aspects of these subunits will be mentioned in the discussion section.

96

```
 1              5                  10                  15
F-Met-Ala-Tyr-Pro-Met-Gln-Leu-Gly-Phe-Gln-Asp-Ala-Thr-Ser-Pro-Ile-Met-Glu-Glu-

 20  21            25                  30                  35
Leu-Leu-His-Phe-Ser-Asp-His-Thr-Leu-Met-Ile-Val-Phe-Leu-Ile-Ser-Ser-Leu-Val-

       40  41          45                  50                  55
Leu-Tyr-Ile-Ile-Ser-Leu(Ile,Leu,Ile,Leu,Ile,Ala,Leu,Thr,His,Trp,Val,Phe,Ser,

          60  61          65                  70                  75
Glx,Asx,Thr,Leu)-Met-Asp-Ala-Gln-Glu-Val-Glu-Thr-Ile-Trp-Thr-Ile-Leu-Pro-Ala-

            80  81          85                  90                  95
Ile-Ile-Leu-Ile-Leu-Ile-Ala-Leu-Pro-Ser-Leu-Arg-Ile-Leu-Tyr-Met-Met-Asp-Glu-

            100 101         105                 110
Ile-Asn-Asn-Pro-Ser-Leu-Thr-Val-Lys-Thr-Met-Gly-His-Gln-Trp-Tyr-Trp-Ser-Tyr-

115                 120                 125                 130
Glu-Tyr-Thr-Asp-Tyr-Glu-Asp-Leu-Ser-Phe-Asp-Ser-Tyr-Met-Ile-Pro-Thr-Ser-Glu-

    135               140 141             145                 150
Leu-Lys-Pro-Gly-Glu-Leu-Arg-Leu-Leu-Glu-Val-Asp-Asn-Arg-Val-Val-Leu-Pro-Met-

    155               160 161             165                 170
Glu-Met-Thr-Ile-Arg-Met-Leu-Val-Ser-Ser-Glu-Asp-Val-Leu-His-Ser-Trp-Ala-Val-

        175               180 181             185                 190
Pro-Ser-Leu-Gly-Leu-Lys-Thr-Asp-Ala-Ile-Pro-Gly-Arg-Leu-Asn-Gln-Thr-Thr-Leu-

        195               200 201             205
Met-Ser-Ser-Arg-Pro-Gly-Leu-Tyr-Tyr-Gly-Gln-Cys-Ser-Glu-Ile-Cys-Gly-Ser-Asn-

210               215               220 221             225
His-Ser-Phe-Met-Pro-Ile-Val-Leu-Glu-Leu-Val-Pro-Leu-Lys-Tyr-Phe-Glu-Lys-Trp-

    230          233
Ser-Ala-Ser-Met-Leu-COOH
```

Fig. 2. The amino acid sequence of bovine heart cytochrome oxidase subunit II.

<u>Predicted</u> <u>Conformation</u> <u>of</u> <u>the</u> <u>Subunits</u>. The Chou-Fasman procedure[27] allows
the prediction of certain protein conformations with about 80% accuracy.
Therefore, the conformation of subunits II (except for 15 residues) and V have
been calculated and the results of the calculation are summarized in Table IV.
Subunit V is predicted to contain 40% α-helix and 15% β-pleated sheet structure,
respectively. It is very much like the 15 proteins that Chou-Fasman used for
their predictive study. On the other hand, subunit II is predicted to contain
about 17% helix and 37% β-pleated sheet structure. The membranous and
amphiphatic cytochrome b_5 has been reported to contain 39% α-helix and 22%
β-pleated sheet conformation.

TABLE IV

SECONDARY STRUCTURE OF THE SUBUNITS OF THE CYTOCHROME OXIDASE

Protein	α-Helix	β-Sheet	Random	β-Turns
1. Chou-Fasman (15 proteins)	36%	17%	47%	
2. Subunit II	17%	37%		10
3. Subunit V	40%	15%		4
4. Cytochrome $\underline{b}_5$	39%	22%	39%	

Cysteine-Cystine Content of Bovine Heart Cytochrome Oxidase Subunits. Thus far, we have only determined the total cysteine plus cystine content of the cytochrome oxidase. Subunit I-VII contained the following number of carboxy-methylcysteinyl-residues: 1, 2, 0, 3, 1, 3, and 0 residues, respectively or a total of 10 such residues. Most of the residues appear to exist as cysteine residues[28,29] as determined by mercurial titration of the native cytochrome oxidase. Yu and Yu[6] have obtained evidence that there is a disulfide bond between subunits IV and VI. As mentioned earlier, subunits IV and VI contain about 3 cysteine residues each. Subunit II has been reported to contain an exposed cysteine residue by Docktor et al[30]. There are two cysteine residues at residues 202 and 206 in subunit II and one of these residues is exposed and reacts with sulfhydryl reagents. Yeast cytochrome oxidase subunit III has been reported to contain a sulfhydryl group which can be -S-S- bonded to a cysteine residue in yeast cytochrome c[31]. Bovine heart cytochrome oxidase subunit II does not contain cystine. Additional experiments are needed to determine the exact amount of cysteine and cystine residues since this data is very important when reconstitution of cytochrome oxidase from the subunits is attempted.

Prosthetic Group Assignments of the Subunits. This is a very controversial topic at the present time which needs to be resolved. Yu and Yu[6] have presented data that subunits I and V are the heme $\underline{a}$ containing subunits. Winters et al[32], on the other hand, report that five labs have obtained evidence that subunits I and III are the heme $\underline{a}$ or $\underline{a}_3$ containing subunits and subunits V and VII are the copper containing subunits. Due to the drastic conditions used for the isolation of the subunits, dissociation of the prosthetic groups is a real danger. Perhaps, a specific method for the isolation of each of the subunits from the native oxidase will be needed to clarify the prosthetic group-subunit relationships. As a matter of fact, whether it is meaningful to specify which subunit links with which prosthetic group (cf. King, et al., this symposium).

DISCUSSION

At the present time, three laboratories are involved in the structure determination of bovine heart cytochrome c oxidase[2,3,25]. The following progress has occurred with respect to the chemical studies of the bovine heart cytochrome c oxidase. The amino acid compositions; the NH_2- and COOH-terminal amino acids have been determined; subunit V has been completely sequenced and nearly the complete sequence of subunit II, except for 15 residues, has been determined; subunits I, III, IV, and VI are currently being sequenced and subunit VII is being stockpiled; the attachment of the prosthetic groups to the subunits are being investigated; the conformation of the subunits is being calculated from the sequence data. The sequence studies are quite involved since the molecular weight of bovine heart cytochrome c oxidase is about 126,000 and some 1100 residues in the seven subunits must be sequenced. However, this data is essential for future crystal X-ray diffraction experiments, for structure-function investigations, for obtaining evolutionary-genetic data as the yeast cytochrome oxidase is also being sequenced by Tsugita.

What has been learned from the sequences studies to date? Subunit V has heme a attached to it when isolated by the procedure of Yu et al[6]. There is 0.5 mole of heme a per mole of subunit V. The heme a is bound very tenaciously to subunit V and the heme a remains bound to the subunit.

However, there are two views concerning the prosthetic group of subunit V. Winter et al[32] report that it is one of the copper containing subunits. What does the sequence investigations indicate about this matter? First of all, the heme a is very tightly bound to subunit V as mentioned earlier and carboxymethylation of the subunit in 8 M urea does not cause the heme a to become liberated. Cleavage of the carboxymethylcysteine derivative by CNBr followed by exclusion-diffusion chromatography on Sephadex G-50 results in the heme being attached to CNBr-I (residues 33-109) while no heme a is found in CNBr-2 (residues 1-33). After tryptic digestion of the Cys-Cm derivative of subunit V and chromatography on Sephadex G-50, the heme is eluted with the largest tryptic peptide (residues 82-109) although at this stage the heme a starts to dissociate from the tryptic peptide. The preliminary evidence thus suggests that heme a is bound very tightly to the COOH-terminal region of the peptide. As reported previously[25], subunit V contains all of the known amino acid side chains which have been shown to chelate with the heme iron. The amino acid sequence of subunit V shows very distant homology with the β-chain of human hemoglobin[25]. However, a drawback concerning this homology is that the calculated conformation of subunit V is different from the β-chain of human hemoglobin although there are some similarities. As stated previously, calculation procedures are only approximate and may not yield the true conformation and therefore the final answer must await crystal X-ray diffraction study of the subunit. On the other hand, if subunit V is a copper containing subunit as reported by Winter et al [32], one must keep one's eyes focused on the sulfhydryl group location in the protein. There is only one sulfhydryl group in subunit V and at position 55. However, the sequence about this cysteine residue is Arg-Ala-Cys-Arg-Arg and is unlike that reported for any copper protein of known sequence. Obviously, further experiments must be conducted to clarify matters.

The sequence determination of subunit II is much more difficult and there is still a stretch of 15 residues which needs to be completed. This includes residues 45-60 which have been extremely difficult to sequence. However, it should be mentioned that the sequence of the rest of subunit II is fairly certain since sufficient types of peptides have been sequenced in order that overlapping sequences exist. Subunit II has been proposed by Yu and Yu[6] to be a copper containing subunit but again, Winter et al[32] disagree and report that this subunit contains no prosthetic group. However, if sulfhydryl groups are copper ligands in cytochrome oxidase, there is a potential copper binding site in this subunit since it contains 2 sulfhydryl groups at residues 202 and 206. This region of the subunit is predicted to contain a β-bend and is a possible copper binding site. However, Dockter et al[30] have recently shown that one sulfhydryl group in subunit II is exposed and reacts with sulfhydryl reagents. Interesting features of this hydrophobic subunit are that it is predicted to contain a considerable amount of β-pleated sheet structure and in addition contains a stretch of many aromatic residues (residues 110, 111, 112, 114, 116, 119, 124 and 127; 8 out of 18 or 44%). Crystal X-ray diffraction studies of proteins have shown that many of the aromatic residues are π-bonded to other aromatic residues and subunit II contains a sufficient number of aromatic residues for π-bonding of this type to occur. Thus, as many as 16 aromatic residues could be localized in this region of the protein subunit. Subunit II also contains many stretches of hydrophobic residues varying from 13 in a row (residues 73-85) to lesser amounts and probably the hydrophobic regions are candidates for the subunit which are localized in the inner membrane. Substantial progress has been made in the sequence determination of other subunits of bovine heart cytochrome oxidase but since the studies are not yet

100

completed, the results obtained will not be discussed here.

It is the hope of our group that these results presented here will be of value to others interested in establishing the molecular mechanism of oxidative phosphorylation at site III in the mitochondrial electron transport system.

The research project was supported in part by Grant GB 22556 from the National Institutes of Health and Grant RIAS SER 77-06923 from the National Science Foundation to KTY, and NIHGM-16767 and HL-12576 to TEK.

REFERENCES

1. Nakashima, T., Higa, H., Matsubara, H., Benson, A. M., and Yasunobu, K. T. (1966) J. Biol. Chem., 241, 1166-1177.

2. Buse, G., and Steffens, G. J. (1976) in Genetics and Biogenesis of Chloroplasts and Mitochondria, Th. Bucherer et al. eds., Elsevier Press, p. 189.

3. Boonman, J. C. P., Muijsers, A. O., and Biersma, R. (1977) 11th FEBS Meeting, Copenhagen.

4. Kuboyama, M., Yong, F. C., and King, T. E. (1972) J. Biol. Chem., 247, 6375-6383.

5. Yu, L., Yu, C. A., and King, T. E. (1977) Biochem. Biophys. Res. Commun., 74, 670-676.

6. Yu, C. A., and Yu, L. (1977) Biochim. Biophys. Acta, 495, 248-259.

7. Tanaka, M., Haniu, M., Yasunobu, K. T., and Mortenson, L. E. (1977) J. Biol. Chem., 252, 7093-7100.

8. Tanaka, M., Haniu, M., Yasunobu, K. T., and Yoch, D. C. (1977) Biochemistry, 16, 3525-3537.

9. Spackman, D. H., Moore, S., and Stein, W. H. (1958) Anal. Chem., 30, 1190-1206.

10. Edman, P., and Begg, C. (1967) Eur. J. Biochem., 1, 80-91.

11. Edman, P. (1956) Acta Chem. Scand., 10, 761-768.

12. Ambler, R. B. (1967) Methods Enzymol., 11, 436-445.

13. Bradbury, J. H. (1958) Biochem. J., 68, 475-486.

14. Kearns, J. J., Yang, C. S., and Gilmour, M. V. (1971) Biochem. Biophys. Res. Commun., 45, 835-841.

15. Rubin, M. (1972) Fed. Proc., 31, 3896.

16. Capaldi, R. A., and Hayashi, H. (1972) FEBS Lett., 26, 261-263.

17. Rubin, M. S., and Tzagoloff, A. (1973) J. Biol. Chem., 248, 4269-4274.

18. Wainio, W. W., Laskowska-Klita, T., Rosman, J., and Grebner, D. (1973) Bioenergetics, 4, 453-467.

19. Takemori, S., Sekuzu, I., and Okunuki, K. (1961) Biochim. Biophys. Acta, 51, 464-472.

20. Tzagoloff, A., Yang, P. C., Wharton, D. C., and Rieske, J. S. (1965) Biochim. Biophys. Acta, 96, 1.

21. Love, B., Chan, S. H. P., and Stotz, E. (1970) J. Biol. Chem., 245, 6664-6668.

22. Orii, Y., and Okunuki, K. (1967) J. Biochem. (Tokyo), 45, 847-854.

23. Matsubara, H., Orii, Y., and Okunuki, K. (1965) Biochim. Biophys. Acta, 97, 61-67.

24. Poyton, R. O., and Schatz, G. (1975) J. Biol. Chem., 250, 752-766.

25. Tanaka, M., Haniu, M., Yasunobu, K. T., Yu, C. A., Yu, L., and King, T. E. (1977) Biochem. Biophys. Res. Commun., 76, 1014-1019.

26. Tanaka, M., Haniu, M., Zeitlin, S., Yasunobu, K. T., Yu, C. A., Yu, L., and King, T. E. (1975) Biochem. Biophys. Res. Commun., 66, 357-367.

27. Chou, P., and Fasman, G. D. (1974) Biochemistry, 13, 222-245.

28. Tsuozuki, T., Orii, Y., and Okunuki, K. (1967) J. Biochem., 62, 37.

29. Kornblatt, J. A., Baraff, G. A., and Williams, G. R. (1973) Canad. J, Biochem., 51, 1417-1427.

30. Dockter, M. E., Steinemann, A., and Schatz, G. (1978) J. Biol. Chem., 253, 311-317.

31. Birchmeirer, W., Kohler, C. E., and Schatz, G. (1976) Proc. Natl. Acad. Sci. U.S.A., 73, 4334.

32. Winter, D. B., Bruyninckx, W. J., and Mason, H. S. (1978) Fed. Proc., 37, 1814, abstract 2984.

NEW METHODOLOGIES APPLIED
TO CYTOCHROME OXIDASE

Cytochrome Oxidase, T.E. King et al. eds.
© 1979 Elsevier/North-Holland Biomedical Press

HEME a MODELS FOR CYTOCHROMES a and a_3 IN CYTOCHROME OXIDASE--EPR, MCD AND RESONANCE RAMAN STUDIES.

G. T. BABCOCK, J. VAN STEELANDT, G. PALMER*, L. E. VICKERY[†] and I. SALMEEN**
Dept. Chemistry, Michigan State Univ., East Lansing, MI 48824; *Dept. Bio-
chemistry, Rice Univ., Houston, TX 77005; [†]Dept. Physiology, Univ. California,
Irvine, CA 92717 and **Ford Motor Company, Dearborn, MI 48121

ABSTRACT

Electron paramagnetic resonance (EPR), magnetic circular dichroism (MCD) and resonance Raman spectroscopies have been used to characterize high- and low-spin heme a compounds. The spectral properties of these models were compared with the corresponding properties previously assigned to cytochromes a and a_3. These comparisons support a model in which cytochrome a is low-spin and cytochrome a_3 is high-spin in resting and fully reduced cytochrome oxidase. The Raman data are interpreted to provide insight into the conformation of heme a in its cytochrome a and a_3 binding sites. In cytochrome a_3^{2+} the heme a formyl group is very nearly in the porphyrin plane and probably not hydrogen bonded; it is thus able to exert both inductive and resonance effects on the ring system. In cytochrome a^{2+} the heme a formyl is much less coplanar with the macrocycle and affects porphyrin and iron behavior in primarily an inductive manner. The modulation of these geometries by, for example, protein conformation or heme redox state, may provide a basis for chromophore interaction in the protein.

INTRODUCTION

Cytochrome oxidase contains two moles of heme a per mole of protein. In the functional enzyme, the two hemes a exist in different local environments and exhibit diverse spectral and ligand binding properties. To describe this situation the cytochrome a/cytochrome a_3 nomenclature, which distinguishes the two hemes a on the basis of their ligand binding properties, has developed: cytochrome a comprises one of the heme a groups and its protein surroundings and does not bind external ligands; cytochrome a_3 includes the second heme a and is accessible to exogenous ligands (e.g., CN^-, N_3^- and CO). Current views of the oxidase generally regard cytochrome a_3 as the site of dioxygen re-duction and cytochrome a as the site of cytochrome c oxidation[1].

An unambiguous determination of the spectral and magnetic properties of cytochromes a and a_3 is complicated by interactions between the metal centers

of the protein. Two kinds of interaction have been recognized. The first
involves an antiferromagnetic coupling between cytochrome a_3 and the EPR undetectable copper (Cu_u) which results in an S=2 ground state for the two metals
in the oxidized enzyme and quenches their expected ferric and cupric EPR
signals[2,3]. We have found that this coupling is quite strong ($2J < -200$ cm^{-1})
and, by analogy with superoxide dismutase, have postulated that the coupling
interaction is mediated by a bridging imidazole ligand as shown in the structure
below[4,5]. The second type of metal-metal interaction involves redox coupling
between cytochrome a and a_3 such that the redox and
ligation state of one of the hemes alters the redox
potential and, to a lesser extent, the spectral
properties of the second heme[6,7]. The molecular
basis for this phenomenon is unknown. Cytochrome
oxidase does undergo substantial conformational changes
during redox cycling[4], however, and it is possible
that the two phenomena are causally related.

   Despite the complications introduced by the metal-metal interactions in the
protein we have been able to interpret our previous MCD, EPR and resonance
Raman results in terms of a model in which cytochrome a is low-spin and cytochrome a_3 is high-spin in both the fully oxidized and fully reduced states of
the enzyme[3,4,5,7,8]. In the experiments reported here we have prepared high-
and low-spin complexes of heme a and have compared the spectral, magnetic and
vibrational properties of these model compounds with the corresponding properties postulated for cytochromes a and a_3. One objective of the present experiments is to confirm and extend the protein assignments; a second and more
important objective is to identify properties of the heme a chromophore which
are especially susceptible to environmental modulation and may consequently
serve as loci for functional control within the protein.

MATERIALS AND METHODS
   Heme a was isolated as described by Takemori and King[9] and handled as des-
cribed previously.[3] MCD and resonance Raman data were obtained with the same
instrumentation used in earlier studies[3,8]. EPR spectra were recorded at
X-band with a Varian E4 EPR spectrometer equipped with an Oxford Instruments
ESR-9 liquid helium cryostat. Optical spectra were recorded using either
Cary 17 or McPherson EU-707D recording spectrophotometers.
   The chemicals used were obtained from standard commercial sources. Imidazole
was recrystallized from benzene or from ethanol. We have found that dithionite

TABLE I

g-VALUES AND LIGAND FIELD PARAMETERS* FOR LOW-SPIN HEME a COMPOUNDS

Species	g-values			$|\Delta/\lambda|$	$|V/\Delta|$
cyt a^{3+}	3.03	2.21	1.45	2.97	0.87
heme a (Im)$_2$	2.96	2.29	1.48	2.88	0.73
heme a (Im)$_2$ + HSO$_3^-$	2.92	2.28	1.52	3.03	0.73
heme a (Im)$_2$ pH 11.3	2.71	2.27	1.74	4.08	0.69
protohemin (Im)$_2$**	3.02	2.24	1.51	3.08	0.86

*Calculated according to the method of Taylor[12] using the axis system of
 ref. 11; $|\Delta/\lambda|$ = tetragonality, $|V/\Delta|$ = rhombicity.
**From ref. 11.

oxidation products enhance the degradation of heme a. To circumvent this
difficulty heme a solutions were deaerated prior to stoichiometric reduction
by aqueous, buffered dithionite. Brij 35 solutions were prepared and stored
under argon.

RESULTS

   Cytochrome a Models. Van Gelder and Beinert[2] originally identified the low-
spin species (g-values: 3.03, 2.21, 1.45) observed in the EPR spectrum of the
oxidized enzyme with cytochrome a. This assignment remained somewhat controversial due to the unusual redox properties of the enzyme but was considerably strengthened by our MCD results which provided strong support for a low-
spin assignment for cytochrome a in both ferric and ferrous states[3]. The axial
ligands to the iron of cytochrome a are unknown; however, Smith and Williams,
arguing from optical data, have suggested bis(imidazole) ligation[10]. From the
g-values for a^{3+}, the ligand field parameters, Δ/λ and V/Δ, can be calculated
(Table I). The numerical values are typical of group B of Peisach, J. et al[11]
again suggesting bis(imidazole) ligation.

   Figure 1 shows EPR spectra we have recorded for various heme a compounds.
The top trace shows bis(imidazole) heme a at neutral pH; the g-values and ligand
field parameters (Table I) are in reasonable agreement with those observed for
cytochrome a^{3+}. The formyl group at the heme a 8 position introduces a strong
perturbation to the π electron system of the porphyrin ring and results in the
characteristic green color of bis(imidazole) heme a complexes. The corresponding protohemin complexes are orange-brown. Surprisingly, the formyl group

108

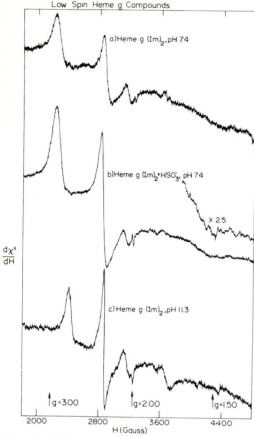

Low Spin Heme a Compounds

a) Heme a (Im)$_2$, pH 7.4

b) Heme a (Im)$_2$+HSO$_3^-$, pH 7.4

x 2.5

c) Heme a (Im)$_2$, pH 11.3

$\frac{d\chi''}{dH}$

g=3.00   g=2.00   g=1.50

2000   2800   3600   4400
H (Gauss)

Fig. 1. EPR spectra of low spin heme a compounds. T = 9.2K, 5 mW microwave power, 20G modulation amplitude. The buffer was 0.1 M sodium phosphate, 0.07 M cetyltetramethyl ammonium bromide and 0.4 M imidazole, and in (b) 0.1 M sodium bisulfite was also present.

appears to have little influence on the ligand field geometry or the electron density at the iron as evidenced by a comparison of the g-values and ligand field parameters in Table I for cyt a^{3+}, heme a (Im)$_2$ and protohemin (Im)$_2$. We have examined this result in more detail by reacting the formyl group of the heme a (Im)$_2$ complex with bisulfite to form the saturated addition product,

$$- \overset{\overset{\displaystyle OH}{|}}{\underset{\underset{\displaystyle H}{|}}{C}} - SO_3^- \; Na^+,$$ at the porphyrin

8 position[13]. The species formed in this reaction is orange-brown. The EPR spectrum of this compound is shown in Figure 1b and the relevant parameters are collected in Table I. As anticipated there is little difference between the ligand field parameters of the bis(imidazole) complexes of heme a and its bisulfite addition compound. We have also prepared the high pH form of bis(imidazole) heme a in which the N1 proton has been removed (Fig. 1c). The ligand field parameters (Table I) allow the assignment of this complex to the H group of Peisach et al which also contains the high pH forms of oxidized cytochrome b$_5$ and protohemin (Im)$_2$. The data of Figure 1 and Table I allow two conclusions: (a) the EPR properties of low-spin heme a^{3+} complexes are similar to those of the corresponding protoheme complexes, and (b) heme a^{3+} (Im)$_2$ at neutral pH reasonably reproduces the EPR properties of cytochrome a^{3+}.

In previous work we were able to decompose MCD spectra of cytochrome oxidase and its derivatives so that Soret spectra of cytochromes a and a$_3$ in both valence states could be calculated[7]. These computed curves are shown in

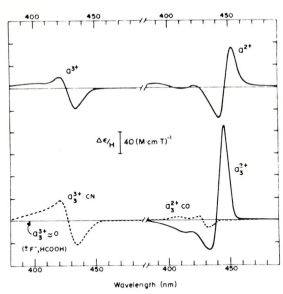

Fig. 2. Computed MCD spectra of cytochrome $\underline{a}$ and $\underline{a}_3$. For details, see ref. 7.

Figure 2. The Soret absorption and MCD spectra of ferric and ferrous heme $\underline{a}$ $(Im)_2$ complexes are shown in Figure 3. The oxidized heme $\underline{a}$ $(Im)_2$ model is in good agreement with cytochrome $\underline{a}^{3+}$ in terms of MCD $\underline{C}$ term band shape and intensity. This is expected, however, since Vickery et al[14] showed that for ferric heme proteins these parameters are primarily determined by the iron spin state and are relatively insensitive to the axial ligands. More striking is the general agreement between the MCD spectrum of cytochrome $\underline{a}^{2+}$ and that of the reduced heme $\underline{a}$ $(Im)_2$ complex. Both show the anomalous -/+ derivative MCD band shape and, in each, the MCD zero crossing is shifted approximately 5 nm to the red of the Soret absorption maximum (recent MO calculations indicate that the "reversed" derivative band shape arises from magnetic field induced mixing of two nearly degenerate excited states[15]). The peak amplitude of the ferrous model compound is less than that of cyt. $\underline{a}^{2+}$ owing to the greater full-width at half-height bandwidth in the former (14.1 nm compared to 8.9 nm). The integrated intensities, obtained as the product of (full-width at half-height)^2 and amplitude, for these two species agree to within 5%. From the combined MCD and EPR results we conclude that the heme $\underline{a}$ $(Im)_2$ complex is a good model for cytochrome $\underline{a}$.

Raman spectra of reduced cytochrome oxidase and a number of low-spin and high-spin model compounds are shown in Figure 4. (Photoreduction of the oxidized enzyme occurs in the laser beam and complicates attempts to record the Raman spectrum of this species[8,16,17]. We have recently circumvented this difficulty and will report our results shortly.) In previous work[8] we showed that both cytochrome $\underline{a}^{2+}$ and $\underline{a}_3^{2+}$ contributed to the observed spectrum but that the bands observed at 215, 364, 1230 and 1670 cm^{-1} were due only to cyt $\underline{a}_3^{2+}$. We assigned the 1670 cm^{-1} band to the formyl group of cyt $\underline{a}_3^{2+}$ and

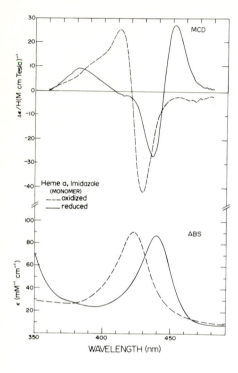

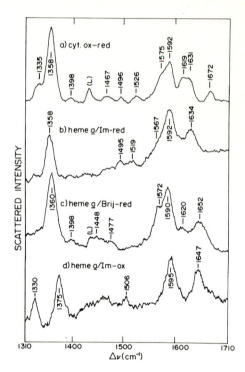

Fig. 3. Absorption and MCD spectra of ferric and ferrous heme a (Im)₂ complexes. The buffer was 0.1 M sodium phosphate, 0.005 M sodium dodecyl sulfate, pH 7.4.

Fig. 4. High frequency Raman spectra (λ_{ex} = 441.6 nm, incident intensity ≈ 30 mW) of reduced cytochrome oxidase and a number of model complexes. Buffers: (a) 0.1 M HEPES, 0.5% Brij 35, pH 7.4; (b) and (d) as in Fig. 1; (c) 0.1 M HEPES, 2% Brij 35, pH 7.4.

suggested that for this species, but not for cyt $\underline{a}^{2+}$, the formyl group is in the porphyrin plane or very nearly so and allows strong coupling between the C=O π electrons and the porphyrin π system. We also speculated that the 215 cm^{-1} band may arise from the mode of coupling between the iron of cytochrome $\underline{a}_3$ and its associated, antiferromagnetically coupled copper.

The Raman spectrum of heme $\underline{a}^{2+}$ (Im)₂ (Fig. 4b) is similar to that expected for cytochrome $\underline{a}^{2+}$ in that bands at 215, 364, 1230 and 1670 cm^{-1} are not observed for this low-spin species. Furthermore the bands at 1634, 1592 and 1358 cm^{-1} are similar in frequency and intensity pattern to lines observed in the oxidase at 1631, 1592 and 1358 cm^{-1}. The similarity between heme

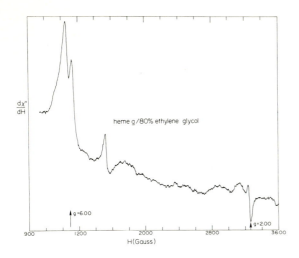

Fig. 5. EPR spectrum of heme $\underline{a}$ in 80%
ethylene glycol/20% 0.1 M sodium phosphate
buffer, pH 8.8. T = 10K, instrumental
conditions as in Fig. 1. The apparent
$\underline{g}$-values are 6.21, 5.57 and 2.0.

$\underline{a}$ (Im)$_2$ and cytochrome $\underline{a}$
established above encourages
us to associate the 1631 cm^{-1}
band seen in the oxidase with
cytochrome $\underline{a}^{2+}$; data are
presented below which indicate
that both cytochromes $\underline{a}$ and
$\underline{a}_3$ contribute intensity to
the 1358 and 1592 cm^{-1} bands.
A comparison of these data
with Raman spectra of low-
spin ferrous heme proteins
and iron porphyrins[18,19]
suggests that the 1631 cm^{-1}
and 1592 cm^{-1} bands are
shifted 6-10 cm^{-1} to higher
frequencies relative to the
corresponding bands in the
protoheme systems. An analogous effect can be seen
for oxidized heme $\underline{a}$ (Im)$_2$ (Fig. 4d): the 1647 cm^{-1} and 1595 cm^{-1} bands occur
at higher frequency than those of low-spin ferric protoheme compounds[18,19].
A possible explanation for these frequency shifts is electron withdrawal from
the porphyrin π system by the heme $\underline{a}$ formyl group. Judging by the absence of
a resonance enhanced formyl vibration this effect would be primarily inductive.

   Cytochrome $\underline{a}_3$ Models. Lemberg[20] originally emphasized the blue shift
that occurs in the cytochrome oxidase α band region upon CO binding to $\underline{a}_3^{2+}$.
Accordingly, our initial work on cytochrome $\underline{a}_3$ models has been with heme $\underline{a}$
species whose ferrous forms show this phenomenon. Thus far we have found
two classes of these compounds: (a) heme $\underline{a}$ in solutions of neutral detergent[20]
and (b) heme $\underline{a}$ in an ethylene glycol/water mixed solvent. In the experiments
described below both types of model systems gave essentially the same results.

   The EPR spectrum of oxidized heme $\underline{a}$ in ethylene glycol/water (4:1), shown
in Figure 5, is that of high-spin ferric iron. Figure 6 shows the Soret
absorption and MCD spectra of ferric, ferrous and carbon monoxy ferrous heme
$\underline{a}$ in the same solvent. For the oxidized species, the MCD intensity is very
weak, as expected from the work of Vickery $\underline{et}$ $\underline{al}$ on ferric high-spin myoglobin
derivatives[14] and the data of Fig. 5. Moreover, the computed MCD spectrum of

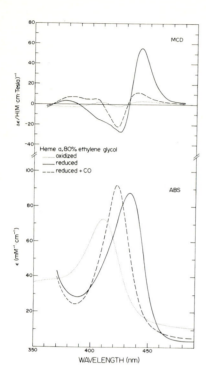

Fig. 6. Absorption and MCD spectra of heme a complexes in 80% ethylene glycol/20% 0.1 M sodium phosphate, pH 8.8.

high-spin cyt. $\underline{a}_{-3}^{2+}$ (Fig. 2) shows the same weak intensity. The ferrous heme $\underline{a}$ compound exhibits the intense MCD spectrum characteristic of S=2, Fe^{2+} hemes[14] and the bandshape agreement with the calculated MCD spectrum of ferrous cyt $\underline{a}_3$ in Fig. 2 is good. In addition, whereas the MCD zero crossing for cytochrome a^{2+} and its heme $\underline{a}^{2+}$ model compound are red-shifted with respect to their Soret optical absorption maxima, both cytochrome $\underline{a}_{-3}^{2+}$ and heme $\underline{a}^{2+}$ in 80% ethylene glycol exhibit zero crossings which are blue-shifted 2-4 nm relative to the absorption maxima (Fig. 6 and ref. 21). The bandwidth of the positive arm of the MCD spectrum of the model compound is about 7 nm greater than that calculated for cytochrome $\underline{a}_{-3}^{2+}$ (this parameter is postulated to be a sensitive function of the relative magnitudes of the Zeeman splitting and spin orbit coupling in the excited state[22]). The MCD spectrum of the heme $\underline{a}^{2+} \cdot$ CO complex is in agreement with that calculated for $\underline{a}_{-3}^{2+} \cdot$ CO except for the low energy, positive feature seen in the model at about 440 nm.

The resonance Raman spectrum of reduced heme $\underline{a}$ in 2% Brij 35 is shown in Figure 4c. Brij 35 was used in these studies rather than, for example, Tween 20 because the former is free of fluorescent impurities. In the high frequency region three features are of interest: (a) the intensity ratio of bands at 1590 and 1572 cm^{-1}, (b) the band at 1620 cm^{-1} and (c) the band at 1652 cm^{-1}. For ferrous protoheme compounds the intensity ratio 1584/1566 is a sensitive function of spin state, being less than one for high-spin species and greater than one for low-spin species[18]. If, as suggested above, the 1590 and 1572 cm^{-1} bands are the analogous ferrous heme $\underline{a}$ bands, then this empirical correlation breaks down for heme $\underline{a}$; while the 1590/1572 intensity ratio is greater for the high-spin heme a/Brij model (Fig. 4c) than for the low-spin heme a (Im)$_2$ species (Fig. 4b), it is still less than one in both cases. This observation may explain why the 1592/1575 intensity ratio is also less than one for the reduced oxidase (Fig. 4a).

In ferrous, high-spin protoheme compounds the highest frequency band occurs in the 1605-1615 cm^{-1} region[18,19]. For the heme a^{2+}/Brij model we suggest that analogous band occurs at 1620 cm^{-1}; the ~5 cm^{-1} shift to higher frequency may again be attributable to the electron withdrawing formyl substituent. The band at 1652 cm^{-1} is anomalously high for ferrous compounds, particularly since heme vibrational bands generally exhibit a shift to lower frequencies upon reduction. For the heme a model compounds of Fig. 4 the 1652 cm^{-1} band is the highest frequency line observed, higher even than the ferric, low-spin line seen at 1647 cm^{-1} in heme a^{3+} (Im)$_2$. The most obvious explanation for the 1652 cm^{-1} band is that it arises from the heme a^{2+} formyl group. The 20 cm^{-1} shift to lower frequency relative to the 1672 cm^{-1} formyl band in the reduced oxidase could be explained if hydrogen bond formation to the formyl occurred for the model but were absent in the protein. The converse situation apparently occurs for chlorophyll b which carries a formyl group at the ring 3 position[23]. Lutz has shown that chlorophyll b in acetone has Raman lines in the 1660-1670 cm^{-1} region which he assigned to the formyl group. In vivo the chlorophyll b formyl C=O group was assigned to vibrations shifted 25-35 cm^{-1} to lower frequencies compared to the monomer. Shifts of this magnitude are typical of hydrogen bonding[24] and consequently Lutz proposed that hydrogen bonds exist between the polypetide and the formyl of chlorophyll b in vivo and these serve to anchor the chromophore to the protein.

The data of Figures 4, 5 and 6 demonstrate that heme a in ethylene glycol or neutral detergent has many features similar to cytochromes a$_3$ but the spectrum of this model system does not show a 215 cm^{-1} band, suggesting again that this vibration is unique to the native reduced oxidase. This reinforces the speculation above concerning the relationship between the 215 cm^{-1} band and the magnetic coupling between cytochrome a$_3$ and copper.

DISCUSSION

The combined use of EPR, MCD and resonance Raman in the study of cytochrome oxidase is an especially useful approach: the magnetic techniques provide spin state and ligand field information on the central iron of heme a while Raman probes electron distribution and geometry for the surrounding porphyrin macrocycle and its substituents.

The results presented above support our earlier assignments of spin states for cytochrome $\underline{a}$ and $\underline{a}_3$ [3,4]. The EPR and MCD properties of the low-spin heme $\underline{a}$ $(Im)_2$ complex are in agreement with those calculated for cytochrome $\underline{a}$; the MCD parameters of high-spin heme $\underline{a}$ in ethylene glycol or Brij 35 compare well with those we have assigned to cytochrome $\underline{a}_3$. In addition, for the heme $\underline{a}$ $(Im)_2$ model compound our combined data present a reasonably strong argument for the presence of bis(histidine) axial ligation in cytochrome $\underline{a}$ in situ. The high-spin heme $\underline{a}$ model for cytochrome $\underline{a}_3$ is much less adequate; neither the proposed histidine axial ligand nor the antiferromagnetically coupled copper has been reproduced.

Hill and Wharton, drawing on results obtained with Pseudomonas cytochrome oxidase, have recently pointed out the functional requirement for an unusual heme (either heme d_1 or heme $\underline{a}$) in the enzyme active site[25]. The Raman results of Figure 4 provide several possible clues to the structural basis for this selectivity. The suggestion, made above, that hydrogen bonding does not occur to the formyl of cytochrome $\underline{a}_3^{2+}$ is noteworthy in that hydrogen bond formation between polypeptide and coenzyme appears to be commonplace in other proteins[23,26,27]. The possible absence of this effect for the formyl of cytochrome $\underline{a}_3^{2+}$ suggests that the function of the carbonyl group is more complex than merely to serve as a site of attachment of the chromophore to the heme pocket. This possibility is made particularly attractive in view of the apparent iron spin state dependence of the formyl geometry discussed in ref. 8 and above. In contrast, the vinyls of the protohemes in hemoglobin and cytochrome $\underline{b}_5$ are conjugated with the porphyrin π system in both iron valence states[28,29]. Geometric modulation of the formyl of heme $\underline{a}$ would control its inductive and resonance effects on the macrocycle π density and could reflect, though not necessarily be the origin of, the electronic basis for redox cooperativity.

ACKNOWLEDGEMENTS

This research was supported by a Cottrell Research Grant from the Research Corporation, a Michigan State University Biomedical Research Support Grant, NIH Grant GM-21337 (GP), Welch Foundation Grant C-636 (GP), and by the U.S. Energy Research and Development Association.

REFERENCES

1. Erecinska, M. and Wilson, D. F. (1978) Arch. Biochem. Biophys., 188, 1-14.
2. Van Gelder, B. F. and Beinert, H. (1969) Biochim. Biophys. Acta, 189, 1-24.

3. Babcock, G. T., Vickery, L. E. and Palmer, G. (1976) J. Biol. Chem., 251, 7907-7919.

4. Palmer, G., Babcock, G. T. and Vickery, L. E. (1976) Proc. Nat. Acad. Sci., 73, 2206-2210.

5. Tweedle, M. F., Wilson, L. J., Garci-Iniguez, L., Babcock, G. T. and Palmer G. (1978) J. Biol. Chem., in press.

6. Wikstrom, M. K. F., Harmon, H. J., Ingledew, W. J. and Chance, B. (1976) FEBS Letts., 65, 259-276.

7. Babcock, G. T. Vickery, L. E. and Palmer, G. (1978) J. Biol. Chem., 253, 2400-2411.

8. Salmeen, I., Rimai, L. and Babcock, G. T. (1978) Biochemistry, 17, 800-806.

9. Takemori, S. and King, T. E. (1965) J. Biol. Chem., 240, 504-513.

10. Smith, D. W. and Williams, R. J. P. (1970) Struct. Bonding (Berlin), Vol. 7, 1-45.

11. Peisach, J. Blumberg, W. E. and Adler, A. (1973) Ann N.Y. Acad. Sci., 206, 310-327.

12. Taylor, C. P. S. (1977) Biochim. Biophys. Acta, 491, 137-149.

13. Yanagi, Y., Sekuzu, I., Orii, Y. and Okunuki, K. (1972) J. Biochem. (Tokyo), 71, 47-56.

14. Vickery, L., Nozawa, T. and Sauer, K. (1976) J. Am. Chem. Soc., 98, 343-350.

15. Kaito, A., Nozawa, T., Yamamoto, T., Hatano, M. and Orii, Y. (1977) Chem. Phys. Letts., 52, 154-160.

16. Kitagawa, T., Kyogoku, K., and Orii, Y. (1977) Arch. Biochem. Biophys., 181, 228-235.

17. Adar, F. and Yonetani, T. (1978) Biochim. Biophys. Acta, 502, 80-86.

18. Yamamoto, T., Palmer, G., Gill, D., Salmeen, I. and Rimai, L. (1973) J. Biol. Chem., 248, 5211-5213.

19. Spiro, T. G. and Burke, J. M. (1976) J. Am. Chem. Soc., 98, 5482-5489.

20. Lemberg, M. R. (1969) Physiol. Rev., 49, 48-121.

21. Vanneste, W. H. (1966) Biochemistry, 5, 838-848.

22. Treu, J. I. and Hopfield, J. J. (1975) J. Chem. Phys., 63, 613-623.

23. Lutz, M. (1977) Biochim. Biophys. Acta, 460, 408-430.

24. Josien, M. L. (1963) Pure Appl. Chem., 4, 33-60.

25. Hill, K. E. and Wharton, D. C. (1978) J. Biol. Chem., 253, 489-495.

26. Salemme, F. R. (1977) Ann. Rev. Biochem., 46, 299-329.

27. Dutta, P. K., Nestor, J. R. and Spiro, T. G. (1977) Proc. Nat. Acad. Sci. U.S.A., 74, 4146-4149.

28. Spiro, T. G. and Strekas, T. C. (1974) J. Am. Chem. Soc., 96, 338-345.

29. Adar, F. (1975) Arch. Biochem. Biophys., 170, 644-650.

Cytochrome Oxidase, T.E. King et al. eds.
© 1979 Elsevier/North-Holland Biomedical Press

MAGNETIC CIRCULAR DICHROISM STUDIES ON CYTOCHROME OXIDASE AND HEME A DERIVATIVES

TSUNENORI NOZAWA, YUTAKA ORII[*], AKIRA KAITO, TAKAO YAMAMOTO AND MASAHIRO HATANO

Chemical Research Institute of Non-aqueous Solutions, Tohoku University, Sendai 980, and *Department of Biology, Faculty of Science, Osaka University, Toyonaka 560 (Japan)

ABSTRACT

Magnetic circular dichroism (MCD) spectra of heme a derivatives were fairly different from those of heme b derivatives whose MCD spectra have been extensively studied. The peculiarities in the MCD spectra of heme a were ascribable to the presence of the formyl substituent at the eight position in the porphyrin periphery. Oxidation and spin dependence of the MCD spectra of heme a derivatives could be attributed to the effect of the iron electronic states just like in those of heme b derivatives. MCD spectra of cytochrome oxidase and its various derivatives were analyzed on the basis of the knowledge obtained from the model systems.

INTRODUCTION

Mammalian cytochrome oxidase contains heme a and copper in a 1 : 1 ratio for about 100,000 g protein[1], and catalyzed the aerobic oxidation of reduced cytochrome c as a natural electron donor. The prosthetic group heme a has a formyl group at the position 8, instead of a methyl group in protoheme (heme b) or heme c, and is responsible for its unique spectral and chemical features. The formyl group so strongly affects the porphyrin π electron system that the absorption bands usually appear at longer wavelength than those of the other two hemes. However, when this group forms addition compounds with carbonyl reagents, the absorption band shifts to shorter wavelength, approaching those of the heme b or heme c derivatives[2,3]. Therefore, it is interesting to see how these characteristics are reflected in MCD spectra, since these are sensitive to differences in electronic structures as well as to changes in the redox and spin states of the hemes[4].

MCD is the optical activity of matters induced by a magnetic field parallel to the incident light. Theories show that the observed MCD can be resolved

into three different terms called Faraday A, B and C terms[5,6]. The A term has
its origin in the Zeeman splitting in the ground or excited state, hence is
nonzero only when the ground or excited state has orbital or spin degeneracy.
The orbital or spin degeneracy in the ground state gives rise to the C term
which comes from the population differences of electrons in the ground state
sublevels separated by the Zeeman effect, hence its magnitude increases with
the decrease of the temperature. The B term originates from the mixing of
other electronic states into the specific ground or excited state by the magnetic field. Usually the S-shaped dispersion for the A term can be distinguished
from the bell shaped ones for the B and C terms. However the distinction among
Faraday parameters from their apparent shapes becomes sometimes superficial.
Thus, in the porphyrin system the chromophore does not have true C_4 symmetry
if the substituents of the porphyrin periphery are rigorously taken into
account. Hence we can not expect Faraday A term for the Soret and Q bands,
because they have no actual degeneracy. However, if the split of the degeneracy is so small that the two split states interact exclusively with
themselves, a couple of Faraday B terms have opposite signs and appear very
close in energy. The resulting spectra will have very similar dispersion shape
to that of Faraday A term. We call these two Faraday B terms as "apparent" A
or A' term which can be treated just like true Faraday A term[7]. When two closely existing Faraday C terms have opposite signs the apparent dispersion shape
will also mimic the A term[4,8].

We investigated the MCD spectra of cytochrome oxidase and heme a derivatives
under several conditions, and observed unique MCD profiles which were ascribed
to the formyl substituent of heme a.

MATERIALS AND METHODS
Cytochrome oxidase was prepared according to the method of Okunuki et al.[9]
with a slight modification; that is, the final ammonium sulfate precipitate
was dissolved in 0.25% Emasol 1130-0.1% (w/v) sodium cholate-0.05 M sodium
phosphate buffer (pH7.4), and dialyzed against the same medium with stirring
at least for 2 hr in a cold room. The concentration was determined by using
a millimolar extinction coefficient difference of $\Delta\varepsilon$ (605 - 630 nm, reduced
form) = 16.5. Heme a was extracted from purified cytochrome oxidase according
to the method of Takemori and King[10], and finally dissolved in 1% (w/v) aqueous
sodium carbonate. The concentration was determined based on a millimolar extinction coefficient of 27.4 at 578 nm for the pyridine ferrohemochrome[2].

The solvent systems for low temperature experiments consist of glycerol (70%) and sodium phosphate buffer (pH = 7.4, 0.015 M) with 2% sodium cholate and 0.5 M potassium chloride.

The MCD magnitudes were expressed in molar ellipticity per gauss on the basis of heme a morality.

RESULTS AND DISCUSSION

Ferroheme a derivatives. Since dithionite-reduced free heme a in SDS failed to give a well defined spectrum, the MCD spectra of imidazole- and pyridine-ferroheme a were recorded (Figure 1). It is noted that these MCD spectra in the Soret and visible regions are reversed in sign compared with those for ferrous low-spin complexes of heme b like cytochrome b_2[8], and b_5[4] and other derivatives of myoglobin and hemoglobin[4]. The monomeric ferroheme a-imidazole complex in the mixture of glycerol (70%) and sodium phosphate buffer (pH 7.4, 0.015 M) with 2% sodium cholate and 0.5 M potassium chloride (the concentration of ferroheme a, 25 μM, imidazole (0.1 M)) did not show any intrinsic temperature variation from 288 to 115 K both in the Soret and Q regions (not shown). This indicates the assignment that the Faraday parameters for the ferroheme a complexes are Faraday A' or a couple of B terms throughout the Soret and Q regions. It may be important to add that sodium cholate titration by absorption spectra in the above conditions implied the aggregate to monomer change of ferroheme a-imidazole complex and completely monomer state in 2% sodium cholate concentration.

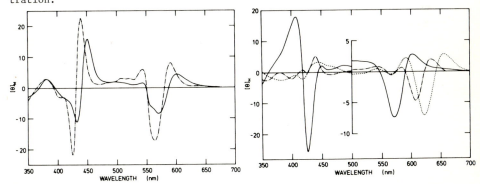

Fig. 1. MCD spectra of imidazole (⟶) and pyridine complexes of ferroheme a. Heme a (20.6 μM) dissolved in 2.5 % (w/v) SDS containing 48 mM imidazole, pH 7.66, or heme a (12 μM) in a mixture of pyridine (20%, v/v) and NaOH (0.1 N) was reduced with sodium dithionite. The concentrations in parentheses represent the final ones.

Fig. 2. MCD spectra of ferriheme a (······) and its complex with imidazole (⟶) or pyridine (⟶·⟶). Heme a (21.6 μM) was dissolved in 2.5% SDS, pH 10.17. The spectra of other complexes were recorded before the reduction of each sample as appeared in the legend to Fig. 1.

120

Ferriheme a derivatives. Figure 2 illustrates the MCD spectra of monomeric
ferriheme a and its derivatives[2]. Since in the visible region the MCD of imidazole-ferriheme a in the low temperature-solvent with 0.5 M potassium chloride
and 2% sodium cholate (Ferriheme a, 25 µM; imidazole, 0.1 M) did not show any
temperature dependence from 291 to 119 K (not shown), this was assigned reasonably to Faraday A' or the two B terms. In the Soret region a less featured
MCD spectrum of free monomeric heme a would be ascribed to its high spin character, and conceivably consists of Faraday C and A' (a couple of B) terms.
Pyridine-ferriheme a exhibited an MCD spectrum characteristic for high spin
heme a. Therefore, it is conceivable that even if pyridine is a strong ligand
to protoheme, it is a weak field-ligand toward ferriheme a. Contrary to free
heme a and its pyridine compound, imidazole-ferriheme a showed a distinct peak
and trough at 486 and 450 nm. The MCD spectrum of the imidazole-ferriheme a
in the low temperature-solvent with 2% sodium cholate and 0.5 M potassium chloride showed temperature dependence as illustrated in Figure 3. In the Figure 3
the MCD magnitude at the peak around 408 nm or at the trough around 430 nm was
plotted against the reciprocal of the absolute temperature. Since the plots
give the linear lines with zero intercept at 1/T = 0, and the absorption spectra suffered no change from 291 K down to 119 K, the Soret MCD peak and trough
of the imidazole-ferriheme a were determined to be composed of Faraday C terms.

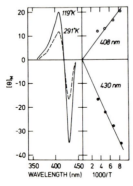

Fig. 3. Effect of temperature on the
MCD spectra of ferriheme a imidazole.
Ferriheme a (25 µM)-imidazole (0.1 M)
was dissolved in the mixture of glyc-
erole (70%) and sodium phosphate
buffer (0.015 M) with sodium cholate
(2%) and potassium chloride (0.5 M).
The right side illustrates the plots
of either the height or trough of the
MCD spectra against the reciprocal of
the absolute temperature.

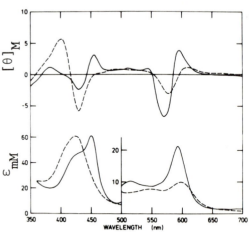

Fig. 4. MCD spectra (above) and absorption spectra (below) of heme a-imidazole.
Heme a was dissolved in 0.05 M sodium
phosphate buffer, pH 7.66, containing 0.1
M imidazole. ----, the oxidized form;
——, the reduced form.

Effect of carbonyl reagents on the MCD of imidazole-heme a. The carbonyl
nature of the formyl side chain in heme a is easily lost once it forms cyan-
hydrin with cyanide or an addition compound with sodium bisulfite, and the nature of heme a derivative is expected to approach that of heme b or heme c derivatives. In the following experiments, as detergents are known to inhibit the
addition compound formation[2], SDS was eliminated from the reaction mixtures.
Consequently the MCD magnitude of imidazole-ferriheme a in the Soret region was
decreased by more than 60%, although it exhibited the same pattern of MCD spectrum as that in the presence of SDS. In Figure 4 the MCD and absorption spectra
for imidazole-heme a in both reduced and oxidized states are illustrated for
comparison. When sodium bisulfite was added to imidazole-ferriheme a, a heme b
type absorption spectrum appeared having the maxima at 533 and 409 nm with a
shoulder around 570 nm and an increased absorption at the shorter wavelength
side of the Soret peak. When reduced with sodium dithionite a complicated spectrum appeared; that is, in the visible region the most intense peak was at 560
nm and this was flanked by a peak at 586 nm and a twin peak with maxima at 531
and 523 nm. The Soret peak was at 413 nm with a slight absorption increase to
the longer wavelength side (Figure 5). The absorption spectrum of the cyan-
hydrin was essentially the same as that described above. In accordance with

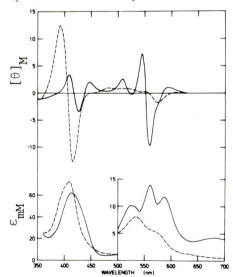
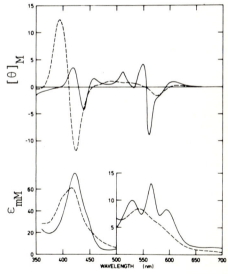

Fig. 5. MCD spectra (above) and absorption spectra (below) of heme a-imidazole
in the presence of sodium bisulfite. To
2.0 ml of a solution of heme a-imidazole
in 0.05 M sodium phosphate buffer which
was prepared as described in the legend
to Fig. 4 was added 100 mg of NaHSO₃.
The final pH was 5.81. ---, the oxidized form; ——, the reduced form.

Fig. 6. MCD spectra (above)and absorption spectra (below) of heme a-imidazole
in the presence of cyanide. Heme a was
dissolved in 0.1 M imidazole-0.2 M KCN-
0.05 M sodium phosphate buffer, pH 7.93.
---, the oxidized form; ——, the reduced form.

these spectral profiles, the MCD patterns of the addition compounds were also
unique. Imidazole-ferriheme a supplemented with sodium bisulfite exhibited the
MCD pattern resembling that of ferric protoheme complexes in a low-spin state.
In the Soret region the magnitude of both the peak at 392 nm and trough at 415
nm was increased about two fold. This intensification must reflect the contribution of Faraday C terms to these extrema which are larger in heme b or heme c
derivatives than that in heme a derivatives. Imidazole-heme a complexes in the
presence of cyanide also gave the MCD spectra of the same pattern (Figure 6).
It is noteworthy that the MCD patterns of the addition compounds in the reduced
state are much more like those of carbonmonoxymyoglobin[4], ferrocytochrome b_5[4,11]
and ferrocytochrome c[4] especially in the visible region. The most characteristic point is the intense peak at 545-549 nm at a shorter wavelength side of a
deep trough at either 559 or 561 nm. In the Soret region the MCD spectra might
be explained by an overlap of a sinusoidal curve due to the addition compound
and an S-shaped curve due to the remaining heme a, since from the absorption
spectra it is apparent that the samples are in fact the mixtures of heme a and
heme b type complexes. From these results the unique features of the MCD pattern of heme a derivatives are ascribed to the existence of a formyl side chain.

Effect of formyl substituent on Faraday parameters

We can summarize the kinds and signs of Faraday parameters of heme a model
complexes as in Table 1 which also includes the data for iron porphin and
heme b. It is noted in Table 1 that kinds of Faraday parameters are same
among three types of heme derivatives over the whole spec-

Table 1
Kinds and signs of Faraday parameters for various hemes

Oxidation State	Fe(II)				Fe(III)			
Spin State	Low		High		Low		High	
Wavelength Region	Soret	Q	Soret	Q	Soret	Q	Soret	Q
Iron Porphin	A	A	C	A	C	A	*	A
Heme b	A'[b]	A'	C	A'	C	A'	*	A'
Heme a	-A'	-A'	C	-A'	C	-A'	*	-A'

[a] At room temperatures. [b] Or two B terms with opposite sings.
* No predominant Faraday parameters. Mixed state of Faraday A (A') and C terms

tral regions, but that the signs are opposite in Faraday A' terms, though they
are same in Faraday C terms between heme a and heme b.

Ferrous low spin case. Since ferrous low spin hemochromes exhibit similar
MCD spectral profiles to those of metal porphyrins without vacant 3d orbitals,
the effect of electronic state of iron on the Faraday parameters can be neglect-
ed in the first approximation. Hence, ignoring orbitals of iron and axial

ligands, we calculated the transition energies, oscillator strengths, and the "actual" and "apparent" Faraday A terms on the basis of the PPP CI approxima- tion[7]. The formal charge of the porphyrin moiety is assumed to be zero. That is, the core charge of the central nitrogen atom is taken to be 1.5. Configura- tion interaction among singly excited configurations below 6 eV was taken into account. The LCAO-MO coefficients, obtained by the PPP method, were deorthogo- nalized by the inverse Löwdin transformation, and the oscillator strength and the "actual" and "apparent" A terms were calculated using Slater orbitals. The electric transition moments were evaluated by the dipole length, r, and by the dipole velocity, ∇, operators.

The calculated transition energy and A/D value of porphin can be reasonably compared with the experimental values of metal octaethylporphyrins (OEP), re- ported by Gale et al.[12], and the result are shown in Table 2. The experimental transition energies of the Q and Soret bands are clearly reproduced by the PPP method using the modified Nishimoto-Mataga equation[13] for the two-center elec- tron repulsion integral. The experimental A/D values are slightly affected by the nature of the central metal. The calculated A/D values are in good agree- ment with the experimental data in sign and in magnitude.

Table 2
Experimental data of metal octaethylporphyrin (OEP) and calculated data of porphin

	$Q_{0\leftarrow0}$		$Q_{v\leftarrow0}$		Soret	
	$\nu \times 10^{-3}$ (cm^{-1})	A/D (β)	$\nu \times 10^{-3}$ (cm^{-1})	A/D (β)	$\nu \times 10^{-3}$ (cm^{-1})	A/D (β)
experimental [22]						
Cu(OEP)	17.78	1.73	18.89	0.59	24.92	0.19
Zn(OEP)	17.54	1.77	18.71	0.9	24.84	0.15
Ag(OEP)	17.91	1.06	19.05	0.36	24.46	0.29
Co(OEP)	18.08	0.93	19.23	0.33	25.4	0.2
calculated						
porphin	16.9	2.25			24.7	0.088

The calculated transition energies, oscillator strengths, and A'/D values of protoporphyrin are compared with the experimental data of low-spin ferrous hemo- globins in Table 3, where the "apparent" A term is designated as A'. The calcu- lated and observed transition energies are slightly smaller than those of metal porphins. The splittings of the Q and Soret bands are calculated to 17 and 73 cm^{-1}, respectively, and are much smaller than the experimental linewidths. For the Q band, two overlapping vibronic bands, $Q_{0\leftarrow0}$ and $Q_{v\leftarrow0}$, were resolved by the gaussian curve fitting procedure, and then the experimental oscillator strengths and A'/D values were obtained by use of the method of moments. The experimental

Table 3
Experimental data of low-spin Fe(II) hemoglobin (Hb) complexes and calculated data of protoporphyrin

	$Q_{0\leftarrow 0}$			$Q_{v\leftarrow 0}$			Soret		
	$\nu \times 10^{-3}$ (cm^{-1})	f	A'/D (β)	$\nu \times 10^{-3}$ (cm^{-1})	f	A'/D (β)	$\nu \times 10^{-3}$ (cm^{-1})	f	A'/D (β)
experimental									
α-HbFe(II) CO	17.5	0.016	1.0	18.5	0.095	0.37	24.2	1.20	0.27
α-HbFe(II) NO	17.5	0.020	2.0	18.6	0.121	0.57	24.3	1.40	0.13
α-HbFe(II) O_2	17.4	0.019	1.3	18.5	0.084	0.10	24.3	1.29	0.23
β-HbFe(II) CO	17.5	0.023	2.3	18.5	0.097	0.59	24.1	1.22	0.20
β-HbFe(II) NO	17.5	0.013	3.0	18.6	0.112	0.33	24.3	1.24	0.08
β-HbFe(II) O_2	17.3	0.020	2.3	18.6	0.094	0.16	24.3	1.18	0.16
calculated									
protoporphyrin									
r a)	16.7	0.018	2.19				23.9	5.53	0.04
$\mathbf{v}$ b)	16.7	0.008	2.19				23.9	1.82	0.04

a) Calculated by the dipole length method.　　b) Calculated by the dipole velocity method.

Table 4
Experimental data of low-spin Fe(II) heme a and calculated data of porphyrin a

	Q			Soret		
	$\nu \times 10^{-3}$ (cm^{-1})	f	A'/D (β)	$\nu \times 10^{-3}$ (cm^{-1})	f	A'/D (β)
experimental						
low-spin Fe(II)						
heme a	16.6	0.13	−0.79	23.0	0.74	−0.46
calculated						
porphyrin a　r a)	16.6	0.039	−1.40	23.7	5.15	−0.05
$\mathbf{v}$ b)	16.6	0.015	−0.55	23.7	1.68	−0.05

a) Calculated by the dipole length method.　　b) Calculated by the dipole velocity method.

A'/D values are somewhat influenced by the axial ligands. Although the dipole length method produces larger theoretical oscillator strengths than the dipole velocity method, the calculated A'/D values reasonably agree with the experimetal values both in sign and in order of magnitude.

The calculated results for porphyrin a and the experimental data of low-spin ferrous heme a are summarized in Table 4. The MCD spectra of heme a have been previously reported. The formyl group in heme a causes the red-shift in the transition energies relative to metal porphin and protoheme. The calculated splittings of the Q and Soret bands in porphyrin a are 73 and 309 cm^{-1}, respectively. The formyl group in heme a splits the degeneracy of the excited states of porphyrin more largely than the vinyl group in protoheme. The calculated oscillator strengths of Q band and the magnitude of the A'/D value of the Soret band are smaller than the experimental values. The PPP method, however, explains the experimental result that the sign of the MCD of low-spin ferrous heme a is opposite to that of metal porphin and low-spin ferrous protoheme.

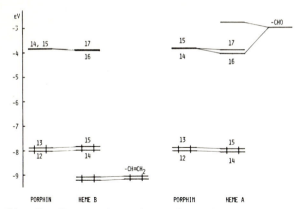

Fig. 7. Interaction scheme of porphyrin HOMOs and LUMOs with orbitals of the substituents.

Table 5
Molecular orbitals of the Soret and Q states

PORPHIN

$\Psi_1 = -0.681\ \Psi_{12 \to 15} + 0.727\ \Psi_{13 \to 14}$

$\Psi_2 = 0.681\ \Psi_{12 \to 14} + 0.723\ \Psi_{13 \to 15}$

$\Psi_3 = 0.727\ \Psi_{12 \to 14} - 0.664\ \Psi_{13 \to 15}$

$\Psi_4 = 0.727\ \Psi_{12 \to 15} + 0.664\ \Psi_{13 \to 14}$

HEME B

$\Psi_1 = 0.321\ \Psi_{14 \to 16} - 0.604\ \Psi_{14 \to 17} + 0.653\ \Psi_{15 \to 16} + 0.308\ \Psi_{15 \to 17}$

$\Psi_2 = 0.606\ \Psi_{14 \to 16} + 0.319\ \Psi_{14 \to 17} - 0.309\ \Psi_{15 \to 16} + 0.652\ \Psi_{15 \to 17}$

$\Psi_3 = 0.548\ \Psi_{14 \to 16} + 0.448\ \Psi_{14 \to 17} + 0.416\ \Psi_{15 \to 16} - 0.517\ \Psi_{15 \to 17}$

$\Psi_4 = -0.450\ \Psi_{14 \to 16} + 0.553\ \Psi_{14 \to 17} + 0.522\ \Psi_{15 \to 16} + 0.418\ \Psi_{15 \to 17}$

HEME A

$\Psi_1 = 0.600\ \Psi_{14 \to 16} - 0.371\ \Psi_{14 \to 17} - 0.437\ \Psi_{15 \to 16} - 0.552\ \Psi_{15 \to 17}$

$\Psi_2 = 0.429\ \Psi_{14 \to 16} + 0.524\ \Psi_{14 \to 17} + 0.623\ \Psi_{15 \to 16} - 0.376\ \Psi_{15 \to 17}$

$\Psi_3 = 0.631\ \Psi_{14 \to 16} - 0.206\ \Psi_{14 \to 17} + 0.180\ \Psi_{15 \to 16} + 0.665\ \Psi_{15 \to 17}$

$\Psi_4 = 0.183\ \Psi_{14 \to 16} + 0.716\ \Psi_{14 \to 17} - 0.606\ \Psi_{15 \to 16} + 0.206\ \Psi_{15 \to 17}$

The reasons why the formyl substituents change the signs of the apparent A (A') terms will be qualitatively explained as follows. The Soret and Q bands are well approximated by considering only 4 orbitals (2 highest occupied (a_{1u} and a_{2u}) and 2 lowest vacant (e_g) orbitals, so called 4 orbital model). The formyl π^* orbital resides very close to the lowest vacant orbitals of porphin π electron system, they interact very strongly. The interactions split the degeneracy of the porphyrin excited state, and also lower the energies. However, the vinyl group orbitals only weakly interact especially with the porphyrin highest occupied orbitals (Figure 7). These make the wavefunctions for the Q (ψ_1, ψ_2) and Soret (ψ_3, ψ_4) change to those shown in Table 5. If care is taken to read that the 16th and 17th orbitals in heme b have similar nature with those of the 17th and 16th orbitals in heme a, respectively, the close examination of Table 5 clarified that the magnitudes of the coefficients in ψ_1 and ψ_2 are reversed, and the nature of ψ_3 and ψ_4 orbitals are exchanged between in heme b and heme a. These make the signs of the A' terms by changing the signs of the electric moments for the transitions to ψ_1 and ψ_2, and the signs of the magnetic moments for the transitions to ψ_3 and ψ_4.

Effect of iron electronic states on Faraday parameters. The simple way to handle the effect of iron electronic states is to construct the Soret and Q wave

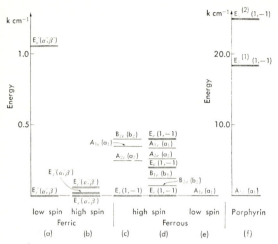

Fig. 8. The representations of the electronic states for the iron and porphyrin unperturbed states necessary for the calculation of the Faraday parameters of the hemoproteins. The symbols in parentheses represent the basis for the configurations. Hole configurations are adopted in (a). In (b), D (the zero field splitting parameter) is taken to be 10 cm^{-1}. For high-spin ferrous iron, (c) and (d) correspond to $^5B_{2g}$ and ^5E_g, and the order and the splitting in the configuration are arbitrary.

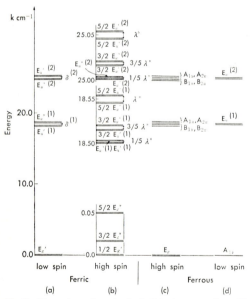

Fig. 9. Electronic configuration in $D_{4h}*$ for heme complexes. The energy differences between E_u' and E_u'' states and their relative energy levels are taken from the results for low-spin ferric hemoproteins (4). The order $(A_{1u}, A_{2u}) > (B_{1u}, B_{2u})$ was determined from the result of the ferrous deoxymyoglobin (4).

function in heme from the products of those of porphyrin and the ground state of iron orbitals, both of which have been extensively explored both theoretically and experimentally[14] (Figure 8). This approximation is valid only when mixing of orbitals between iron and porphyrin does not significantly affect the discussion. Previous papers showed that this method could explain Faraday parameters of various derivatives of iron porphin and heme b. The group theoretical representations of the electronic configurations of iron and porphyrin, and heme as the product of those of iron and porphyrin are shown in Figures 8 and 9[14]. In this approximation the Soret and Q states are split into two nearly degenerate states designated as E_u' and E_u'' in the D_{4h} double group. Inserting these representations to the equations of Faraday parameters gives the results shown in Table 6. Sum of the Faraday parameters to E_u' and E_u'' states gives the Faraday parameters for the Soret and Q bands. The details how these calculated data predict the results

TABLE 6

Faraday Parameters for Low-spin Ferric Derivatives

Faraday parameter (transition)[a]	$E_g' \rightarrow E_u'$ [c]	$E_g' \rightarrow E_u''$ [c]
$A(S)$	$1/2\,a_2{}^2b_{22}$	$1/2\,a_2{}^2b_{22}$
$B(S)$	$a_1a_2(b_{12}-b_{00})S_{0j}{}^2/W(Q,S)$	$a_1a_2(b_{12}+b_{00})S_{0j}{}^2/W(Q,S)$
$C(S)$	$-1/2\,a_2{}^2b_{00}$	$1/2\,a_2{}^2b_{00}$
$D(S)$[b]	$a_2{}^2$	$a_2{}^2$
$A(Q)$	$1/2\,a_1{}^2b_{11}S_{0j}{}^2$	$1/2\,a_1{}^2b_{11}S_{0j}{}^2$
$B(Q)$	$a_1a_2(b_{12}-b_{00})S_{0j}{}^2/W(S,Q)$	$a_1a_2(b_{12}+b_{00})S_{0j}{}^2/W(S,Q)$
$C(Q)$	$-1/2\,a_1{}^2b_{00}S_{0j}{}^2$	$1/2\,a_1{}^2b_{00}S_{0j}{}^2$
$D(Q)$	$a_1{}^2$	$a_1{}^2$

[a] The S and Q indices represent the Soret and Q transitions, respectively. [b] D is the dipole strength, defined in Stephens' paper (4). [c] $W=h\nu$ is the energy of the state: $W(Q,S)=W(Q)-W(S)$, $W(S,Q)=W(S)-W(Q)$.

listed in Table 1 for heme b have been discussed already[14]. Here we wish to stress only the point that the signs of Faraday C terms will be determined by those of the ground state magnetic moment (b_{00}). Since the porphyrin ground state has no magnetic moment (Figure 8), the magnetic moment of the heme ground state comes exclusively from that of the iron ground electronic state. This is the reason why the sign of the Faraday C terms does not differ between heme b and heme a, provided that the oxidation and spin states of the iron are similar.

Effect of the formyl substituent on MCD of cytochrome oxidase

Figures 10A, B and 11 exhibit MCD spectra for reduced cytochrome oxidase and its derivatives. The MCD spectra of the cyanide and CO complexes are quite similar to the MCD spectra of imidazole-ferroheme a in the absence of SDS. This may suggest some kind of interactions including heme in these cytochrome oxidase

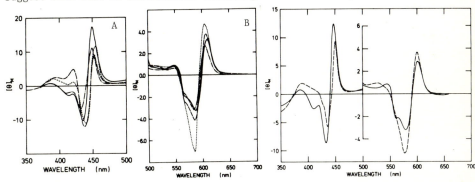

Fig. 10. MCD spectra of reduced cytochrome oxidase compounds in the Soret (A) and visible (B) regions. Cytochrome oxidase in emasol -cholate-phosphate buffer was reduced with sodium dithionite at least for 25 min, the pH being 7.32 (——). To the reduced cytochrome oxidase was added cyanide to a final concentration of 48 mM, pH 7.34 (----), or azide at 48 mM and pH 7.21 (-·-→). Alternatively, cytochrome oxidase-cyanide (48 mM) at pH 7.33 was reduced with sodium dithionite (←--→).

Fig. 11. MCD spectrum of cytochrome oxidase-CO. Through a solution of reduced cytochrome oxidase in emasol -cholate-phosphate buffer, pH 7.40 (——) was bubbled a stream of CO gas for 2 min.(---).

128

complexes. Instead of adding cyanide to the reduced enzyme (Type A complex), a cyanide complex of the oxidized enzyme was reduced with sodium dithionite (Type B complex)[15]. In this case, the MCD spectrum in the visible region was almost identical to that of reduced cytochrome oxidase. In the Soret region an enhanced peak at 420 nm was noted. Figure 12A and B show the MCD spectra of oxidized cytochrome oxidase complexes. Different effects of azide and cyanide may indicate that the sites of attack by these ligands are different. The intensification due to the cyanide complexing might be ascribed to an increase of the low-spin content as has been discussed by Babcock et al.[16-18]. The MCD in the visible region would be composed of Faraday A' (a couple of B) terms as with those of imidazole-ferriheme a.

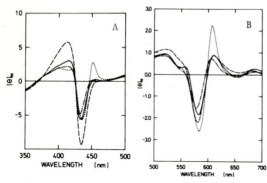

Fig. 12. MCD spectra of oxidized cytochrome oxidase compounds in the Soret (A) and visible (B) regions. ——, oxidized enzyme in an emasole-cholate-phosphate buffer; ———, plus either 48 mM KCN at pH 7.34; —·—, or 48 mM NaN₃ at pH 7.21.

REFERENCES

1. Orii, Y. and Yoshikawa, S. (1973) in Oxidase and Related Redox Systems, King, T. E., Mason, H. S. and Morrison, M. eds., University Park Press, 11, p 639.
2. Yanagi, Y., Sekuzu, I., Orii, Y. and Okunuki, K. (1972) J. Biochem., 71, 47.
3. Orii, Y. and Iizuka, T. (1975) J. Biochem., 77, 1123.
4. Vickery, L., Nozawa, T. and Sauer, K. (1976) 98, 343 and 351.
5. Buckingham, A. D. and Stephens, P. J. (1966) Ann. Rev. Phys. Chem., 17, 399.
6. Schatz, P. N. and McCaffery, A. J. (1969), Quart. Rev. Chem. Soc., 23, 552.
7. Kaito, A., Nozawa, T., Yamamoto, T., Hatano, M. and Orii, Y. (1977) Chem. Phys. Lett., 52, 154.
8. Briat, B., Berger, D. and Lelibous, M. (1972) J. Chem. Phys., 57, 5606.
9. Okunuki, K., Sekuzu, I., Takemori, S., and Yonetani, T. (1958) J. Biochem., 45, 847.
10. Takemori, S. and King, T. E. (1965), J. Biol. Chem., 249, 504.
11. Vickery, L., Salmon, A. and Sauer, K. (1975) Biochim. Biophys. Acta, 386, 87.
12. Gale, R., McCaffery, A. J., and Row, M. D. (1972) J. Chem. Soc., Dalton Trans., , 596.
13. Tomono, K. and Nishimoto, K. (1977) Bull. Chem. Soc. Japan, 49, 1179.
14. Hatano, M. and Nozawa, T. (1978) Adv. Biophys., 11, 95.
15. Orii, Y. and Okunuki, K. (1964) J. Biochem., 55, 37.
16. Palmer, G., Babcock, G. T. and Vickery, L. E. (1976) Proc. Natl. Acad. Sci., 73, 2206.
17. Babcock, G. T., Vickery, L. E. and Palmer, G. (1976) J. Biol. Chem., 251, 7909.
18 Babcock, G. T., Vickery, L. E. and Palmer, G. (1978) J. Biol. Chem., 253, 2400.

Cytochrome Oxidase, T.E. King et al. eds.
© *1979 Elsevier/North-Holland Biomedical Press*   129

TWO KINDS OF THE OXIDIZED STATE OF CYTOCHROME OXIDASE STUDIED BY RESONANCE
RAMAN SPECTRA

TEIZO KITAGAWA and YUTAKA ORII*

Institute for Protein Research, Osaka University, Suita, Osaka, 565, Japan.
*Department of Biology, Faculty of Science, Osaka University, Toyonaka, Osaka
560, Japan.

ABSTRACT

   The resonance Raman spectra of cytochrome oxidase excited at 514.5 nm are
reported. The ferricyanide-oxidized and dithionite-reduced forms gave a single
line in the Band IV region whereas the resting enzyme as purified gave two lines.
These two lines were replaced by the line of the dithionite-reduced oxidase
after anaerobic laser illumination and this photoreduction was confirmed by the
appearance of two absorption bands at 440 and 602 nm. The Raman spectra characterized by the two Band IV lines were also observed upon aerobic laser illumination of the ferricyanide-oxidized oxidase and upon bubbling oxygen through
the dithionite-reduced one. One of the two lines disappeared when the resting
enzyme was monomerized by a SDS treatment, and the whole spectrum resembled that
of ferricyanide-oxidized enzyme. Therefore the oxidase which gives two Band IV
lines is inferred to stay in an oxidized state specific to some dimer structure
but conformationally different from the ferricyanide-oxidized state.

INTRODUCTION

   Purified preparations of cytochrome oxidase usually exist as dimers containing two heme a molecules and two copper atoms. With sodium dodecyl sulfate
(SDS), guanidine hydrochloride or alkali under mild conditions,[1-3] these dimers
are dissociated into active monomers containing presumably one heme and one
copper.[3,4] Therefore, the monomer can be regarded as a minimal functional unit
of this enzyme. Elucidation of interactions between the two monomers in the
dimeric unit as well as those between heme a and copper in the monomer has
currently been a subject of spectroscopic investigations.[5,6]
   The reaction mechanism of reduced cytochrome oxidase with molecular oxygen
and the accompanying changes of oxidation states of heme iron have been studied
in detail (Ref. 7 and references cited therein). It was noted that the final
product of oxygenated oxidase [Comp. III] gives an absorption spectrum different
from the fully oxidized oxidase.[7] Previously Orii and Okunuki had already
noticed that cytochrome oxidase as purified does not always show the absorption

spectrum of fully oxidized form[8] but the manifold spectra of oxidized oxidase
have not been studied in detail.

Resonance Raman scattering from hemoproteins provides selectively the vibrational spectra of iron-porphyrin interacting in situ with immediate environment, reflecting a state of heme (Refs. 9 and 10 and references cited therein).
The Raman spectral changes of isolated heme a derivatives upon axial ligation
and reaction at the peripheral formyl group were studied previously.[11] Thus
the resonance Raman spectra of cytochrome oxidase may now reveal structural
details of the heme moieties of the enzyme. The first resonance Raman study of
cytochrome oxidase[12] emphasized, for the ferricyanide oxidized form, coexistence
of two "oxidation state marker" (Band IV)[13] lines characteristic of ferric and
ferrous iron-porphyrins. Subsequent studies[14-16] clarified that the lower-
frequency line out of the two emerged from the reduced oxidase which resulted
from photoreduction by laser illumination. We present here Raman evidence for
the photoreduction and also for existence of two kinds of oxidized state, discussing structural features of hemes in these two states as well as in the
reduced and cyanide-complexed states.

EXPERIMENTAL PROCEDURE

Cytochrome oxidase was purified from bovine heart muscle by the method of
Okunuki et al.[17] with some modification,[1] and was dissolved in a mixture of
0.25% Emasol 1130, 0.1% cholate and 0.05M sodium phosphate buffer at pH 7.5.
For a SDS treatment, 0.82 mM cytochrome oxidase solution was diluted with an
equal volume of SDS-buffer mixture of appropriate SDS concentration. The enzyme
was reduced with powdered dithionite and oxidized with concentrated solution of
potassium ferricyanide.

The Raman spectra were measured with an argon ion laser (Spectra Physics,
model 164) and a JEOL 400D Raman spectrometer equipped with an HTV-R649 photomultiplier. Frequency of the spectrometer was calibrated with indene as standard.[18] Uncertainty of the peak frequency is as large as ± 2 cm^{-1} for ordinary
Raman lines of cytochrome oxidase. For the measurement of Raman spectra a 30 μl
portion of 0.41 mM oxidase solution was placed in a cylindrical cell with a
thermostatted cell holder and temperature of the solution was kept at ca. 10°C.
After the measurement, an absorption spectrum of diluted sample was recorded
on a Hitachi 124 spectrophotometer. Usually no spectral alteration due to laser
illumination was observed.

For the examination of photoreduction, a 10 μl portion of cytochrome oxidase
solution at 160 μM was placed in a melting point capillary tube (Fisher Scienti-

fic Corp., 2 mm). Before the sample injection, nitrogen gas was flushed through the tube and was also bubbled into the sample solution without foaming. After being sealed, it was kept in a water bath at 5°C perpendicularly to the laser beam and subjected to laser illumination. Since the size of a laser spot was so small, the capillary was advanced every 30 min by a distance equal to the spot diameter, thus ensuring irradiation of the whole portion (488.0 nm, 400 mW at the laser exit). The absorption spectrum of the illuminated sample in the capillary was measured with a device constructed by assembling Union Giken spectrophotometer modules and a light guide.

Cytochrome oxidase cyanide complexes in the oxidized and reduced states were prepared in a Raman cell by adding solid potassium cyanide to the enzyme solution and by reducing it afterwards, respectively. Since direct pH measurement of the sample was impossible due to a small volume, the pH of a buffer solution which did not contain the enzyme but which was otherwise treated in the same way was measured separately. Thus the pH of the oxidase solution was estimated to be ca. 12 in the presence of KCN and ca. 9 upon further addition of dithionite.

RESULTS

The resonance Raman spectra of reduced, resting, and oxidized cytochrome oxidase under air are shown in Fig. 1, where "resting" signifies the enzyme preparation as purified. A single Band IV line is seen at 1362 cm^{-1} for the reduced enzyme and at 1372 cm^{-1} for oxidized one, whereas two lines are observed at 1370 and 1357 cm^{-1} for the resting enzyme.

When the ferricyanide-oxidized enzyme was exposed to laser illumination aerobically, the Raman spectrum changed with time as illustrated in Fig. 2, where recording of spectra 1, 2, 3, 4, 5, and 6 were started at 0, 12, 24, 36, 48, and 60 min, respectively, after addition of ferricyanide. Spectrum 7 was obtained by computer-averaging the spectra accumulated in a period of 100 min from 2 hr after the ferricyanide addition. In spectrum 1, there is no trace of Raman line between the 1375- and 1340-cm^{-1} lines but in spectrum 4 a Raman line is discernible at 1358 cm^{-1} and is prominent in the computer-averaged spectrum. The presence of two Raman lines in this frequency region resembles the Raman spectrum of the resting enzyme.

On the other hand, when the resting enzyme was anaerobically laser-illuminated, a different spectral change took place; in 1 hr two lines appeared at 1362 and 1372 cm^{-1} and finally (ca. 4 hr) the 1372-cm^{-1} line disappeared completely. Furthermore, when this solution was exposed to air for 24 hr, the 1362-cm^{-1} line was replaced by two lines at 1372 and 1356 cm^{-1} and the whole spectrum resembled that of the resting enzyme. The 1356-cm^{-1} line differed

132

distinctly from the 1362-cm^{-1} line in its shape and intensity. Therefore it is evident that under anaerobic conditions cytochrome oxidase is reduced completely and reversibly by laser illumination at 488.0 nm and that the photoreduced species differs from the resting and the aerobically laser-illuminated enzymes.

The absorption spectra of cytochrome oxidase in a capillary tube are shown in Fig. 3. Spectrum A was recorded immediately after the addition of ferricyanide and was unaltered for 24 hr. Spectrum B was observed for the laser illuminated resting oxidase (see Experimental Procedure). This exhibits a broad Soret peak with a plateau from 428 to 440 nm and an α band at 602 nm. The absorption bands at 440 and 602 nm indicate the formation of reduced oxidase. After the sample had been left in contact with air by breaking the seal of the capillary, the plateau was replaced by a single peak at 428 nm and the α band

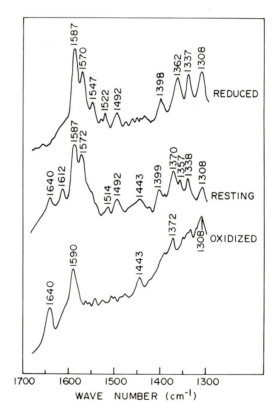

Fig. 1. Resonance Raman spectra of reduced, resting, and oxidized cytochrome oxidase. "Resting" means the enzyme as purified, and this was excited at 488.0 nm. The top and bottom spectra were obtained by excitation at 514.5 nm for the dithionite-reduced and ferricyanide-oxidized enzymes, respectively.

was shifted to 596 nm as shown in spectrum C. When the resting enzyme was laser illuminated in the presence of air, the absorption spectrum was almost unaltered.

The cyanide complex of the resting oxidase gave a Raman spectrum similar to that of the ferricyanide oxidized one, although when it was reduced the resultant Raman spectrum was clearly different from that of the intact reduced enzyme. To examine a possibility that the dimeric structure of cytochrome oxidase yields Raman spectrum characterized by the two Band IV lines, we tried to prepare monomeric cytochrome oxidase. An alkali treatment[3] resulted in strong fluorescence, thus preventing recording of Raman spectra. Accordingly the SDS treatment was adopted. As illustrated in Fig. 4, the 1356-cm^{-1} line remained when the resting enzyme was treated with 0.31 or 1.6% (not shown) SDS. However, at 3.2% SDS, which was high enough to induce monomerization of dimeric cytochrome oxidase at 410 µM, the 1356-cm^{-1} line disappeared almost completely and the whole spectrum looked much the same as that of the intact ferricyanide-oxidized oxidase shown at the bottom of Fig. 1. Therefore, it is conceivable that a certain type of quaternary structure specific to oxidized dimeric oxidase is necessary for the appearance of the 1356-cm^{-1} line and that the quaternary

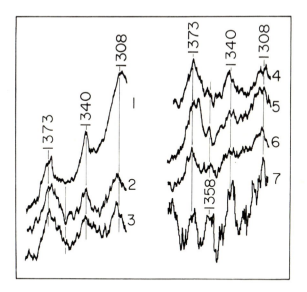

Fig. 2. Raman spectral change of the ferricyanide-oxidized enzyme with time. Scanning of spectra 1, 2, 3, 4, 5, and 6 were started at 0, 12, 24, 36, 48, and 60 min after the ferricyanide treatment, respectively. Spectrum 7 was obtained by computer-averaging the spectra accumulated in a period of 100 min. The accumulation was initiated at 2 hr after the ferricyanide addition. Laser, 514.5 nm; power, 100 mW.

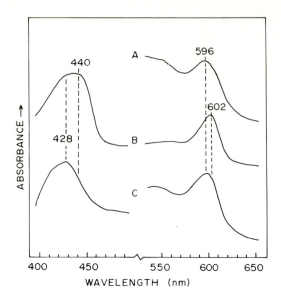

Fig. 3. Absorption spectra of cytochrome oxidase in a capillary tube.
A, ferricyanide-oxidized; B, after laser illumination of the resting
enzyme in a sealed tube (see Experimental Procedure); C, one day
after breaking the seal of the laser-illuminated sample.

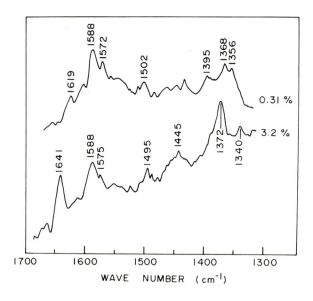

Fig. 4. Resonance Raman spectra of the resting enzyme in the presence of SDS.
The concentration of SDS are specified beside the spectrum. Laser,
488.0 nm; power, 100 mW.

structure is destroyed upon addition of ferricyanide.

Dithionite-reduced cytochrome oxidase was exposed to laser illumination for initial 2 hr and was allowed to stand under air for 48 hr. The Raman spectra were recorded intermittently in the region between 1300 and 1400 cm^{-1}. As illustrated in Fig. 5, the Raman lines at 1398, 1362, and 1308 cm^{-1} remained almost unchanged although intensity of the 1337 cm^{-1} line diminished with time. The last spectrum resembles that of reduced cytochrome oxidase-cyanide shown below. Since the sample gave a normal absorption spectrum of the reduced form after the last recording of Raman spectrum, the possibility of irreversible protein denaturation during this experiment can be ruled out. Thus the disappearance of the 1337 cm^{-1} line was interpreted in terms of a change of state of the heme formyl group.[16]

When molecular oxygen was bubbled into a solution of dithionite-reduced cytochrome oxidase, a Raman spectrum with two characteristic lines at 1372 and 1357 cm^{-1} was observed. This Raman spectrum presumably stands for Comp. III of the oxygenated preparation.[7] Recent studies[19,20] indicate that Comp. III is in the same oxidation state as the oxidized enzyme. Accordingly it is concluded that there are two types of Raman spectra for oxidized enzyme; one corresponds to the spectrum without the 1357-cm^{-1} line as shown at the bottom of Fig. 1 and the other to the middle spectrum in Fig. 1 with two Band IV lines. The interrelation of the reduced and two oxidized states is summarized in Table 1.

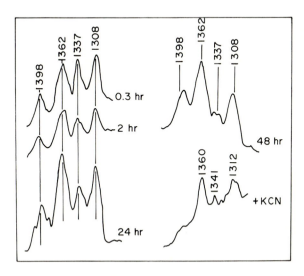

Fig. 5. Raman spectral change of the dithionite-reduced cytochrome oxidase with time. Scan starting time measured from addition of dithionite is specified beside the spectrum. The right bottom is for cyanide complex.

136

Table 1. Raman characteristics of three states of cytochrome oxidase and their interrelation.

	fully-oxidized	resting	fully-reduced
Band IV	one line $\xleftrightarrow[b]{a}$	two lines $\xleftrightarrow[e]{c,d}$	one line
ν (cm^{-1})	1372	~1370 ~1357	1362

a) ferricyanide   b) aerobic laser-illumination
c) anaerobic laser-illumination   d) dithionite
e) oxygen bubbling

DISCUSSION

The resonance Raman spectra of ferrous and ferric heme a bis-imidazole complexes [a·Im$_2$] (low spin) and that of ferric heme a observed previously[11] are shown in Fig. 6. Band IV of a(II)·Im$_2$ and a(III)·Im$_2$ are seen at 1360 and 1375 cm^{-1}, respectively, at almost the same frequencies of Band IV lines of the dithionite-reduced and ferricyanide-oxidized enzymes, respectively. Therefore, two heme a molecules of dimeric oxidase are equally reduced by dithionite or oxidized by ferricyanide. It is noted that a prominent difference between high- and low-spin ferric heme a's is found in the Raman spectra around 1600-1650 cm^{-1} region but not in the Band IV region.

Adar and Yonetani[15] observed that intensity of the 1358-cm^{-1} line gradually increased as the oxidized oxidase was exposed to laser illumination at 413 nm. Since this spectral change was accompanied by intensity decrease of the 1372-cm^{-1} line and was more effective with higher laser power, they explained this phenomenon as due to photoreduction of cytochrome oxidase by electrons supplied from contaminated flavins. This photoreduction was confirmed in this study by both Raman and absorption spectroscopies. However, upon aerobic laser illumi-nation we could not see the photoreduction. The discrepancy between Adar and Yonetani's and our results may be attributed to a different rate of photo-reduction either due to the different excitation-wavelength or a different amount of contaminated flavins or both.

The ferricyanide oxidized enzyme exhibits the characteristic Raman line of ferric low spin state at 1640 cm^{-1} as a(III)·Im$_2$ does. The resting enzyme gave several Raman lines in addition to those of the ferricyanide oxidized one, suggesting that the two hemes present in the dimeric unit provide different sets of Raman lines. Although some of frequencies of the Raman lines specific to the resting enzyme are close to those of the reduced enzyme, the resting oxidase

---

showed no trace of the 440 nm band even after laser illumination in the presence of air. Therefore, it is unlikely that the resting enzyme is a mixture of the oxidized and reduced oxidases.

One of the two hemes in the resting enzyme must be in the ferric low spin state, giving rise to the Raman lines at 1640, 1587, 1443, 1370 and 1308 cm^{-1}. The heme a moiety of this kind seems to be affected little by the ferricyanide addition or monomerization. Another set of the Raman lines at 1612, 1492, 1397, and 1357 cm^{-1}, on the other hand, is assignable to the second heme a moiety. Since the frequencies of the latter Raman lines are significantly different from

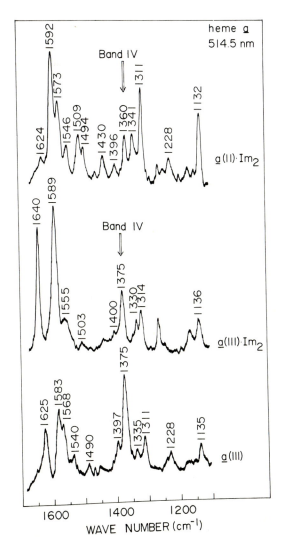

Fig. 6. The resonance Raman spectra of isolated heme a derivatives in an aqueous solution containing 0.1 M Na$_2$CO$_3$. Top, ferrous heme a bis-imidazole; middle, ferric heme a bis-imidazole; bottom, free ferric heme a. Laser, 514.5 nm; power, 200 mW.

138

those of either high- or low-spin ferric heme $\underline{a}$ derivatives, it is conceivable that a certain kind of strain is imposed on the second heme $\underline{a}$ through the immediate environment. In this context, the 1357-cm^{-1} line is regarded as representing a conformation specific to the dimer structure. In fact, the Raman lines of the latter group disappeared as the enzyme was monomerized.

In conclusion the resting, the reduced and oxygen bubbled [Comp. III], and the aerobically laser-illuminated oxidases gave similar Raman spectra characterized by the presence of two Band IV lines. These are distinctly different from those of the monomerized, the ferricyanide treated oxidase or the isolated heme $\underline{a}$ derivatives. All these hemes are considered to stay in the oxidized state, but in the former group of oxidized oxidase one of the two hemes in the dimeric unit is under stronger influence of the immediate environment specific to the dimeric oxidase than the other.

REFERENCES

1. Orii, Y., Manabe, M. and Yoneda, M. (1977) J. Biochem. 81, 505-517.

2. Orii, Y. and Okunuki, K. (1967) J. Biochem. 61, 388-403.

3. Orii, Y., Matsumura, Y. and Okunuki, K. (1973) in Oxidase and Related Redox Systems, King, T. E., Mason, H. S. and Morrison, M. eds. University Park Press, Baltimore, pp.666-672.

4. Love, B., Chen, S. H. and Stotz, E. (1970) J. Biol. Chem. 245, 6664-6668.

5. Malmström, B. G. (1973) Q. Rev. Biophys. 6, 389-431.

6. Nicholls, P. and Chance, B. (1974) in Molecular Mechanism of Oxygen Activation, Hayaishi, O. ed. Academic Press, New York, pp.479-534.

7. Orii, Y. and King, T. E. (1976) J. Biol. Chem. 251, 7487-7493.

8. Orii, Y. and Okunuki, K. (1965) J. Biochem. 57, 45-54.

9. Spiro, T. G. (1975) Biochim. Biophys. Acta, 416, 169-187.

10. Kitagawa, T., Ozaki, Y. and Kyogoku, Y. (1978) Adv. Biophys. 11, 153-196.

11. Kitagawa, T., Kyogoku, Y. and Orii, Y. (1977) Arch. Biochem. Biophys. 181, 228-235.

12. Salmeen, I., Rimai, L., Gill, D., Yamamoto, T., Palmer, G., Hartzell, C. R. and Beinert, H. (1973) Biochem. Biophys. Res. Commun. 52, 1100-1107.

13. Kitagawa, T., Iizuka, T., Saito, M. and Kyogoku, Y. (1975) Chem. Lett. 849-852.

14. Salmeen, I., Rimai, L. and Babcock, G. (1978) Biochemistry, 17, 800-806.

15. Adar, F. and Yonetani, T. (1978) Biochim. Biophys. Acta, in press.

16. Kitagawa, T. and Orii, Y. (1978) J. Biochem. in press.

17. Okunuki, K., Sekuzu, I., Yonetani, T. and Takemori, S. (1958) J. Biochem. 45, 847-854.

18. Hendra, P. J. and Loader, E. J. (1968) Chem. Ind. 718-719.

19. Kornblatt, J. A., Kells, D. I. C., and Williams, G. R. (1975) Canad. J. Biochem. 53, 461-466.

20. Dasgupta, U. and Wharton, D. C. (1977) Arch. Biochem. Biophys. 183, 260-272.

Cytochrome Oxidase, T.E. King et al. eds.
© 1979 Elsevier/North-Holland Biomedical Press

ALTERNATIVE OXIDIZED STATES OF CYTOCHROME c OXIDASE. A NOVEL EPR SIGNAL OF
THE "OXYGENATED" FORM

HELMUT BEINERT, ROBERT W. SHAW and RAYMOND E. HANSEN
Institute for Enzyme Research, University of Wisconsin, Madison, Wisconsin
53706 (U.S.A.)

ABSTRACT

EPR and light absorption characteristics are described of three transient alternative oxidized states of cytochrome c oxidase different from the aerobic,
resting one. These are, 1) a form generated by anaerobic reoxidation of the
completely reduced enzyme by ferricyanide or porphyrexide; 2) a form obtained
by exposing this product of anaerobic oxidation to oxygen; and 3) a form generated by rapid reoxidation with oxygen of the completely reduced enzyme. Some
reactions of these forms with substrates and ligands are reported.

The product of anaerobic reoxidation is characterized by a strong rhombic,
high spin ferric heme signal at g $\sim$ 6, which disappears rapidly on exposure to
oxygen. The product of rapid reoxidation by oxygen shows an intense, thus far
unrecognized EPR signal with g=5; 1.78; 1.69. Only this product of direct reoxidation with oxygen shows a pronounced absorption at 655 nm. It is suggested
that this form is identical with or related to what has collectively been
called the "oxygenated" form and to "activated" forms obtained under similar
conditions.

INTRODUCTION

In the past few decades it has been amply documented that active site structure and therewith activity of enzymes can be modulated by substrates, reaction
products, effectors as well as other conditions. In many instances, the different forms of enzymes so generated are of transient nature and are thus difficult to characterize, particularly with enzymes that do not possess specific
prosthetic groups. One would think, however, that with an enzyme such as cytochrome c oxidase with four metals as active components, which lend themselves
to a variety of spectroscopic approaches, intermediate and transient alternative forms should be more readily detected and characterized.

Thus, ever since the discovery of what is generally known today as the "oxygenated" form of cytochrome c oxidase by Okunuki and his collaborators[1], spectrophotometry has given evidence that alternative oxidized forms of the enzyme

of transient nature exist. More recently these observations have assumed new significance as it has become apparent that an oxidized form or forms of the enzyme other than the resting one, viz. that generally obtained on purification and after storage, is the species active in catalysis of oxidation-reduction[2,3] and recent work on the reoxidation of the reduced enzyme by O_2 after flash photolysis of the reduced CO-compound has produced evidence for early intermediates representing transient states from the initial O_2-complex of the reduced enzyme to states reached after (partial) transfer of electrons to O_2[4].

Although spectrophotometry as a means of observing such forms of cytochrome c oxidase has the advantage of sensitivity and convenience, the extensive overlap of the spectra of the various components as well as the difficulty of a quantitative evaluation in the case of forms of unknown absorption spectra and extinction coefficients, is a distinct disadvantage. We have, therefore, attempted to apply our experience with EPR spectroscopy of the enzyme and with rapid freeze-quenching techniques to the study of transient forms of cytochrome c oxidase. We will be concerned here with transient forms which appear to represent oxidized states of the enzyme and will consider their relationship to alternative oxidized states previously described by other workers.

MATERIALS AND METHODS

These have been described in recent publications from our laboratory[5-9]. As in these reports, all concentrations given refer to the state after mixing reactant solutions, unless it is specifically stated that a solution of a certain concentration was mixed with another solution of a specified concentration. Enzyme concentrations are expressed in terms of total heme a, neglecting for this purpose differences between the heme components. The reaction times are given as calculated from the dimensions of the mixing apparatus and the speed of the syringe ram, not considering the effective quenching time of the reaction(s) involved.

RESULTS

From their distinct EPR features we can distinguish 3 alternative oxidized states, including the resting one, and a fourth such state from its EPR plus optical characteristics. These forms are the aerobic, resting one; a form obtained on rapid reoxidation by O_2 of the reduced enzyme, related to or identical with what has collectively been called the "oxygenated" form; a form obtained on anaerobic reoxidation of the reduced enzyme by electron acceptors other than O_2, and finally a form produced from this anaerobic oxidation pro-

duct by reaction with O_2. In addition, of course, we have observed forms arising by reaction of the mentioned ones, with ligands such as CO, cyanide, sulfide, azide etc.

I. The aerobic resting form. The EPR spectra of the resting oxidized state
have been extensively described and quantitatively evaluated[5,10]. It will suf-
fice, therefore, to refer here to this previous work. One set of observations,
however, which we have made in our studies on the other, transient forms de-
serves mentioning. Namely, during a detailed study of the time course of the
reoxidation of the reduced enzyme we have observed slow but significant changes
in the high spin iron signals. There are also indications that the copper signal decreases slowly but we have not convinced ourselves of the significance of
these changes, which are of the order of 10-20%. The high spin ferric heme
signal at g=6 as well as the signal at g=4.3, however, increase steadily, immediately after reoxidation by O_2. The increase of the signal at g=4.3 is presumably due to a slow reoxidation of contaminating iron compounds, which are
not in efficient electronic communication with the active metal components of
the enzyme. This explanation may, however, not be correct for the ferric heme
signals at g=6. Since these signals actually arise on partial reduction of
cytochrome c oxidase, we are rather inclined to attribute the appearance of
signals at g=6 to a slow autoreduction. That cytochrome c oxidase preparations, possibly by additions made during preparation, have a tendency toward
autoreduction is well known[11]. A procedure for production of the "mixed
valence" state starting with the oxidized resting form in the presence of CO
and absence of O_2 is based on this phenomenon[12]. We mention these observations
and conclusions here, inasmuch as they were of importance to us in the characterization of the transient "oxygenated" form and also to point out that even
in the resting oxidized state, it cannot be assumed that all components are
100% oxidized at all times.

II. The anaerobically oxidized form. This form is readily obtained on anaerobic reoxidation of reduced enzyme by such oxidants as ferricyanide or porphyrexide[8]. The procedure used for reduction is of no consequences. This form
is characterized by the presence of an intense, largely rhombic, high spin
ferric signal at g=6 (Fig. 1, A) and by the absence of a significant absorption
at 655 nm. Whatever absorption at 655 nm is observed, is presumably due to
contaminating O_2. In this form both hemes as well as the EPR detectable copper
are largely oxidized, a_3 70-80%, a 80-100% and the copper signal is mostly 10-
25% stronger than in the resting form, where it is generally found to represent
35-40% of the chemically determined copper. Since there is no reliable measure

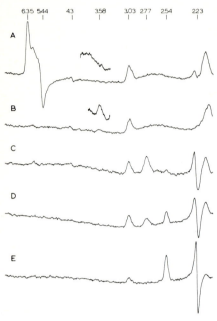

Fig. 1. EPR spectra of cytochrome c oxidase recorded in experiments in which the transient high spin species was exposed to cyanide or sulfide. The enzyme was initially reduced anaerobically by an excess of NADH[9] in the presence of a small quantity of cytochrome c. A, enzyme in the transient state as produced by oxidation with porphyrexide at 820 msec and then mixed with anaerobic buffer (at a 2:1 ratio) for 6 msec; B, as A but mixed anaerobically with a 50 mM solution of KCN (neutralized) instead of buffer; C, D and E, as in A but mixed anaerobically with a 40 mM solution of NaHS (pH 7.4) for 6, 200 and 6600 msec, respectively. The final enzyme concentrations were 205 μM in A-C and 192 μM in D and E. The conditions of EPR spectroscopy were: microwave power and frequency, respectively, 2.7 mwatt and 9.2 GHz; modulation frequency and amplitude, respectively, 100 KHz and 0.8 mT; scanning rate 50 mT/min and temperature 13K. For the inserts in A and B, an amplification was used fivefold higher than for the main spectra and the time constant and scanning time were increased.

of the oxidation state of the EPR-undetectable copper, we are not certain of the state of this copper in the anaerobically reoxidized form of the enzyme. We would, however, assume that an oxidant such as porphyrexide, with a midpoint potential of +725 mV[13], should have reoxidized this copper component, unless some kinetic inhibition exists. The anaerobically reoxidized form is optimally developed at 1 to 3 sec after mixing the oxidant with the reduced enzyme and has largely decayed in 1 min.

Although cytochrome c oxidase is not likely to be 100% converted into this form under its normal operating conditions, the transient anaerobically reoxidized form is nevertheless of interest for several reasons. First, it is entirely possible that a fraction of enzyme molecules may be converted into this or a closely related state during the functioning of the enzyme. Second, it is a quantitative study of the EPR signals of this form that has produced the most substantial evidence that the rhombic high spin ferric heme signal at g=6 - at least in this form of the enzyme - is due to a_{-3}^{3+} [8]. Third, in contrast to the resting oxidized form, the anaerobically reoxidized form, or more precisely the a_3^{3+} in this form, reacts rapidly with ligands typical for a_{-3} such as cyanide and sulfide and it responds rapidly to O_2, ferrocytochrome c and to CO, even in the oxidized state. Although CO does presumably not bind to a_{-3}^{3+}, the oxidized heme nevertheless senses the presence of this gas and responds to it by an increase in rhombicity (Fig. 2). If this response of the high spin ferric heme signal at g=6 to CO is taken as a diagnostic for a_3^{3+}, then the almost identical highly rhombic signal observed on photodissociation of the reduced a_3^{2+}-CO complex in the presence of ferricyanide[9] is also due to a_3^{3+} and the same holds for the rhombic signal at g=6 observed on partial reduction of the oxidase by ferrocytochrome c[7,9].

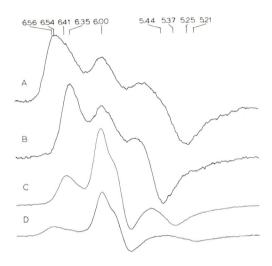

Fig. 2. EPR spectra of cytochrome c oxidase samples exposed to the following conditions: A, 0.8 mM enzyme reduced anaerobically with a slight excess of ascorbate in the presence of an equimolar quantity of cytochrome c, reoxidized anaerobically with 6 mM porphyrexide for 1.2 sec and then exposed anaerobically to a saturated solution of CO in 10 mM cacodylate, pH 7.2, for 100 msec at 15°. B, (control) as A except that CO was replaced by N_2. C, cytochrome c oxidase partly reduced anaerobically with cytochrome c to maximal development of high spin signals at g=6. D, as C, but then mixed with a saturated solution of CO for 6 msec. Equal volumes were mixed and rapidly frozen. The concentrations given refer to the initial concentrations in the syringes in terms of total heme a. The conditions of EPR spectroscopy were those of Fig. 1 except that the scanning rate was 20 mT/min.

The reactions of the anaerobically reoxidized form with cyanide and sulfide occur within ≤ 5 msec and with ferrocytochrome c within ~ 20 msec. The conversion of a_3^{3+} into the low spin a_3^{3+}-CN^- and a_3^{3+}-S^{2-} complexes occurs quantitatively without reduction (Fig. 1, B-E), whereas it is known that the resting oxidized form reacts with these reagents only slowly and after partial reduction. Sulfide will of course also reduce the anaerobically reoxidized form, but this occurs much more slowly. Azide, known to have a lower affinity for $\underline{a}_3^{3+}$ than cyanide or sulfide[14] reacts more slowly with $\underline{a}_3^{3+}$ in the anaerobically reoxidized form. With both sulfide and azide, low spin species are formed at the earliest reaction times, which are different from those described in experiments carried out on a slower time scale. These are then converted within seconds to the species previously described[9].

We had previously observed that on partial reduction of cytochrome $\underline{c}$ oxidase at pH > 8 no rhombic high spin ferric heme signals are observed. Instead low spin signals arise at g=2.6, 2.2 and 1.84. It was, therefore, of interest to subject the anaerobically reoxidized form, which exhibits a strong high spin ferric signal at g=6, to a pH-jump to the alkaline region. Under these conditions the high spin ferric signal disappeared in $\leq$ 5 msec, whereas the low spin signal at g=2.6, 2.2 and 1.84 did not appear immediately. The signal arose, however in the course of seconds to minutes. This indicates that the rhombic high spin ferric heme species at g=6 is not in direct equilibrium with the low spin species at g=2.6, 2.2 and 1.84. An EPR silent state obviously intervenes. Nevertheless, this experiment lends support to the idea that the high spin ferric heme signal at g=6 and the low spin signal at g=2.6, 2.2 and, 1.84 represent the same species, namely $\underline{a}_3^{3+}$.

III. The product of exposure to O_2 of the anaerobically reoxidized enzyme. This species has been observed on exposure of the anaerobically reoxidized form to O_2 but has not been further characterized. It should be recalled that a study of the properties of this latter form already involved 3-syringe-double mixing experiments, so that a further study of the reactions of a form arising in such a system would call for 4 syringe - triple mixing, which is laborious and expensive in terms of materials. The "oxidation" product of the anaerobically oxidized form is characterized by the absence of a significant signal at g=6 and at the same time the absence of a fully developed 655 nm band. It is this latter feature which distinguishes it from the resting oxidized form as well as the "oxygenated" form, which arises on treatment of the reduced form with O_2, without prior use of a chemical oxidant (there are additional differences to the "oxygenated" form, see below). The 655 nm absorption therefore, appears to be a characteristic of direct oxidation with O_2.

IV. An early product of aerobic oxidation related to the "oxygenated" form. Time ranges used for study of reoxidation. In discussing our observations on this form, it may be useful to consider the time ranges involved in experimentation on "early" oxidation products. With the techniques employed by us, viz., rapid mixing close to room temperature and freeze-quenching, we can cover the range from $\sim$ 5 msec on, similar to what is possible by stopped-flow spectro-photometry. Much of the work on the "oxygenated" species has been carried out in the range of seconds to minutes. The recent work on low temperature trapping of early reaction products after photodissociation of the a_3^{2+}– CO compound[4], when extrapolated to room temperature conditions, probably extends into the μsec range. We are aware of the limitations of our approach, but in view of the time ranges covered, must assume that we have observed a species which must also have been present in most previous experiments dealing with the "oxygenated" form.

EPR signal of early oxidation product. We had observed in our previous EPR studies on cytochrome c oxidase that immediately - on the time scale of our ex-periments - after reoxidation of the completely reduced enzyme by O_2, and only by O_2, two resonances appeared at g=1.78 and 1.69 [5] which largely decayed with-in a few minutes. However, we did not find any clues as to other resonances which could be part of such a signal. The shape of the two resonances indicated that other lines belonging to the same signal, would have to be at lower field. The unidentified lines were of sufficiently low intensity that there was little hope to find additional lines of the same signal in the area around g=2 where the copper and heme absorptions are dominant. Not knowing the extent of the signal, we were unable to estimate, even approximately, how much material this signal might represent.

We have now found a broad resonance at g=5 which in all experiments carried out to date (see below) has shown behavior parallel to the two resonances at g=1.78 and 1.69 (ref. 14a). The three resonances of the signal, g=5; g=1.78 and g=1.69 exhibit the following properties:

1.) They are maximally developed after reaction of the fully reduced enzyme by O_2 at the earliest times accessible by our technique and decay with a $t_{1/2}$ at 16°C of $\sim$ 100 sec.

2.) The method used for reduction of the enzyme (dithionite, ascorbate, cyto-chrome c, NADH) makes no difference.

3.) Oxidation by electron acceptors other than O_2 did not produce these signals.

4.) Below saturating concentrations of O_2, the signal intensity is proportional to O_2 concentration.

5.) The use of the O_2 enriched 90% with ^{17}O does not change the line width (at high field) which is $\sim$ 13 mT with $^{16}O_2$.

6.) The signal is only slowly ($\sim$ 1 sec) abolished on addition of ferrocyto-chrome c, whereas the signal at g=3;2.2 and 1.5, which represents cytochrome <u>a</u>, disappears within ∿ 5 msec.

7.) The signal is abolished within < 5 msec on addition of cyanide or sulfide.

8.) The signal appears on reoxidation at pH 6 and 7.2, but not at pH 9.

9.) The maximal signal has never been seen in the absence of a fully developed 655 nm absorption or in the presence of a strong high spin signal at g=6; both of these features are thought to be characteristic of cytochrome a_3^{3+} [5-9,15].

10.) The signal is not readily saturated. It is not saturated at 90 mW at 13K or at 25 mW at 6.6K. It is saturated at 25 mW at 5.3K.

11.) The signal shape is not significantly influenced by temperature over the range 5-50K. However, the signal intensity decreases below 13K when observed at nonsaturating powers.

Figs. 3-6 present EPR spectra of experiments aimed at points 1 and 4 above. Figs. 3A and 4A show the high field and low field spectra, respectively, of a sample reoxidized in 8 msec, Figs. 3B and 4B and 3C and 4C show the corresponding signals at 220 sec and 13.5 min, respectively. Figs. 5A and 6A again show high and low field spectra, respectively, of another sample, reoxidized with saturating O_2, while Figs. 5B and 6B show a sample in which the ratio of O_2 molarity to heme concentration was about 1:10, i.e., the total electron accepting capacity in the O_2 available was 1/5 that required for the electron donors in the oxidase.

We would like to draw attention to the following features in the signal: Any changes in intensity of the two lines at g=1.69 and 1.78 occur in parallel and are proportional — within error — to the intensity changes of the signal at g=5. At maximum development of the signal at g=5; 1.78 and 1.69 the signal of cytochrome <u>a</u> at g=3.0; 2.2 and 1.5 shows two components of about equal intensity and has ∿ 70% of its maximal intensity. Fig. 3B and C shows that, as the new signal disappears, the signal of cytochrome <u>a</u> rises to full intensity with the more rhombic component at g=1.46 only gaining additional intensity.

<u>Interpretation of EPR signal</u>. The temperature behavior reported in item 11.) above excludes the possibility that we are dealing with a simple spin 1/2 system. A spin system of higher multiplicity is, therefore, indicated. However, if the line at g=5 is indeed a derivative line (as it seems, Fig. 4,6) and represents g_x and g_y of the compound that gives rise to it, this would exclude a spin 3/2 system from consideration and would indicate that the 2 lines at high field must arise from a splitting due to spin-spin interaction. In view of the size of the splitting (19 mT) an interaction of 2 electron spins is likely.

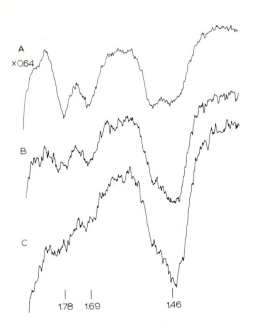

Fig. 3. EPR spectra (high field) from a study of the time course of decay of the transient oxidized form. Cytochrome c oxidase, 1.82 mM, was reduced as for Fig. 1, A and reoxidized at 16°C. A, exposed to tricine-cacodylate buffer, pH 7.2, saturated with O_2, for 8 msec; B, for 218 sec; C, for 13.5 min. The conditions of EPR spectroscopy were those of Fig. 1, except that the scanning rate was 100 mT/min. Spectra B and C were recorded at 1.56 times the amplification used for A; 4 scans were averaged.

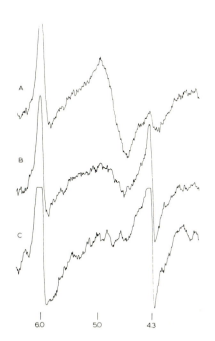

Fig. 4. EPR spectra (low field) from a study of the time course of decay of the transient oxidized form. The samples and the conditions of EPR spectroscopy were those of Fig. 3, except that 9 mW microwave power was used and the scanning rate was 40 mT/min. The amplification used was 1.25 times that used for recording Fig. 3A. (With permission from Ref. 14a).

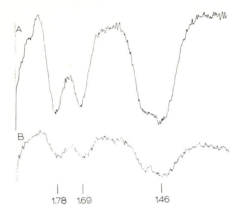

Fig. 5. EPR spectra (high field) from samples showing the dependence on oxygen concentration of the signals at 1.78 and 1.69. Cytochrome c oxidase, 1.58 mM, was reduced as for the experiment of Fig. 1. The enzyme was then rapidly mixed with 0.5 times its volume of tricine-cacodylate buffer of pH 7.2, saturated with oxygen (A) or with air (B) at 16°C. The conditions of EPR spectroscopy were those of Fig. 3; 8 scans were averaged.

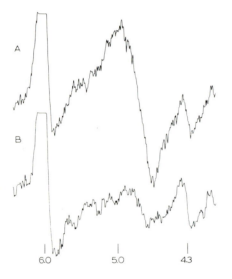

Fig. 6. EPR spectra (low field) from samples showing the dependence on oxygen concentration of the signal at g=5. The samples and the conditions of EPR spectroscopy were those of Fig. 5, except that the amplification was twice that used for Fig. 5 and the scanning rate was 40 mT per min.

Because of the width of the signal and its component resonances and is consequently rather indistinct features we have not been able to date to record spectra at a higher frequency in order to verify this last point. We are, therefore, unable to give this signal a unique interpretation. Consideration must be given to the question whether it could arise by spin-spin interaction of a high spin ferric heme of spin 5/2 with another paramagnet or by mixing of spin 5/2 and 3/2 states (cf. 16).

If we assume that we are dealing with a system of effective spin 1/2 and compare the intensity of the signal to that of the low spin ferric heme signal of cytochrome a present in the same spectra, by application of the procedures of Aasa and Vänngård[17], we come to a concentration of unpaired spins approximately equal to 50% of the heme in the enzyme. Thus, unless this approach is entirely inapplicable, we are dealing here with a major species which might well represent a large fraction, if not all, of one of the metal components of the enzyme.

Assignment of EPR signal. Obviously, without an understanding of the nature of the observed signal, any assignment of this signal must remain tentative. Under conditions when the new signal is observed, the EPR detectable Cu is fully oxidized, within error, and represented in its characteristic signal at g=2. This and the location of g-values exclude the EPR detectable Cu from consideration. The argument concerning the g values would also exclude the undetectable Cu, except that this Cu might qualify as the paramagnet interacting with the species represented in the new signal, if spin-spin interaction will have to be invoked in order to explain the signal. Cytochrome a is 70% represented in the low spin ferric heme signal at g=3.0;2.2;1.5 at the time of maximal development of the new signal. Cytochrome a can, therefore, not be excluded from consideration, as long as we have no reliable quantitative measurements of the unpaired spins represented in the new signal. The following arguments can be made in favor of the suggestion that cytochrome a_3 is the species represented in the new signal: 1.) cytochrome a_3 is not represented in any other signal in samples exhibiting the new signal, although optical data (Soret and 655 nm absorption) indicate that it is oxidized. 2.) There is now good evidence that cytochrome a_3 is in a high spin state[8,9,18], that is, it could represent the spin 5/2 system that is presumably involved in the new signal. 3.) The new signal is eliminated instantly ($\leq$ 5 msec) by cyanide or sulfide, which are both known ligands to cytochrome a_3^{3+}. 4.) The signal is eliminated within a few hundred msec on addition of ferrocytochrome c, whereas the signal of cytochrome a is abolished within $\leq$ 5 msec. 5.) The new signal is never observed in the presence of strong rhombic signals at g=6 and vice versa. Strong rhombic signals at g=6 in the presence of close to maximal signals at g=3 have been shown to originate from cytochrome a_3[8]. 6.) Strong rhombic signals at g=6, representing cytochrome a_3, are only observed on anaerobic reoxidation of reduced oxidase (see above, Results, II) at pH 6 to 7.5 but not at pH 9. The same is true for the formation of the new signal in the rapidly reoxidized oxidase.

For these reasons we favor the idea that the new signal represents cytochrome a_3^{3+} in a transient form, which only exists for a short period < 10 min following reoxidation.

Relationship to other described oxidized forms of the enzyme. According to the time range after exposure of the reduced enzyme to O_2, over which we observe the EPR signal of the transient oxidized form, we would expect it to be identical or related to forms described in spectrophotometric studies of the oxidation of the reduced enzyme by a number of workers[1, 19-23]. Following the original observations of Okunuki and his collaborators[1] such form(s) are generally referred to as "oxygenated" oxidase and are characterized by a Soret peak at 428 nm at room temperature and a slight shift of the α-band toward longer wavelengths together with a small increase in the height of this band. Although evidence for several discrete states of the enzyme in the range from msec to minutes has been reported[19,20] we find it difficult to identify these states, characterized only by slight changes of light absorption features, with the state we have observed through a discrete EPR signal, whose presence or absence provides an unambiguous criterion. The species we have observed comes closest to compounds I or II of the "oxygenated" form described by Orii and King[19] or the intermediates decaying in reaction 2 or 3 of Tiesjema et al.[20]. There is no doubt that the intermediate we observe by EPR must have been present in the reaction mixtures used by previous workers and may have been the principal species in some instances. According to reflectance spectra recorded at 100 K, the Soret peak is at 427-428 nm during the life-time of our transient intermediate and this intermediate shows the small changes in the α-peak described for the "oxygenated" form by others. According to a comparison of our low temperature reflectance spectra with the EPR spectra of the same samples, however, the decay of the EPR signal of the transient form is more rapid than the shift of the Soret peak toward 422-424 nm, where it is observed with the aerobic resting form of our preparations. From the time range in which activated forms of the enzyme have been observed[2,24] which react more rapidly with ferrocytochrome c or ligands, we would also conclude that the species we have identified must have been present under the conditions of those experiments and may well be identical to the activated species observed there. This is particularly likely since we have found our transient species to react rapidly with cyanide and sulfide, in contrast to the aerobic resting enzyme. We have also found the reduction by ferrocytochrome c of cytochrome a in our transient form to be very rapid and beyond resolution by our rapid freezing technique.

We cannot readily make a comparison between the species we have observed and early states in the reaction with O_2 of the photolyzed CO compound of the reduced enzyme as they have been described from trapping experiments in aqueous-organic solvents[4]. These states are thought to include forms in which electrons from the enzyme have not or only partly been transferred to O_2. From optical and EPR criteria and also from its relatively slow rate of decay, we conclude that the transient oxidation product, characterized by the EPR signal we described, is a fully oxidized form. Our experiments with ^{17}O enriched O_2 have given no evidence that a species of oxygen is part of the structure from which the EPR signal originates. However, in view of the linewidth of our signal ($\Delta H_{1/2} \sim 13$ mT at high field), it is possible that a weak interaction with oxygen might have escaped detection.

We expect that progress in the interpretation of the signal we have observed will contribute to our understanding of the nature and particular features of early transient and apparently activated states of reoxidized cytochrome c oxidase.

ACKNOWLEDGEMENTS

This work was supported by the National Institute of General Medical Sciences of the National Institutes of Health by a research grant (GM12394) and a Research Career Award (5-K06-GM-18492) to H. B. and a postdoctoral fellowship (F-32-GM-05772) to RWS.

REFERENCES

1. Sekuzu, I., Takemori, S., Yonetani, T., and Okunuki, K. (1959) J. Biochem., 46, 43-49.

2. Antonini, E., Brunori, M., Colosimo, A., Greenwood, C., and Wilson, M. T. (1977) Proc. Natl. Acad. Sci. USA, 74, 3128-3132.

3. Rosén, S., Bränden, R., Vänngård, T., and Malmström, B. G. (1977) FEBS Letters, 74, 25-30.

4. Chance, B., Saronio, C., and Leigh, J. S., Jr. (1975) J. Biol. Chem., 250, 9226-9237.

5. Hartzell, C. R. and Beinert, H. (1974) Biochim. Biophys. Acta, 368, 318-338.

6. Hartzell, C. R. and Beinert, H. (1976) Biochim. Biophys. Acta, 423, 323-338.

152

7. Beinert, H., Hansen, R. E. and Hartzell, C. R. (1976) Biochim. Biophys. Acta 423, 339-355.

8. Beinert, H. and Shaw, R. W. (1977) Biochim. Biophys. Acta 462, 121-130.

9. Shaw, R. W., Hansen, R. E., and Beinert, H. (1978) Biochim. Biophys. Acta, in press.

10. Aasa, R., Albracht, S.P.J., Falk, K.-E., Lanne, B. and Vänngård, T. (1976) Biochim. Biophys. Acta, 422, 260-272.

11. Tzagoloff, A., and Wharton, D. C. (1965) J. Biol. Chem. 240, 2628-2633.

12. Greenwood, C., Wilson, M. T., and Brunori, M. (1974) Biochem. J., 137, 205-215.

13. Kuhn, R. and Franke, W. (1935) Ber.Dtsch. Chem. Ges., 68, 1528-1536.

14. Wever, R., Van Gelder, B.F., and DerVartanian, D. V. (1975) Biochim. Biophys. Acta, 387, 189-193.

14a. Shaw, R. W., Hansen, R. E. and Beinert, H., J. Biol. Chem., in press.

15. Hartzell, C. R., Hansen, R. E. and Beinert, H. (1973) Proc. Nat. Acad. Sci. (USA), 70, 2477-2481.

16. Maltempo, M. M. and Moss, T. H. (1976) Quarterly Reviews of Biophysics, 9, 181-215.

17. Aasa, R., and Vänngård, T. (1975) J. Mag. Res., 19, 308-315.

18. Babcock, G. T., Vickery, L. E. and Palmer, G. (1976) J. Biol. Chem., 251, 7907-7919.

19. Orii, Y. and King, T. E. (1976) J. Biol. Chem., 251, 7487-7493.

20. Tiesjema, R. H., Muijsers, A. O. and Van Gelder, B. F. (1972) Biochim. Biophys. Acta, 256, 32-42.

21. Muijsers, A. O., Tiesjema, R. H., and Van Gelder, B. F. (1971) Biochim. Biophys. Acta, 234, 481-492.

22. Wharton, D. C. and Gibson, Q. H. (1968) J. Biol. Chem., 243, 702-706.

23. Lemberg, R. and Mansley, G. E. (1966) Biochim. Biophys. Acta, 118, 19-35.

24. Brittain, T. and Greenwood, C. (1976) Biochem. J., 155, 453-455.

Cytochrome Oxidase, T.E. King et al. eds.
© *1979 Elsevier/North-Holland Biomedical Press*

SOME POSSIBLE CHEMICAL AND ELECTRONIC STATES OF
CYTOCHROME c OXIDASE AND ITS INTERMEDIATE REDOX STATES

W. E. BLUMBERG

Bell Laboratories, Murray Hill, N. J. 07974, USA.

J. PEISACH

Departments of Molecular Pharmacology and Molecular Biology,
Albert Einstein College of Medicine of Yeshiva University,
Bronx, N. Y. 10461, USA.

ABSTRACT

The various oxidation states of cytochrome c oxidase are discussed with reference to their observed magnetic and EPR properties as well as to their chemical structures. The structure of heme a in oxidized cytochrome a is shown to be a *bis*-imidazole compound, thereby resembling cytochromes b_2 and b_5. The heme a in partially or fully reduced cytochrome a_3 has a single imidazole ligand and in this way resembles myoglobin and hemoglobin. A μ-oxo structure is discussed as a possible bridge between iron and copper in fully oxidized cytochrome a_3.

INTRODUCTION

Cytochrome c oxidase is a multisubunit protein[1,2] which contains both iron[3,4] and copper[5,6] as redox active components, the former being incorporated into heme a.[7] The heme a is bound to two different polypeptides, cytochrome a and cytochrome a_3,[8] which are associated with several other non-metal containing proteins to complete the cytochrome c oxidase molecule. Nicholls and Chance[9] have recently reviewed the optical, magnetic and ligand binding properties of cytochromes a and a_3.

Cytochrome c oxidase, when fully oxidized, requires four electron equivalents in order to become fully reduced.[10] It is usually stated that the hemes a are both ferric and the copper atoms both cupric in a formal valence sense. The EPR spectrum of only two paramagnetic species, one heme a, that in cytochrome a, and one copper atom are observed.[11] The paramagnetic components of cytochrome a_3 are not observed in EPR experiments because of strong antiferromagnetic coupling, leading to an even spin state.[12,13] The spectrum of the EPR observable copper lacks resolved nuclear hyperfine structure[14,15,16] leading Peisach and Blumberg[17] to suggest that the unpaired spin is not principally localized on the metal atom.

STRUCTURE OF IRON IN CYTOCHROME a

The EPR spectrum of the heme a observed in the fully oxidized protein was first observed by van Gelder and Beinert.[18] The spectrum, now generally ascribed to cytochrome a, is that of a low spin ferric heme compound ($S = \frac{1}{2}$). Blumberg and Peisach[19] showed that one can analyze EPR parameters for low spin heme complexes using a crystal field approach so as to obtain information concerning the structure of the axial ligands. The EPR parameters provide signatures of various combinations of these ligands. Since chemical substitution of side chains of porphyrin molecules only shift the EPR parameters slightly,[20] one may make ligand assignments for low spin compounds of heme a using data obtained for corresponding complexes of heme.

Peisach[21] has recently summarized the EPR g values of a number of low spin heme compounds and heme proteins. Both on the basis of the g values and the nuclear modulation effect in EPR,[22] one can determine that the heme a component of cytochrome c oxidase observed by EPR has two imidazole groups as axial ligands. Thus structurally it resembles cytochromes b_2[23] and b_5.[24] This suggests that the function of heme a in cytochrome a is electron transfer, as in the cases of the cytochromes b.

STRUCTURE OF COPPER IN CYTOCHROME a

The EPR customarily ascribed to copper is observed in fully oxidized cytochrome c oxidase and is generally believed to arise from the copper of cytochrome a.[11,15] The spectrum is axial ($g_{\parallel} = 2.17$, $g_{\perp} = 2.03$) and has no resolved hyperfine structure. The lack of hyperfine pattern in the EPR spectrum ascribed to copper would suggest that the unpaired spin of the formally cupric copper is largely transferred to the ligand or ligands. Indeed, the ratio of $\Delta g_{\parallel}$ to $\Delta g_{\perp}$ ($0.17/0.03 = 5.7$) also indicates this, as a d electron on copper has a maximum Δg ratio of 4,[25] and this ratio usually decreases upon tetrahedral distortion. An unpaired spin on a sulfur atom may have a Δg ratio approaching infinity.[26] Further evidence for this suggestion is derived from nuclear modulation effect experiments carried out by Mims and Peisach.[27,28] In the case of Cu(II) complexes, the nuclear modulation produces a pattern which depends on the type and number of interacting nuclear spins and their distances. For example, these authors have shown that the copper atom in a variety of mononuclear copper proteins all have imidazole ligation. For cytochrome c oxidase the pattern characteristic of imidazole ligation was not observed. The most prominent features in the modulation pattern arose from interactions of the paramagnetic center with protons. Thus, either the copper binding site of cytochrome c oxidase is unique in that Cu(II) is not ligated to imidazole or the unpaired spin largely resides on a ligand other than imidazole. If the copper atom is a part of the paramagnetic center, then the best electronic description is Cu(I)·SR where a ligand to metal charge transfer equivalent to almost one complete electron has taken place. This charge transfer would place almost all the spin density on the ligand.

Neither the low spin ferric EPR spectra for heme *a* of cytochrome *a* nor the "copper" spectrum show prominent effects of dipolar coupling. If these effects are present, they are within the natural linewidths of a few gauss, leading to the conclusion that the two paramagnetic centers are no closer than about 6-8 Angstroms. Even at this distance, however, electron redistribution after oxidation or reduction could take place very rapidly.

STRUCTURE OF IRON IN CYTOCHROME a_3

When cytochrome *c* oxidase is partially reduced, new EPR spectra can be seen which probably arise from the heme *a* of cytochrome a_3.[29] In addition to the EPR of several high spin ferric heme compounds,[18,30,31] one can also observe a spectrum which is almost identical to that of ferric myoglobin hydroxide. It is interesting to speculate that this hydroxide form of ferric cytochrome a_3 is in internal pH equilibrium with one or more of the high spin forms observed. If azide,[18] cyanide,[32] or hydrosulfide[33] are added, the high spin and hydroxide forms are converted to low spin forms which are very similar in their EPR properties to myoglobin azide,[34] cyanide,[35] and hydrosulfide.[36] Thus an axial ligand to the ferric heme *a* of partially reduced cytochrome a_3 is an imidazole group. The ligand *trans* to the imidazole is freely exchangeable with exogenous ligands and is probably oxygenous in nature, perhaps even water under certain circumstances.

In fully reduced cytochrome *c* oxidase, NO binds to the ferrous heme *a* of cytochrome a_3. This NO-bound form has an EPR spectrum with a nine-line superhyperfine pattern consistent with imidazole as the axial ligand *trans* to the NO.[13] Thus the imidazole group can be a ligand to heme *a* of cytochrome a_3 in both oxidation states of the iron. There is no direct evidence, however, that imidazole is a ligand to the heme *a* of fully oxidized cytochrome a_3, but by inference it seems likely to be so.

COUPLING BETWEEN IRON AND COPPER IN CYTOCHROME a_3

From the magnitude of the antiferromagnetic coupling of the heme *a* and copper in cytochrome a_3,[13] one concludes that the iron and copper atoms are separated by no more than a few atoms. One can envision three hypothetical cases: *(1)* the imidazole group bridges the two metal atoms as suggested by Palmer;[37] *(2)* an endogenous group from the protein bridges the two metal atoms on the side of the heme *a* distal to the imidazole; *(3)* an exogenous group performs the same function as suggested by Peisach.[21]

One can discount hypothesis *(1)* on several counts. The *J* coupling transferred by an imidazole group (based on Cu−Cu bridging with a coupling of 26 cm[-1] as determined by Kolks and Lippard[38]) would be insufficient to account for the magnitude of *J* in cytochrome *c* oxidase, although it is conceivable that Cu−Fe coupling via an imidazole group might be somewhat larger. More direct evidence, however, comes from the EPR of the partially reduced intermediates mentioned above. The proximal imidazole group of heme in myoglobin (as well as in hemoglobin) is deprotonated or hydrogen bonded, that is, the axial ligand resembles imidazolate in contrast to cytochrome P-450 where the axial ligand is believed to be neutral imidazole.[39] Copper ligated to imidazolate would mimic neutral imidazole, and the hydroxide, azide, cyanide, and hydrosulfide compounds observed for partially reduced cytochrome a_3 would not have the same EPR properties as the analogous myoglobin compounds. Likewise in the reduced state, the nine-line superhyperfine pattern of the NO adduct indicates that the imidazole ligand *trans* to NO binds neither a proton nor a metal atom, a conclusion based on an analogy to low spin ferric hemoproteins.[39]

Case *(2)* would hypothesize an imidazole or carboxyl function to bridge the two metal atoms. An imidazole group would probably provide insufficient *J* coupling as discussed above. A carboxyl function ligated to heme *a* of cytochrome a_3 after the associated copper atom is reduced would give high spin EPR spectra consistent with those observed in partially reduced cytochrome *c* oxidase. No low spin ferric heme compound is known which involves a carboxyl group as an axial ligand to heme iron. Therefore, one must postulate that this group would be readily dissociable to form the hydroxide compound and could be displaced by azide, cyanide, and hydrosulfide. Of course, in the fully reduced state one does not know whether there is a bridge or not, but, if there is, it must be easily displaced by O_2, CO, and NO.

Based on the *J* coupling observed in crystalline copper acetate,[40] where four carboxyl bridges provide a *J* of about 300 cm[-1], one would also expect that the *J* coupling provided by a carboxyl function would be somewhat too small to account for the antiferromagnetic coupling in cytochrome *c* oxidase. Just for completeness, it might be mentioned that mercaptide sulfur can theoretically act as a bridge, but this has neither been suggested nor observed in multimetal protein complexes.

An exogenous bridge, case *(3)*, poses problems of its own. It is likely that, if such a group arose from a small organic molecule, it would not have escaped the notice of the myriad of biochemists who have made the various preparations of cytochrome *c* oxidase. On the other hand, a single oxygen atom, $O^=$, would be undetectable as a separate chemical entity. Such an atom could make a μ-oxo bridge between the two metal atoms. Although this particular mixed metal μ-oxo structure is not known in inorganic chemistry, there are many examples of Fe−O−Fe[41] and

$Cu-O-Cu^{42}$ bridges. When a μ-oxo bridge is displaced, it would become lost, but it could be almost instantaneously replaced from water as needed. It would be jettisoned and regenerated during each turnover cycle of the oxidase.

One should examine the chemistry of the reduction of O_2 with this μ-oxo bridge in mind. In this case one starts with $Fe^{3+}-O^=-Cu^{2+}$ in the fully oxidized protein. After a one electron reduction, the oxygen atom would no longer bridge and would cling to the metal atom still oxidized. The spin coupling would be broken. Since one observes transient EPR signals from ferric heme a but not from copper,[18] one would conclude that the copper atom must always be reduced first. As soon as the copper atom is reduced, it would no longer bind to the oxygen atom, leaving it free to gain one proton (forming ferric heme a hydroxide) or two protons (forming aquo ferric heme a). Indeed the EPR spectra one observes are consistent with the formation of these species. Complete reduction of the ferric complex would dissociate the oxygen atom from the presumed bound water or hydroxide, leaving a suitable binding site for O_2 between the two reduced metal atoms consisting of high spin ferrous heme a, analogous to the heme of deoxyhemoglobin, and Cu^{1+}, analogous to the copper at the O_2 binding site of hemocyanin. The oxygenated complex, $Fe^{2+}-{\cdot}OO^{\cdot}-Cu^{1+}$, could undergo electron tautomerization to the oxidized metal-peroxide complex, $Fe^{3+}-O^-O^--Cu^{2+}$, or the ferryl structure, $Fe^{4+}-O^=-O^--Cu^{2+}$. The latter would probably spontaneously cleave the $O-O$ bond either before or immediately after undergoing a one-electron reduction. After this the oxygen atom associated with the copper atom would be converted to water and could exchange with the bulk solvent. This would leave $Fe^{4+}-O^= + Cu^{2+}$, which could reform the bridge, or the bridge would reform after another single electron reduction to yield $Fe^{3+}-O^=-Cu^{2+}$, the antiferromagnetically coupled oxidized state. There are, of course, other pathways of electron tautomerization leading to different electronic descriptions of the intermediates. Whether the individual steps in this illustrative scheme occur sequentially or in pairs is not essential to the overall mechanism. What we have tried to do is to present a dissected picture of a possible chemical course of events.

Several experiments, both physical and chemical, come to mind to test the μ-oxo hypothesis. Infrared stretching frequencies for both $Fe-O$ and $Cu-O$ should be observed in the resting oxidized enzyme. An experiment using isotopically labelled O_2 under *single turnover* conditions would leave a labelled bridge which would appear in the infrared stretching spectrum and would best be observed by difference spectroscopy. Analysis of the extended fine structure of the iron and copper x-ray absorption edges would show an iron–copper distance of a little over 3 Angstroms. These experiments would also ascertain whether the imidazole observed to be bound to heme *a* in partially or fully reduced cytochrome a_3 is indeed bound in the fully oxidized protein.

Acknowledgement. That portion of this work carried out at Albert Einstein College of Medicine was supported in part by Research Grant HL-13399 from the Heart and Lung Institute and as such is Communication 386 from the Joan and Ester Avnet Institute of Molecular Biology.

REFERENCES

1. Chuang, T. F., and Crane, F. L. (1971) *Biochem. Biophys. Res. Commun.* **42,** 1076.
2. Shakespeare, P. G., and Mahler, H. R. (1971) *J. Biol. Chem.* **246,** 7649.
3. Dannenberg, H., and Kiese, M. (1952) *Biochem. Z.* **322,** 395.
4. Person, P., Wainio, W. W., and Eichel, B. (1953) *J. Biol. Chem.* **202,** 369.
5. Keilin, E., and Hartree, E. F. (1938) *Nature* **141,** 870.
6. Keilin, D., and Hartree, E. F. (1939) *Proc. Roy. Soc. (London)* **B127,** 167.
7. Lemberg, M. R. (1969) *Physiol. Rev.* **49,** 48.
8. Keilin, D. (1925) *Proc. Roy. Soc. (London)* **B98,** 312.
9. Nicholls, P., and Chance, B. (1974) in *Molecular Mechanisms of Oxygen Activation* (O. Hayaishi, ed.), Academic Press, New York. p. 479.
10. van Gelder, B. F., and Muijsers, A. O. (1964) *Biochim. Biophys. Acta* **81,** 405.
11. Hartzell, C. R., and Beinert, H. (1976) *Biochim. Biophys. Acta* **423,** 323.
12. Babcock, G. T., Vickery, L. E., and Palmer, G. (1976) *J. Biol. Chem.* **251,** 7907.
13. Palmer, G., Antolis, T., Babcock, G. T., Garcia-Iniguez, L., Tweedle, M., Wilson, L. J., and Vickery, L. E. (1978) in *Mechanisms of Oxidizing Enzymes* (T. P. Singer and R. N. Ondarza, eds.), Elsevier Press, New York. p. 221.
14. Beinert, H., and Palmer, G. (1964) *J. Biol. Chem.* **239,** 1221.
15. Beinert, H. (1966) in *The Biochemistry of Copper* (J. Peisach, P. Aisen, and W. E. Blumberg, eds.), Academic Press, New York. p. 213.
16. Beinert, H., Griffiths, D. E., Wharton, D. C., and Sands, R. H. (1967) *J. Biol. Chem.* **237,** 2337.
17. Peisach, J., and Blumberg, W. E. (1974) *Arch. Biochem. Biophys.* **165,** 691.
18. van Gelder, B. F., and Beinert, H. (1969) *Biochim. Biophys. Acta* **189,** 1.
19. Blumberg, W. E., and Peisach, J. (1971) in *Probes of Structure and Function of Macromolecules and Membranes* (B. Chance, T. Yonetani, and A. S. Mildvan, eds.), Academic Press, New York. p. 215.

20. Peisach, J., Blumberg, W. E., and Adler, A. (1973) *Ann. New York Acad. Sci.* **206,** 310.

21. Peisach, J. in *Frontiers of Biological Energetics* (L. P. Dutton, J. S. Leigh, Jr., and A. Scarpa, eds.), Academic Press, New York (in press).

22. Peisach, J., and Mims, W. B. (1977) *Biochemistry* **16,** 2795.

23. Watari, H., Groudinsky, O., and Labeyrie, F. (1967) *Biochim. Biophys. Acta* **131,** 592−594.

24. Bois-Poltoratsky, R. and Ehrenberg, A. (1967) *Eur. J. Biochem.* **2,** 1511−1533.

25. Griffiths, J. S. (1961) *The Theory of Transition Metal Ions,* Cambridge University Press, Cambridge. pp. 341−345.

26. Akasaka, K. (1965) *J. Biol. Chem.* **43,** 1182−1183.

27. Mims, W. B., Peisach, J., and Davis, J. L. (1977) *J. Chem. Phys.* **66,** 5536.

28. Peisach, J. (1978) in *Mechanisms of Oxidizing Enzymes* (T. P. Singer and R. N. Ondarza, eds.), Elsevier Press, New York. p. 285.

29. Beinert, H., and Shaw, R. W. (1977) *Biochim. Biophys. Acta* **462,** 121.

30. Peisach, J., Blumberg, W. E., Ogawa, S, Rachmilewitz, E. A., and Oltzik, R. (1971) *J. Biol. Chem.* **246,** 3342.

31. Beinert, H., Shaw, R. W., and Hansen, R. E. (1978) in *Mechanisms of Oxidizing Enzymes* (T. P. Singer and R. N. Ondarza, eds.), Elsevier Press, New York. p. 239.

32. Dervartanian, D. V., Lee, I. Y., Slater, E. C., and van Gelder, B. F. (1974) *Biochim. Biophys. Acta* **347,** 321.

33. Wever, R., van Gelder, B. F., and Dervartanian, D. V. (1975) *Biochim. Biophys. Acta* **387,** 189.

34. Helckè, G. A., Ingram, D. J. E., and Slade, E. F. (1968) *Proc. Roy. Soc. (London)* **B169,** 275.

35. Hori, H. (1971) *Biochim. Biophys. Acta* **251,** 227.

36. Berzofsky, J. A., Peisach, J., and Blumberg, W. E. (1971) *J. Biol. Chem.* **246,** 3367.

37. Palmer, G., Babcock, G. T., and Vickery, L. E. (1976) *Proc. Natl. Acad. Sci. USA* **73,** 2206−2210.

38. Kolks, G., and Lippard, S. J. (1977) *J. Am. Chem. Soc.* **99,** 5804−5806.

39. Chevion, M., Peisach, J., and Blumberg, W. E. (1977) *J. Biol. Chem.* **252,** 3637−3645.

40. Bleaney, B., and Bowers, K. D. (1952) *Proc. Roy. Soc. (London)* **A214,** 451.

41. Fleischer, E., and Srivastava, T. S. (1969) *J. Amer. Chem. Soc.* **91,** 2403.

42. Kato, M., Jonassen, H. B., and Fanning, J. C. (1964) *Chem. Rev.* **64,** 99−128.

Cytochrome Oxidase, T.E. King et al. eds.
© *1979 Elsevier/North-Holland Biomedical Press*

CYTOCHROME OXIDASE VESICLES WITH TWO-DIMENSIONAL ORDER

WILLIAM GOLDFARB, JOACHIM FRANK AND MARTIN KESSEL

New York State Department of Health
Division of Laboratories and Research, Albany, NY 12201

J. C. HSUNG, CHONG H. KIM AND TSOO E. KING

Laboratory of Bioenergetics, State University of New York at Albany, Albany,
New York, U. S. A.

ABSTRACT

 The artificially formed liposomal membrane of highly purified cytochrome
oxidase prepared by a cholate method yields crystals revealed by electron
microscopy. These crystals are suitable for low dose high resolution electron
microscopy and image reconstruction. Our results show the cytochrome oxidase
crystal lattice of pgg symmetry and the unit vector of $\underline{a}$ = 91 ± 4 Å and
$\underline{b}$ = 117 ± 5 Å. A computer reconstruction of the two-dimensional projection is
also presented.

INTRODUCTION

 About 10 years ago Oda[1] first reported two-dimensional crystalline order in
cytochrome oxidase membranes. Subsequent papers[2-5] showed extended crystalline
order in two dimensions under electron microscopic examination. Oda et al.[1,2]
used deoxycholate for solubilization of the enzyme whereas Vanderkooi et al.[3-5]
employed Triton.

 Cytochrome oxidase vesicles which have regular order in two dimensions,
termed two-dimensional crystals, promise to be a form of the enzyme particularly suited for direct methods of structural investigation using computer
reconstruction[6,7]. These thin crystals are good objects for the recently
developed technique of low dose high resolution electron microscopy applied by
Henderson and Unwin to give a 7 Å three-dimensional density map of the
purple membrane of *Halobacterium halobium*. However, enzymes solubilized by
deoxycholate or Triton show very low respiratory control when reconstituted
into liposomes. No report on a cytochrome oxidase study has appeared where
enzymes other than those solubilized by those detergents were used in the formation of crystalline sheets. It has been shown by Hinkle et al.[9] and by Racker
and Kandrach[10] that cytochrome oxidase prepared with cholate has high respiratory control and oxidative phosphorylation in the presence of other factors.

We failed to see any reason why this technique[8] could not be applied to
highly purified cytochrome oxidase solublilized by cholate in well-defined,
artificial vesicular systems. These considerations prompted us to investigate methods for producing crystalline vesicles of cytochrome oxidase solubilized by the cholate method. We have succeeded in making such a preparation,
and have used the crystals to produce computer filtered reconstructions from
electron micrographs of the negatively stained enzyme (Fig. 1,2).

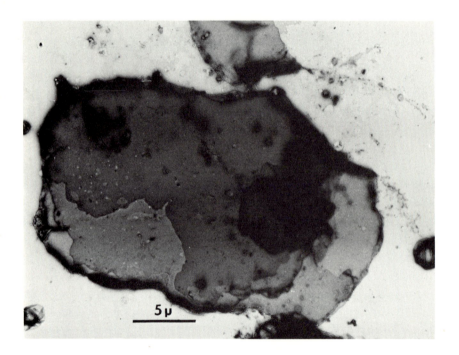

Fig. 1. An electron micrograph of low magnification of cytochrome oxidase
vesicles. Optical diffraction shows the whole vesicle to be a single
ordered lattice.

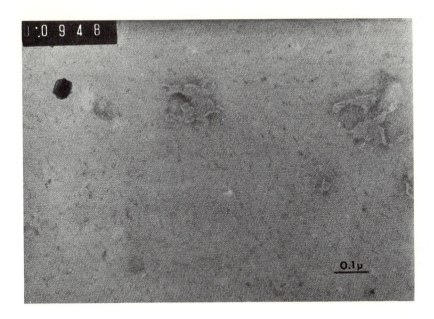

Fig. 2. A view of higher magnification of a thin part of the vesicle as shown in Fig. 1.

MATERIALS AND METHODS

Liposomes were formed from phospholipid-depleted cytochrome oxidase[11] by an extension of the method[12] based on sequential fragmentation of the respiratory chain[13]. The oxidase thus prepared was devoid of even electron transport activity but resumed fully[11] in the presence of azolectin, a soybean phospholipid mixture. The reconstituted system was made from the cytochrome oxidase prepared and mixed with purified azolectin[14] in 1:1 ratio (w/w) in the presence of cholate. The mixture was then sonicated, diluted, dialyzed, and finally fractionated after density gradient centrifugation. We used a similar, but not identical, centrifugation method employed by Drs. H. Griffith and P. Jost (personal communication, 1978). They have used their method for spin-labelled EPR spectroscopy in studies of interactions of different pure phospholipids with cytochrome oxidase.

One drop of the liposome sample at approximately 1.2 mg oxidase per ml in 50 mM phosphate buffer, pH 7.4, with about 45% sucrose was applied to a grid coated with carbon film. The grid was washed 3 times with buffer, then stained with one drop of 1% uranyl acetate and the excess stain was blotted off. The dried grid was examined in a Philips electron microscope EM 301

operated at 100 KeV with a 40 μ objective aperture. Micrographs used for
computer filtering were recorded at 43,000 magnification. The best micrographs
were selected by examination of optical diffraction patterns and scanned in
512 x 700 arrays with a Perkin Elmer PDS 1010 automatic microdensitometer at
25 μ intervals corresponding to a sampling of 5.8 Å.

The computer generated diffractogram and filtered reconstruction (Fig. 3)
were obtained using the SPIDER image processing system[15] on a PDP 11/45.
From the 512 x 700 array of input data, a 256 x 256 area representing about
165 unit cells was selected for filtering (cf. Fig. 3a). In order to avoid
overlaps in the averaging, the 256 x 256 picture was inserted into a blank
512 x 512 array which had a gray level equal to the average density value of
the picture being filtered. From the Fourier transformed data (Fig. 3b) an
optimum filter mask was calculated which included all reflections out to
28 Å. A 3 x 3 area around each Fourier maximum was passed by the filter
mask. The filtered Fourier data were back-transformed to produce the filtered
images, as shown in Figs. 3c and 4.

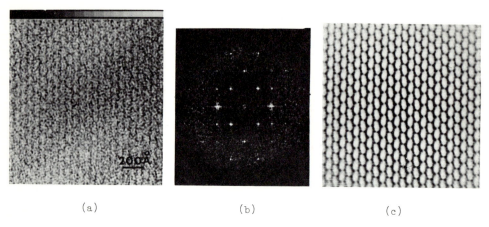

| (a) | (b) | (c) |

Fig. 3. Negatively stained membranous crystals of cytochrome oxidase.
(a) Digitally sampled micrograph before the filtering operation. (b) Computed optical diffraction pattern of the area shown in (a). The b* direction
is horizontal. (c) Computer filtered image reconstruction produced from
an 11 by 15 unit cell area of the original area.

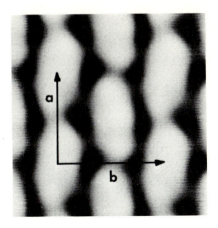

Fig. 4. A unit cell taken from the larger area of filtered cytochrome
oxidase crystal sheet. The white areas are the stain excluding or protein
regions.

RESULTS AND DISCUSSION

In their study of cytochrome oxidase crystals prepared with the Triton
method[16], Henderson et al.[6] have concluded that the two-dimensionally
ordered sheets are formed when the vesicles collapse and enzyme molecules on
opposite sides of the membrane interdigitate. For our cholate enzyme the
osmotic action in centrifugation may cause the collapse of the vesicles in a
manner favorable to the formation of crystals. In many preparations, we
found nearly all the membranous cytochrome oxidase visible on the grid to be
in an ordered condition. Sometimes the vesicles appeared as isolated, large
sheets of crystals but frequently as stacks of two, three or more vesicles.
These stacks might show only a single lattice in projection or a moiré
pattern of multiple lattices. In the first case the stack of vesicles apparently formed with three-dimensional order. Cytochrome oxidase examined before
the step of density gradient centrifugation rarely showed any order within
the vesicles; only very few highly overlapped vesicles showed crystallinity in
this earlier step. The cholate prepared sheets are frequently seen to be
relatively large and well ordered over areas greater than four square microns
(see Figs. 1 and 2). This means that our preparation should be suitable for
high resolution electron diffraction study from both stained and unstained
samples.

It may be mentioned that several samples nearly 3 months old stored at
0 - 4° in about 45% sucrose showed practically the same microscopic crystals
but retained only about 20% of the original electron-transport activity.

The significance of this observation is not known. However, we did not analyze these crystals in detail to ascertain any difference from the fresh samples. Similar results but of much shorter time in storage have been observed[7].

From an analysis of the optical diffraction patterns we found that the crystal lattice had the same pgg symmetry and other features as reported previously[6,17] while differing from the reconstruction of some other investigators[7] in the apparent degree of symmetry.

An average of measurements from 30 optical diffraction patterns gave the unit vectors of $a = 91 \pm 4$ Å, $b = 117 \pm 5$ Å in good agreement with the previous results on Triton preparations. Based on these results, we found, at a resolution of 28 Å, no significant difference from the reconstruction of Maniloff et al.[17] and Henderson et al.[6]. Low dose microscopy and three-dimensional analysis should enable us to probe the structure of cytochrome oxidase at higher resolution.

SUMMARY

The artificially formed liposomal membranes of highly purified cytochrome oxidase prepared by a cholate method yield crystals. These crystals observed in an electron microscope are suitable for low dose high resolution electron microscopy and image reconstruction. The results show the crystal lattice of pgg symmetry and the unit vector of $a = 91 \pm 4$ Å and $b = 117 \pm 5$ Å. The computer-reconstructed image in two-dimensional patterns is also given.

ACKNOWLEDGEMENT

This research was supported by grants GM-24412, GM-16767 and HL-12576 of the NIH, U. S. A. One of the authors (TEK) acknowledges stimulating discussions with Drs. H. Griffith and P. Jost.

REFERENCES

1. Oda, T. (1968) in Structure and Function of Cytochromes, K. Okunuki, M. D. Kamen, and I. Sekuzu, eds. University of Tokyo Press, Tokyo, pp. 500-515.

2. Seki, S., Hayshi, H. and Oda, T. (1970) Arch. Biochem. Biophys. 138,110-121.

3. Vanderkooi, G., Senior, A. E., Capaldi, R. A. and Hayshi, H. (1972) Biochim. Biophys. Acta 274, 38-48.

4. Vanderkooi, G. (1972) Ann. N.Y. Acad. Sci. 195, 6-15.

5. Vanderkooi, G. (1974) Biochim. Biophys. Acta 344, 307-345.

6. Henderson, R., Capaldi, R. A. and Leigh, J. S. (1977) J. Mol. Biol. 112, 631-648.

7. Frey, T. G., Chan, H. P. and Schatz, G. (1978) J. Biol. Chem. 253, 4389-4395.

8. Henderson, R. and Unwin, P. N. T. (1975) Nature 257, 28-32.

9. Hinkle, P. C., Kim, J. J. and Racker, E. (1972) J. Biol. Chem. 247, 1338-1339.

10. Racker, E. and Kandrach, A. (1971) J. Biol. Chem. 246, 7069-7071.

11. Yu, C. A., Yu, L. and King, T. E. (1975) J. Biol. Chem. 250, 1383-1392.

12. Kuboyma, M., Yong, F. C. and King, T. E. (1972) J. Biol. Chem. 247, 6375-6383.

13. King, T. E. (1966) Adv. Enzymol. 28, 155-236.

14. Kagawa, Y. and Racker, E. (1971) J. Biol. Chem. 246, 5477-5487.

15. Frank, J. and Shimkin, B. (1978) 9th Intl. Congr. El. Micros., Toronto, Vol. I, pp. 210-211.

16. Sun, F. F., Prezbindowski, K. S., Crane, F. L. and Jacobs, E. E. (1968) Biochim. Biophys. Acta 153, 804-818.

17. Maniloff, J., Vanderkooi, G., Hagashi, H. and Capaldi, R. A. (1973) Biochim. Biophys. Acta 298, 180-183.

Cytochrome Oxidase, T.E. King et al. eds.
© *1979 Elsevier/North-Holland Biomedical Press*

QUANTUM-MECHANICAL TUNNELING IN BIOMOLECULES

HANS FRAUENFELDER

Department of Physics, University of Illinois at Urbana-Champaign, Urbana,
Illinois 61801

Some physicists are astonished that no vital force is necessary to explain
biological phenomena; some biologists are amazed that molecules can tunnel
through classically impenetrable barriers. These notes are not concerned with
the first problem, but are an attempt to explain quantum-mechanical tunneling
in simple terms.

It is not clear yet if molecular tunneling plays a role in biology at ordinary temperatures. However, many biological processes do not cease at the
freezing point. Some phenomena can be observed even at 4 K. These phenomena
can be important for understanding biological problems, from a possible cold
prehistory of life to enzyme activity.

The present notes are brief and incomplete; they do not contain references.
Extensive treatments and references can be found in the proceedings of a recent
tunneling conference[1].

1. CONCEPTS

1.1 Waves and Particles. In order to understand tunneling, some quantum-
mechanical background is needed. The fundamental fact is contained in the
deBroglie relation: any material particle has wavelike properties. The wavelength associated with a particle of mass m moving with velocity v is given by
the deBroglie relation,

$$\lambda = h/mv \ .\tag{1}$$

Here h is Planck's constant. Eq. (1) has been tested in countless experiments.
Every electron diffraction picture is a verification of Eq. (1). As a consequence of the wave properties, the Heisenberg uncertainty principle holds. In
particular, if the measurement of the energy of a particle takes a time Δt, the
energy is uncertain by an amount ΔE given by

$$\Delta E \ \Delta t \ \geq \ \hbar = h/2\pi \ .\tag{2}$$

1.2 Tunneling of Free Particles. Assume that a particle of mass m and energy
E moves against a classically impenetrable barrier, as in Fig. 1. Classically,
the particle will be reflected and not pass through the barrier. Quantum
mechanically, however, there is a probability P that the particle will <u>tunnel</u>
through the barrier. There are two ways to look at tunneling.

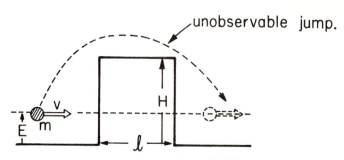

Fig. 1. Tunneling of a free particle through a classically forbidden barrier.

(i) The wave properties of the particle allow it to leak through the barrier.
Light can similarly leak through barriers.

(ii) According to Eq. (2), the particle can have, within a time Δt, an energy
$E + \Delta E$, where $\Delta E \lesssim \hbar/\Delta t$. Such energy fluctuations are well known and their
effects have been seen in many experiments. If the fluctuation ΔE is large
enough, the particle can jump over the barrier without the violation of
classical laws being noticed. The uncertainty relation protects the jumping
particle from observation!

The two ways (i) and (ii) are of course two versions of the same quantum
mechanical fact. In either case, the result is the same – the particle appears
on the right side of the barrier, violating classical laws, but fully in agreement with quantum mechanics and with all experiments.

To calculate the probability of tunneling, the Schroedinger equation is
solved for the situation of Fig. 1. Details are given in nearly all texts on
quantum theory; the result is

$$P = \exp\{-2\ell[2m(H-E)]^{1/2}/\hbar\} \ . \tag{3}$$

The tunneling probability decreases exponentially with increasing barrier width
ℓ and with $(2m(H-E))^{1/2}$. Heavy particles tunnel much less easily than light ones.

It should be stated emphatically that tunneling is nothing strange. If
$m(H-E)\ell^2 \lesssim \hbar^2$, tunneling <u>must</u> occur with a sizable probability.

1.3 Phonons. As we will see below, the interaction of the tunneling particle with its surrounding is essential for biological processes. In the language of physicists the interaction involves emission and absorption of phonons. Phonons are not real particles but quantized lattice vibrations. The concept goes back to Einstein, Debye, Born and von Karman, and was elucidated in studies of the interaction of radiation with matter and of specific heat. The basic fact is that a solid cannot emit or absorb energy in a continuous manner, but only in quanta. In the simplest model, the Einstein solid, the quanta are given by $k_B T_E$, where k_B is the Boltzmann constant and T_E the Einstein temperature. If such a hypothetical solid absorbs an energy of $3k_B T_E$ and as a result is heated slightly, we say that three phonons have been created in the solid.

1.4 Transitions. Consider the two-well system shown in Fig. 2. Assume that a molecule is initially in the higher well B. How can the molecule make the transition to the lowest (bound) state in well A? The energy levels in the system are quantized and common to the two wells. We can distinguish a low- and a high-temperature situation. At very low temperatures, the molecules in wells B will all be in the lowest state. At higher temperatures, the various levels in well B will be populated according to a Boltzmann distribution. Most will still be below the barrier between B and A but some will be above.

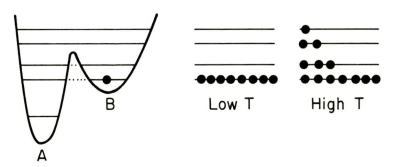

Fig. 2. Transitions between two unequal wells. At time $t = 0$ all molecules are in well B. At very low temperatures $(T \to 0)$, all molecules are in the ground state of well B. At high temperatures levels above the top of the barrier between B and A can be populated.

In Fig. 3a, we look at a molecule in a level above the barrier between B and A, and assume that the walls of the wells are rigid. No energy can be exchanged between the molecule in B and the surrounding. The molecule can then easily make the transition from B to A, but cannot get rid of its excess energy to reach the ground state in A. It therefore must return to B and it will forever oscillate between B and A.

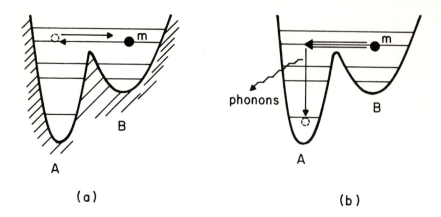

Fig. 3. (a) In a rigid system, with no interaction between the molecule m and the surrounding, m cannot reach the ground state in A, but will forever oscillate between B and A. (b) If the molecule can exchange energy with the surrounding, it can first move from B to an excited state in A. From there it can reach the ground state by emission of one or more phonons.

If, as in Fig. 3b, the molecule in transition can interact with its surrounding, it can transfer energy to the surrounding lattice and drop to the ground state in well A. The transition from B to A involves emission of one or more phonons.

Similarities and differences between Arrhenius and tunneling transitions can now be explained easily (Fig. 4). Below the top of the barrier, the process is called tunneling, above Arrhenius. The two are thus just two different cases of

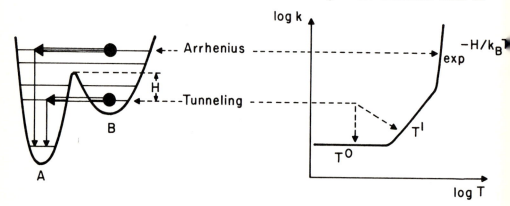

Fig. 4. If the initial state in well B is above the barrier, the transition B→A is called an Arrhenius process. If the initial state is below the barrier, transitions B→A must proceed by tunneling. The temperature dependence of the two processes is shown at right.

the same quantum mechanical transition from B to A. In both cases, the final step to the ground state in A cannot occur without emission of phonons. The differences occur in the dependence of the transition rate k on barrier width and temperature. The Arrhenius process does not depend on the barrier width, ℓ, while ℓ appears in Eq. (3) for the tunneling probability. The Arrhenius rate depends exponentially on T while the tunneling rate is temperature independent at the lowest T, becomes proportional to T at some higher T, and finally may depend on a higher power of T.

Occasionally one hears the remark "tunneling occurs at low, Arrhenius motion at high temperature". This statement is not literally correct. Tunneling takes place at all temperatures, because the levels below to barrier are always occupied. However, the Arrhenius transition increases exponentially with T and thus, once it sets it, will rapidly swamp tunneling.

2. MOLECULAR TUNNELING IN HEME PROTEINS

As an example of molecular tunneling, we discuss the binding of carbon monoxide to heme proteins.

2.1 A Model. The active center in a heme protein is sketched in Fig. 5. The heme group is fixed to the protein mainly at the periphery. The fifth position of the central iron atom is covalently linked to a residue of the backbone,

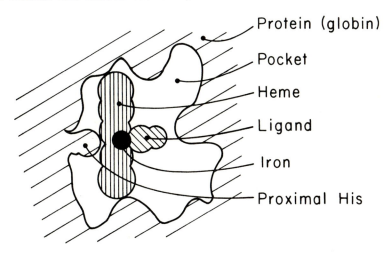

Fig. 5. Heme is embedded in a protein. The fifth position of the iron is covalently bonded to a residue of the protein backbone. The ligand, for instance dioxygen or carbon monoxide, binds at the sixth position. A prominent pocket on the distal side permits access from the outside; at low temperatures, the pocket is closed.

usually the proximal histidine. In myoglobin and hemoglobin, the ligand binds
at position 6. The protein forms a hydrophobic pocket on the distal side. At
temperatures below about 200 K, the pocket is closed off and ligands cannot
enter or exit. Once inside, they are trapped.

Two states are involved in the binding of ligands to ferroheme. Without
ligand ("deoxy state"), the iron is slightly out of the mean heme plane, has
spin $S = 2$, and the heme is probably domed. We call this state B. With bound
CO ("carbon monoxide state"), the heme is planar and the iron is in the heme
plane and has spin 0. We denote this state with A. The two states are sketched
in Fig. 6 and we identify them with states A and B in Figs. 2-4.

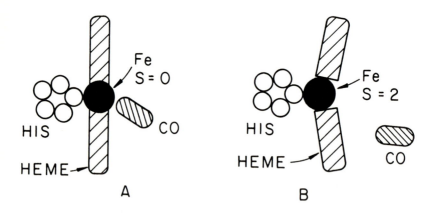

Fig. 6. In state A, CO is bound to the heme iron, the heme is planar, the iron
is in the heme plane and has spin $S = 0$. In state B, CO has moved away, the heme
may be domed, and the iron is slightly out of the heme plane and has spin $S = 2$.

2.2 Experimental Study of Tunneling. The situation shown in Fig. 2, with
all molecules at time $t = 0$ in well B, can be realized with photodissociation.
The ferrous heme protein with bound ligand, for instance MbCO, is prepared in a
proper solvent and cooled to the desired temperature. The bond between Fe and
CO is broken by a light pulse, the free CO can move away from the heme and later
rebind. Rebinding can be followed optically because Mb and MbCO have different
absorption spectra. Below about 200 K, the CO molecules remain in the pocket
(Fig. 5) and single-step rebinding is observed. Fig. 7 gives the rate of
binding of CO to the separated β chain of hemoglobin as a function of temperature between 2 and 50 K. Two regions are evident: Above about 30 K, k is
approximately exponential in 1/T, below about 10 K nearly temperature independent. To interpret this observation, we assume that the CO remains close to
the heme iron after photodissociation at very low temperatures, and the iron

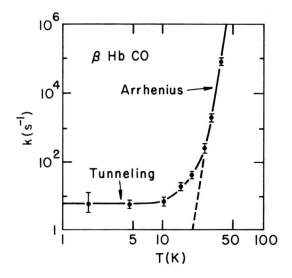

Fig. 7. The rate k of binding of CO to the β chain of hemoglobin after photo-dissociation.

moves out of the heme plane and changes spin from $S = 0$ to $S = 2$. We assume that well A in Fig. 4 represents the $S = 0$ bound state, well B the $S = 2$ state with CO close by. The barrier between B and A can be caused by the motion of the Fe into the heme plane and the concurrent change in heme geometry. Photodissociation moves the system from A to B, rebinding corresponds to the transition $B \rightarrow A$. Fig. 7 implies that the system moves over the barrier by classical Arrhenius motion above 30 K, and quantum-mechanical tunneling through the barrier below about 20 K.

3. OUTLOOK

Tunneling of CO in the separated β chain of hemoglobin, shown in Fig. 7, is not an isolated case. Tunneling has been observed in many heme proteins and also in polymer chains[1]. What can we learn from such experiments? Is tunneling of importance to biological phenomena? The first question is easier to answer. Arrhenius transitions give information about the barrier height, but not on barrier width. With the height known from the Arrhenius regime, the width can be determined and this parameter can help understanding the last binding step in a biological reaction. Probably equally important is the information that can be obtained about the phonon spectrum at the active center. Biological discussions usually do not mention the word "phonon". As we have pointed out earlier, however, the interaction of the ligand with the system is crucial for

the actual binding process. A more thorough understanding of biological reactions will require a better knowledge of the coupling between ligands and
proteins. Tunneling may be one way to gain this knowledge.

The question of the importance of tunneling for biological phenomena can
only be partially answered at the present time. Electron tunneling, not treated
here, is possibly the main mechanism by which electrons are transported in
biological systems. Molecular tunneling may contribute to some processes but is
unlikely to be dominant in normal situations. There may be exceptions, however.
Goldanskii, for instance, has pointed out that crucial steps in evolution may
have taken place at very low temperatures ("storage of order") and that
tunneling could have played a crucial role. In many reactions at biological
temperatures, proton tunneling may be present and important. The experimental
work in the field of atomic and molecular tunneling is only at a beginning and
the relevance may well become clear only in a few years.

REFERENCE

1. Tunneling in Biological Systems, B. Chance et al., eds. Academic Press,
   in press.

Cytochrome Oxidase, T.E. King et al. eds.
© 1979 Elsevier/North-Holland Biomedical Press

THE NATURE OF THE "VISIBLE" COPPER IN CYTOCHROME c OXIDASE

SUNNEY I. CHAN, DAVID F. BOCIAN, GARY W. BRUDVIG, RANDALL H.
MORSE, AND TOM H. STEVENS
Department of Chemistry, California Institute of Technology, Pasadena,
California 91125 USA

ABSTRACT

There are two non-equivalent coppers in cytochrome c oxidase, only one of
which is epr visible. The epr spectrum of the "visible" copper center is unusual for copper(II) in that no copper hyperfine interaction is observed and one
g-value is below that of the free electron. It has been suggested that the epr
spectrum is more characteristic of a sulfur radical than of a copper(II). We
present new epr data which are consistent with this hypothesis and which
support our recently proposed model for the "visible" copper center in which a
sulfur radical is adjacent to a copper(I). In particular, these epr data indicate
that the "visible" copper center is not a Type 1 copper. We also present the
results of resonance Raman studies which further demonstrate that neither
this site nor the epr "invisible" copper in cytochrome c oxidase can be characterized as a Type 1 copper.

INTRODUCTION

Mitochondrial cytochrome c oxidase catalyzes the four-electron reduction of
oxygen to water in the last step of cellular respiration[1]. The functioning enzyme contains four non-equivalent metal centers: one low-spin heme and one
copper, which are each magnetically isolated, and a strongly magnetically
interacting heme-copper pair[2,3]. In the oxidized enzyme only the magnetically non-interacting heme and copper exhibit electron paramagnetic resonance
(epr) signals[3]. However, the epr spectrum attributed to the "visible" copper
is not characteristic of normal Cu(II). No copper hyperfine splitting is re-
solved ($A_{||} \leq 0.003$ cm^{-1}) and one of the g-values ($g_x = 1.99$) is below that of
the free electron. Because of the unusual epr spectrum it has been suggested
that the observed epr signals are not due to copper, but could be better attributed to a sulfur radical[4].

We recently proposed a model (Figure 1) for the "visible" copper center in
cytochrome c oxidase in which a copper-sulfur radical complex is, indeed, the
source of the observed epr signal[5]. In this model two histidines and two cysteines are ligated to the copper, with electron density delocalized from the

cysteines onto the copper, rendering it effectively Cu(I). This structure, in which the unpaired spin density resides primarily on sulfur, provides an explanation for the unusual epr spectrum and also explains recent x-ray absorption results[6] which indicate that one of the coppers is a Cu(I) even in the fully oxidized enzyme. Note that our proposed structure for the "visible" copper center in cytochrome c oxidase differs from that of the Type 1 copper in plastocyanin only in that a cysteine ligand has been substituted for the methionine ligand.

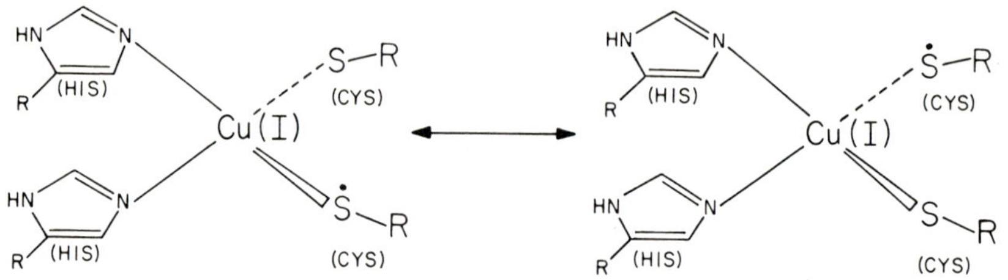

Fig. 1. Proposed structure for the "visible" copper center in cytochrome c oxidase.

In order to better characterize the nature of the "visible" copper center we have investigated its epr saturation behavior. We expect the delocalization of unpaired electron spin density into ligand orbitals to be directly reflected in the saturation behavior of the epr signals. Since we proposed that a sulfhydryl group is an integral part of the "visible" copper center, we also examined the effect of the sulfhydryl binding reagents Ag^+ and p-mercuribenzoate (p-HMB) on cytochrome c oxidase. In addition, we have obtained resonance Raman spectra with excitation in the 600 nm region. It has been suggested[7] that the "visible" copper center in cytochrome c oxidase contains a strong absorption in this region which is buried under the intense α-bands of cytochromes a and a_3. The above studies will be shown to provide further support for our model of the "visible" copper center.

MATERIALS AND METHODS

Beef heart cytochrome c oxidase was isolated by the procedures of Yu et al[8]. and Hartzell and Beinert[9]. The purified enzyme was stored at -80°C until use. The preparations contained 9-11 nmole heme A/mg protein as measured by the pyridine hemochromagen assay[10]. Protein concentrations were measured by the method of Lowry et al.[11]. Reduced cytochrome c oxidase was prepared by the addition of excess sodium dithionite to the protein under an atmosphere of nitrogen. Cytochrome c oxidase activities were measured[12] with a YSI oxygen electrode at 30°C. All work with the enzyme was carried out at 0-4°C unless otherwise specified. The Type 1 copper proteins used in the epr saturation studies were the generous gift of Professor H. B. Gray.

For the Ag^+ binding studies cytochrome c oxidase was dissolved in 50 mM Tris-HNO_3 and 0.5% Tween 20 at pH 7.4 and dialyzed against a solution of $AgNO_3$ dissolved in the same buffer as the protein. Ag^+ was added to cytochrome c oxidase by dialysis, since direct addition of Ag^+ was found to cause significant precipitation of the enzyme. Control samples were prepared as above except KNO_3 was substituted for $AgNO_3$. Individual samples of cytochrome c oxidase were removed at various times during the dialysis, and their activity was measured immediately. An aliquot of the sample was then removed for a protein assay and the remainder of the sample was placed in an epr tube and frozen at -80°C. For the mercurial binding studies, a solution of p-HMB was added directly to cytochrome c oxidase. No precipitation of the enzyme was evident, even when a large excess of p-HMB was added to the protein solution.

EPR spectra were recorded on a Varian E-line Century Series X-band spectrometer equipped with an Air-Products Heli-Trans low temperature system. Resonance Raman spectra were recorded in the laboratory of Professor A. Lewis at Cornell University with the aid of Drs. A. Lemley and N. Petersen. The experimental conditions for the epr and resonance Raman experiments are described in the legends of the appropriate figures.

RESULTS

EPR Saturation Measurements

Type 1 copper proteins exhibit epr spectra with small copper hyperfine splittings. In this regard, the "visible" copper center in cytochrome c oxidase more closely resembles the Type 1 copper center than the copper site in any other known copper protein. We have investigated the epr saturation behavior of the "visible" copper center in cytochrome c oxidase as well as a number of Type 1 copper proteins, inorganic copper complexes, and a cysteine sulfur radical, and these results[13] are shown in Figure 2. It is evident that the saturation behavior of the "visible" copper center in cytochrome c oxidase is different from that of all the Type 1 copper proteins examined. On the other hand, it is similar to that of a sulfur radical, a result which is expected on the basis of our model for the "visible" copper center.

Resonance Raman Studies

The resonance Raman spectra of oxidized and reduced cytochrome c oxidase obtained with excitation at 600 nm are shown[13] in Figure 3. The spectrum of the oxidized enzyme contains a number of intense low frequency bands (< 500 cm^{-1}). Type 1 copper proteins exhibit similar low frequency resonance Raman bands when the incident excitation wavelength falls within the intense 600 nm absorption band which is characteristic of these proteins[14, 15]. It is known that the reduction of Type 1 copper results in the disappearance of the 600 nm absorption as well as the low frequency bands in the resonance Raman spectrum. However, the low frequency bands observed in the resonance Raman spectrum of oxidized cytochrome c oxidase do not disappear upon reduction of the protein. Inasmuch as all copper associated vibrations are not resonance enhanced upon reduction of the copper, we conclude that our resonance Raman spectra of cytochrome c oxidase obtained with excitation at 600 nm are

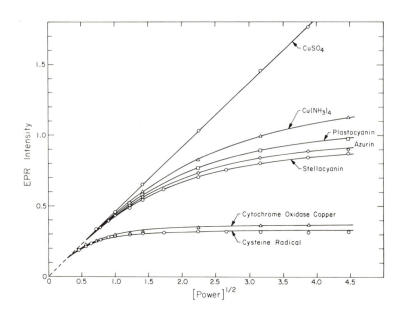

Fig. 2. The saturation behavior of the epr signals of various copper proteins, inorganic copper complexes, and a cysteine sulfur radical. The sulfur radical was generated by UV irradiation of an aqueous solution of cysteine frozen at 77°K. The epr intensities were normalized to give an initial slope of 1/g. The epr spectra were recorded at 10°K in all cases.

entirely due to the hemes.

Sulfhydryl Binding Studies

In our model for the "visible" copper center, two cysteine sulfurs are ligated to the copper. Sulfhydryl binding reagents, such as Ag^+ and p-HMB, would be expected to disrupt the integrity of the copper site if the "visible" copper center is accessible to these reagents.

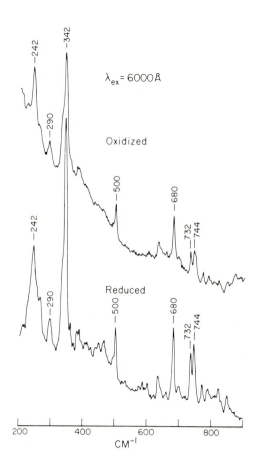

Fig. 3. The resonance Raman spectra of oxidized and reduced cytochrome c oxidase in the low frequency region obtained with excitation at 600 nm. The spectra were obtained using photon counting at 2 cm^{-1} intervals with a count time of 30 seconds per channel. The actual spectral resolution was 2.3 cm^{-1}. The laser power at the sample was 100 mW for the oxidized sample and 150 mW for the reduced sample. Both spectra are unpolarized and were recorded at 120°K.

The addition of p-HMB to cytochrome $\underline{c}$ oxidase produces no change in the epr spectrum of the "visible" copper center or in the enzymatic activity, even after long exposure. However, dialysis of the enzyme against Ag^+, a much smaller sulfhydryl binding reagent, results in the appearance of a Type 2-like copper epr signal (Figure 4) and a concomitant decrease in the activity of the enzyme (Figure 5). This suggests that Ag^+ binds a sulfhydryl group which is inaccessible to p-HMB, and is essential to the integrity of the "visible" copper center.

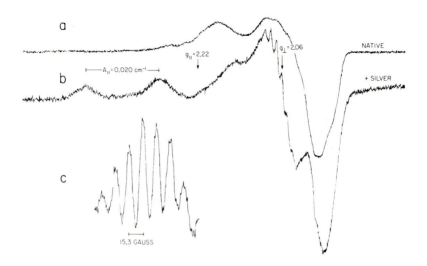

Fig. 4. EPR spectra of a) native cytochrome $\underline{c}$ oxidase, b) cytochrome $\underline{c}$ oxidase plus 16 equivalents of Ag^+, and c) the hyperfine structure in the $g_\perp$ region of b. The temperature was $10°K$, microwave power was 0.2 mW, and microwave frequency was 9.16 GHz for all spectra.

The dialysis of cytochrome $\underline{c}$ oxidase against Ag^+ results in a biphasic decrease in the enzyme's activity which can be fitted to the sum of two exponentials. A scheme such as that shown below will result in the observed activity decrease.

$$A \xrightarrow{k_1} B \xrightarrow{k_2} C$$

Here A represents native cytochrome $\underline{c}$ oxidase; B, a less active state of the enzyme which is rapidly formed upon the addition of Ag^+; and C, an inactive state which is formed very slowly.

The initial rapid loss of activity is dependent on the Ag^+ concentration and probably is due to a Ag^+ induced aggregation of the protein. Indeed, it has been observed that the activity of cytochrome c oxidase is dependent on the state of aggregation of the enzyme[16]. However, the rate of protein aggregation in our experiment depends on a number of factors including the rate at which Ag^+ diffuses across the dialysis membrane as well as the number of Ag^+ binding sites available on the protein. Thus the kinetics of the initial rapid loss of activity are complex.

Cytochrome c oxidase has a pI of 4.0-5.0 which indicates that an excess of negatively charged residues is present under the conditions of our experiments[17] (pH 7.4). The binding of Ag^+ to these negatively charged residues (carboxyl, histidine, and phosphate) would reduce the electrostatic repulsion between protein molecules and lead to protein aggregation. Over the range of

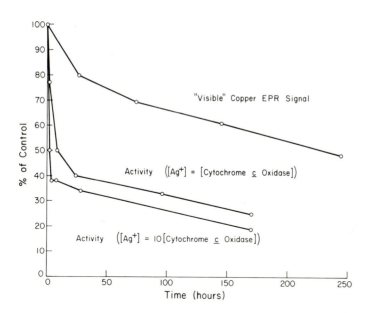

Fig. 5. The activity of cytochrome c oxidase and the epr intensity of the native "visible" copper center during dialysis against Ag^+. The concentration of cytochrome c oxidase was 7.2×10^{-5} M (assuming two heme A/enzyme). The activity points are the average of three measurements and the error is estimated to be ±20%. The epr intensity points are the average of two samples and the error is estimated to be ±10%. The method of Aasa and Vänngård[18] was used to integrate the overlapping epr signals. The temperature was 10°K, microwave power was 0.2 mW, and microwave frequency was 9.16 GHz for all epr spectra.

Ag^+ concentrations examined (7×10^{-4} - 7×10^{-5} $\underline{M}$), the binding of Ag^+ to carboxyl and phosphate groups would be concentration dependent. At the higher Ag^+ concentrations more Ag^+ ions would be bound per protein molecule and consequently larger aggregates would be formed. Thus we expect the activity of state B to depend on the Ag^+ concentration, and this is observed in Figure 5.

The slow loss of activity observed at long times in the biphasic Ag^+ dialysis activity curve occurs at a rate which is equal to the rate of disappearance of the epr signal of the "visible" copper center. This step occurs at a rate which is independent of the Ag^+ concentration over the range from 7×10^{-4} - 7×10^{-5} $\underline{M}$, with a first-order rate constant $k_2 = 0.003$ hr^{-1}. Since the epr signal of the "visible" copper center disappears at the same rate as the loss of activity, it appears that this slow step is a disruption of the "visible" copper center by Ag^+ which produces an inactive enzyme.

As the native "visible" copper epr signal disappears, a Type 2-like Cu(II) epr signal appears with no change in the total epr intensity of the copper signals. This Type 2-like copper epr signal exhibits superhyperfine structure in the $g_\perp$ region due to three nitrogens. This result suggests that in the native enzyme the "visible" copper is ligated to one or more nitrogens. ENDOR results[19] have also shown that one or more nitrogens are associated with the "visible" copper center. Furthermore, electron spin echo experiments suggest that any nitrogens associated with the "visible" copper center must be at least one atom removed from the atom on which the electron spin density is localized[20]. All of the above observations are consistent with our model for the "visible" copper center.

DISCUSSION

Since the epr signal from the "visible" copper center in cytochrome $\underline{c}$ oxidase contains no resolved copper hyperfine splitting, this site has often been grouped with Type 1 coppers, which also exhibit small copper hyperfine splittings. It is believed that the reduced copper hyperfine interaction in Type 1 copper proteins is the result of distorting a Cu(II), which prefers a square-planar geometry, into a nearly tetrahedral environment. In a tetrahedral geometry the anisotropic copper hyperfine interaction is reduced. However, as long as the unpaired electron resides predominantly in a copper $\underline{d}$-orbital, spin polarization of copper $\underline{s}$ electrons will always result in some isotropic copper hyperfine splitting[21]. In the "visible" copper center in cytochrome $\underline{c}$ oxidase, the lack of resolved copper hyperfine splitting places an upper limit on $|A_{\parallel}|$ of 0.003 cm^{-1}, from which the isotropic hyperfine interaction can be determined to be less than that observed for any of the Type 1 copper proteins. In addition to exhibiting no resolved copper hyperfine interaction, the epr spectrum for the "visible" copper center has g-values which are unusual for Cu(II), whereas the g-values for Type 1 coppers are typical of Cu(II) complexes. This implies that the spin-orbit coupling is different in the "visible" copper center from that found in other Cu(II) complexes. Taken together, the unusual g-values and unresolved copper hyperfine splitting as well as our epr saturation data indicate that the "visible" copper center in cytochrome $\underline{c}$ oxidase is best characterized as a Cu(I)-sulfur radical complex and not a Type 1 copper.

Further evidence that the "visible" copper center is not a Type 1 copper is provided by the resonance Raman results which we presented here. Characteristic of all Type 1 copper proteins is an intense absorption near 600 nm which has been assigned to a σ (thiolate) $\rightarrow d_x{}^2{}_{-y}{}^2$ charge transfer transition[22]. A σ (thiolate) $\rightarrow d_x{}^2{}_{-y}{}^2$ charge transfer transition would not occur in our model for the "visible" copper center. A copper$\rightarrow$sulfur radical charge transfer transition has not previously been observed and its position and intensity are unknown.

Cytochrome $\underline{c}$ oxidase has an absorption band at 830 nm which has been associated with the "visible" copper center[23,24]. An absorption near 800 nm of similar intensity is also observed in Type 1 copper proteins. This absorption has been assigned to a π (thiolate) $\rightarrow d_x{}^2{}_{-y}{}^2$ charge transfer transition[22]. In view of our conclusion that the "visible" copper center is not a Type 1 copper, it seems unlikely that the 830 nm band in cytochrome $\underline{c}$ oxidase has the same origin as the near IR band found in Type 1 copper proteins. If, indeed, the 830 nm band is due to the "visible" copper center, then a new assignment consistent with a Cu(I)-sulfur radical structure should be found. However, low-spin hemes are also known to have an absorption in the near IR[25] and it is possible that the 830 nm band in cytochrome $\underline{c}$ oxidase is due to cytochrome a.

The ligation for the "visible" copper center in cytochrome $\underline{c}$ oxidase which we have suggested is similar to that of the Type 1 copper in plastocyanin, differing only in that a cysteine has been substituted for the methionine. Binding of the second cysteine sulfur to the copper in cytochrome $\underline{c}$ oxidase would facilitate the reduction of Cu(II) to Cu(I). It is known[26] that the addition of an excess of cysteine to Cu(II) in solution causes a spontaneous reduction of the copper to Cu(I) with a concomitant oxidation of cysteine to cystine. The oxidation of cysteine to cystine is a two electron process and requires two Cu(II) ions. Presumably this process cannot go to completion in the "visible" copper center, since this site is isolated in the protein matrix, thereby stabilizing a Cu(I)-sulfur radical complex.

Our Ag⁺ dialysis results indicate that the "visible" copper center is quite inaccessible to water soluble species. This result is expected in view of the above description of the "visible" copper center. A major consideration for a copper ion buried in a medium of low dielectric constant is charge balance, and in our model the electrostatic energy is minimized by delocalization of a negative charge from two cysteines onto a Cu(II).

Reduction of the "visible" copper center results in an isolated negative charge within a region of low dielectric constant. This situation should result in a large increase in potential energy of the "visible" copper center which could be coupled to the conservation of energy in cytochrome $\underline{c}$ oxidase. One scheme for the utilization of the increase in potential energy upon reduction of the "visible" copper center involves a proton pumping mechanism which is depicted in Figure 6. In this scheme, the increased potential energy which is generated by reduction of the "visible" copper center is utilized by pulling a proton from the matrix solution to balance the charge. Upon transfer of an electron away from the "visible" copper center, this proton would leave an isolated positive charge buried in a hydrophobic environment, which could be expelled from the "visible" copper center to the opposite side of the mitochondrial

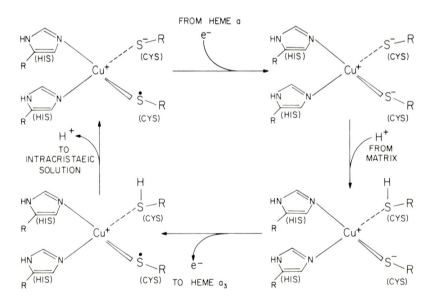

Fig. 6. A possible proton pumping mechanism in cytochrome $\underline{c}$ oxidase.

membrane. Through a mechanism such as this the "visible" copper center could be intimately involved in a proton pumping mechanism.

It is interesting that the mitochondrially synthesized subunit II of cytochrome c oxidase appears to exhibit sequence homology with members of the azurin/ plastocyanin protein family[27]. As we have previously discussed, the single substitution of a cysteine for a methionine as ligands to a copper buried in a hydrophobic environment may be all that is necessary to transform a Type 1 copper center into the "visible" copper center. It may well be that the unusual nature of the "visible" copper center is directly related to its role of coupling electron transfer to proton pumping in cytochrome c oxidase.

CONCLUSIONS

We have presented new evidence that the epr spectrum of the "visible" copper center is best interpreted in terms of a Cu(I)-sulfur radical complex and not a Type 1 copper. This result together with the model which we have recently proposed for this copper center suggests a possible role for this site in a proton pumping process in cytochrome c oxidase.

ACKNOWLEDGMENTS

This work was supported in part by grant GM 22432 from the National Institute of General Medical Sciences, U.S. Public Health Service and by BRSG grant RR07003 awarded by the Biomedical Research Support Grant Program, Division of Research Resources, National Institutes of Health. We thank Drs. A. Lemley and N. Petersen for aid in obtaining the resonance Raman spectra and Professor A. Lewis for the use of the Raman equipment in his laboratory. The copper proteins were the generous gift of Professor H. B. Gray. GWB and THS are recipients of National Institutes of Health predoctoral traineeships. This is contribution No. 5856 from the Division of Chemistry and Chemical Engineering.

REFERENCES

1. Lemberg, M. R. (1969) Physiol. Rev., 49, 48-121.
2. Falk, K., Vänngård, T. and Ångström, J. (1977) FEBS Lett., 75, 23-27.
3. Aasa, R., Albracht, S. P. J., Falk, K. E., Lanne, B. and Vänngård, T. (1976) Biochim. Biophys. Acta, 422, 260-272.
4. Peisach, J. and Blumberg, W. E. (1974) Arch. Biochem. Biophys., 165, 691-708.
5. Chan, S. I., Bocian, D. F., Brudvig, G. W., Morse, R. H. and Stevens, T. H. (1978) in Frontiers of Biological Energetics: From Electrons to Tissues, Dutton, P. L., Leigh, J. S. and Scarpa, A. eds., Academic

Press, New York, N. Y., in press.

6. Hu, V. W., Chan, S. I. and Brown, G. S. (1977) Proc. Nat. Acad. Sci. USA, 74, 3821-3825.

7. Gibson, Q.H. and Greenwood, C. (1965) J. Biol. Chem., 240, 2694-2698.

8. Yu, C., Yu, L. and King, T. E. (1975) J. Biol. Chem., 250, 1383-1392.

9. Hartzell, C. R. and Beinert, H. (1974) Biochim. Biophys. Acta, 368, 318-338.

10. Takemori, S. and King, T. E. (1965) J. Biol. Chem., 240, 504-513.

11. Lowry, O. H., Rosebrough, N. J., Farr, A. L. and Randall, R. J. (1951) J. Biol. Chem., 193, 265-275.

12. Hu, V. W. and Chan, S. I., Biochemistry, in press.

13. Brudvig, G.W., Bocian, D. F. and Chan, S.I., details to be published.

14. Miskowski, V., Tang, S. P. W., Spiro, T. G., Shapiro, E. and Moss, T. H. (1975) Biochemistry, 14, 1244-1250.

15. Siiman, O., Young, N. M. and Carey, P. R. (1976) J. Amer. Chem. Soc., 98, 744-748.

16. Robinson, N. C. and Capaldi, R. A. (1977) Biochemistry, 16, 375-381.

17. Capaldi, R.A. and Briggs, M. (1976) in The Enzymes of Biological Membranes, Vol. 4, Martinosi, A. ed., Plenum Press, New York, N. Y., pp. 87-102.

18. Aasa, R. and Vänngård, T. (1975) J. Mag. Res. 19, 308-315.

19. Van Camp, H. L., Wei, Y. H., King, T. E. and Scholes, C. P. (1978) Biophys. J. 21, 33a.

20. Peisach, J. (1978) in Frontiers of Biological Energetics: From Electrons to Tissues, Dutton, P. L., Leigh, J. S. and Scarpa, A. eds., Academic Press, New York, N. Y., in press.

21. Watson, R. E. and Freeman, A. J. (1961) Phys. Rev., 123, 2027-2047.

22. Solomon, E. I., Hare, J. W. and Gray, H. B. (1976) Proc. Nat. Acad. Sci. USA, 73, 1389-1393.

23. Beinert, H., Hansen, R. E. and Hartzell, C. R. (1976) Biochim. Biophys. Acta, 423, 339-355.

24. Yong, F. C. and King, T. E. (1972) J. Biol. Chem., 247, 6384-6388.

25. Smith, D. W. and Williams, R. J. P. (1969) Structure and Bonding, 7, 1-45.

26. Hemmerich, P. (1966) in The Biochemistry of Copper, Peisach, J., Aisen, P. and Blumberg, W. E.eds., Academic Press, New York, N. Y. pp. 15-34.

27. Buse, G., Steffens, G. J., Steffens, G. C. M. and Sacher, R. (1978) in Frontiers of Biological Energetics: From Electrons to Tissues, Dutton, P. L., Leigh, J. S. and Scarpa, A. eds., Academic Press, New York, N. Y., in press.

Cytochrome Oxidase, T.E. King et al. eds.
© 1979 Elsevier/North-Holland Biomedical Press

THE NATURE AND FUNCTION OF COPPER AND HEME IN CYTOCHROME OXIDASE

LINDA POWERS and W.E. BLUMBERG

Bell Laboratories, Murray Hill, NJ

BRITTON CHANCE, CLYDE H. BARLOW, J.S. LEIGH, Jr., J. SMITH and T. YONETANI

Johnson Research Foundation, University of Pennsylvania, Philadelphia, PA

STEVEN VIK

Institute of Molecular Biology, University of Oregon, Eugene, OR

JACK PEISACH

Department of Pharmacology, Albert Einstein College of Medicine, Bronx, NY

ABSTRACT

New data are presented on the nature and function of the copper atoms of cytochrome oxidase as studied by optical and X-ray edge absorption spectroscopy. A binuclear spin paired Fe-Cu complex is proposed as a possible structure for cytochrome $\underline{a}_3$ and its associated copper.

INTRODUCTION

The nature and function of copper and heme $\underline{a}$ groups in cytochrome $\underline{c}$ oxidase have been intensively investigated for many years[1]. The two heme $\underline{a}$ groups are functionally nonequivalent and associated with the two copper sites. Magnetic properties suggest one of the heme $\underline{a}$ groups (in cytochrome $\underline{a}_3$) is spin-paired to its associated copper ($Cu_{\underline{a}_3}$), and optical properties indicate absorption of the hemes and coppers overlap in the visible region[2-4].

Recently three types of copper sites in proteins have been identified by their optical and magnetic properties. Type I or "blue" copper sites have absorption bands near 450, 600 and 800 nm with a narrow nuclear hyperfine EPR structure ($A_{||} \sim 4$ to 9 mK), and, while Type II sites absorb near 600 to 700 nm, the extinction coefficient is at least an order of magnitude smaller and a broadly spaced nuclear hyperfine EPR structure ($A_{||} \sim 12$ to 22 mK) is observed. Type III sites are diamagnetic; either two Cu^{1+} or spin-paired $Cu^{2+} - Cu^{2+}$ atoms which absorb near 340 nm[5].

X-ray absorption spectroscopy can be used to investigate these sites in different oxidation states. Absorption edge measurements contain information about the charge density of the absorber, degree of covalency of bonding and

190

coordination geometry since the edge measures one electron (ls) transitions to bound states[6]. The achievement of concentrated, well characterized samples and on-line optical assay techniques have permitted monitoring of the copper sites in cytochrome $\underline{c}$ oxidase with X-rays[7].

EXPERIMENTAL METHODS

Preparative procedures that obtain a maximal concentration of cytochrome oxidase in chemically defined states are prerequisite for the edge measurements. At maximal concentration, the samples are highly viscous and are difficult materials for initiating chemical reactions. The 20 × 2 × 2 mm sample holders are closed on the bottom with Mylar tape and filled with 1 to 2 mM cytochrome oxidase. The fully oxidized state may be ensured by addition of 1 equivalent of ferricyanide. The fully reduced state is obtained by first gassing with carbon monoxide for 10 min. Reductants are added serially to the sample in a small chamber containing carbon monoxide and are further flushed with CO for at least 1 hr. This produces a reduced state with no detectable infrared absorption band. If the mixed valence state is desired, the sample is chilled to -25° and ferricyanide in molar excess of the reductants is added. All samples are freeze-trapped immediately after the preparative procedures and maintained at -73°

A second mixed valence state is obtained by binding the iron with cyanide, formate, or azide and then and then adding an excess of reductant for a short time or a stochiometrically equivalent reductant for a longer time. NADH, PMS, or DAD, plus a molar excess of NADH are effective reductants (20 min at 23°) and cytochrome a_3 and Cu_{a_3} become reduced in this time. Intermediate compounds of the reaction of cytochrome oxidase and oxygen are obtained by oxygenating the fully reduced CO compound at -28°, freeze-trapping and flash photolysing at -60 to 100°. This is a difficult task because of the limit of solubility of oxygen, and H_2O_2 is therefore used in order to saturate the cytochrome oxidase with oxygen.

Assay Method. Infrared absorption spectra are taken through the 20 2 2 mm sample holder and absorptions in the vicinity of OD = 0.5 for a 2 mm path are obtained. The wavelength scanning technique is used with RCA type 7102 photomultiplier as a detector and a scan from 1000 to 540 nm is possible. The infrared absorption has inflections at 790 and 804 nm at -100° which may ultimately be attributable to the two copper components of the oxidase.

Less than half of the absorption band disappears when Cu_{a_3} is reduced and Cu_a is oxidized. Not only is some estimate afforded of the degree of reduction of the oxidase, but also the mixed valence states can be evaluated.

Since optical assays are made before and after irradiation, controls are thereby afforded as to the possibility that the redox state is altered by the irradiation. No evidence of alterations has been obtained when the irradiation occurs in the vicinity of -100°, as is our customary experimental procedure. However, a number of tests at room temperature suggest that runs lasting between 20 and 30 min at 23° may indeed involve reduction of the oxidase presumably by hydrated electrons formed from the high energy radiation. Thus, caution must be exercised in the interpretation of any edge data of cytochrome oxidase recorded at room temperature.

Enzymatic activity determinations before and after irradiation at room temperature suggest no detectable loss of activity under conditions of assay ("reconstruction" of the activity by lipid, etc. is not employed).

Adventitious Iron and Copper. Since edge studies deal with total population of iron and copper atoms, it is essential to identify whether adventitious iron and copper are present. Atomic absorption assays indicated that the iron and copper contents agree with 10%. This encourages us to believe that the samples are pure in this respect. However, undetermined contributions of iron-sulfur proteins contained in the DPNH dehydrogenase, together with small amounts of adventitous copper may make reliance upon the ratio of the two metal atoms somewhat tenuous.

Sample Holders. In order to afford temperature regulation with minimal sample volume and optimal ease of adjustment, we have constructed special sample holders for edge and optical observations. The temperature of the sample can be regulated from the range of -104 to -30°. The solid angle which the fluorescence signals can be viewed from the sodium iodide counters is optimized.

RESULTS

Since both copper atoms are observed by absorption edge spectroscopy, the spectra for the various oxidation states of cytochrome oxidase are the sum of the respective copper edges. In order to determine the contribution of each, a wide variety of model compounds were also studied for which EPR measurements have been made[8]. The edge spectrum of one model compound, stellacyanin, was strikingly similar to the higher energy portion of oxidized cytochrome oxidase (Figure 1a). In the reduced states, stellacyanin and cytochrome oxidase are very similar to the lower energy region (Figure 1b). Studies of model compounds (Figure 2) indicate this is expected behavior if cytochrome oxidase were to contain a copper site similar to the Type I or "blue" copper site of stellacyanin.

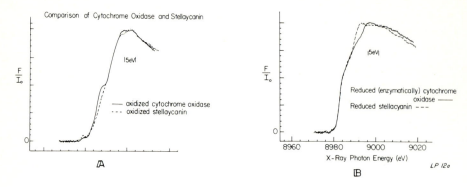

Figure 1. Comparison of X-ray absorption edges in the oxidized and reduced
states of cytochrome oxidase with those of stellacyanin. (Courtesy of Biochim.
Biophys. Acta).

COMPARISON OF COPPER (I) AND COPPER (II) MODEL COMPOUNDS

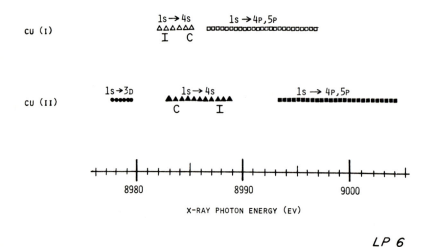

Figure 2. Comparison of absorption energy ranges observed for transitions
in Cu(I) and Cu(II) model compounds. I and C denote ionic and covalent,
respectively. (Courtesy of Biochim. Biophys. Acta).

The site is considered to be Cu_{a_3} because the observed EPR signal for cyto-chrome oxidase is different from that of steallacyanin, and Cu_{a_3} is spin-paired to the heme of cytochrome $\underline{a_3}$. The contribution of Cu_a in the oxidized and reduced states can be found by subtraction of the appropriate stellacyanin edges from that of cytochrome oxidase. This model was further tested since mixed valence states with CO, formate and cyanide were also studied. Figure 3 shows a comparison for sums of the appropriate model edges and the mixed valence states studied (formate and cyanide states gave identical spectra). The broader high energy region observed for mixed valence state + CO implies that this state may be geometrically stressed compared to the oxidized and reduced states. However, the agreement is quite good and the model edges deduced for Cu_a are similar to those observed for the covalent compounds shown in Figure 4. The smaller change in energy upon reduction also indicates that

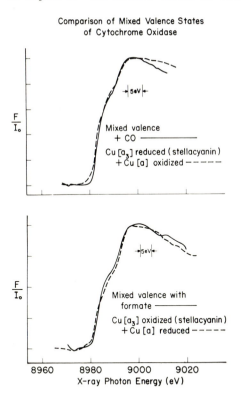

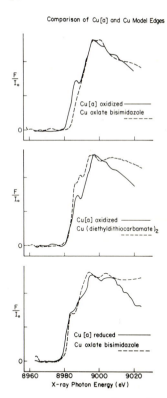

Figure 3. Comparison of the mixed valence states of cytochrome oxidase and the respective sums of stella-cyanin (Cu_{a3} model edge) and Cu_a model edge.

Figure 4. Comparison of model edges for Cu_a in cytochrome oxidase with other Cu model compounds.

Cu_a is the more covalent of the two coppers. This model suggests that Cu_{a_3} is a Type I or "blue" copper site having a local electronic environment similar to that of stellacyanin and that Cu_a is the more covalent of the two sites.

Hu, et al.[6] first reported edge spectra for cytochrome oxidase in the oxidized and reduced states and concluded one copper site was Cu^{1+} in the oxidized state. Although our data are qualitatively similar, the transition (1s → 4s) at 8983.7 eV in the oxidized state they observed is not reproducible in our three different preparations. In addition, our model studies (Figure 2) indicate this transition cannot be unambiguously identified as Cu(I) or Cu(II). Since Cu/Fe ratios were ∿1.1 for our preparation, the slight inflection observed at 8983.7 eV in our data is attributed to "adventitious copper" (not functional).

CONCLUSION

In conclusion we have demonstrated that the local environments of the Cu atoms of cytochrome-c oxidase differ from one another in both the oxidized and reduced states. This is not surprising since one heme a environments in cytochromes a + a_3 also differ from one another and the functions of the two cytochromes are different. It is suggested then that the Cu atoms play just as important a role as heme a in the reactions catalysed by the cytochrome-c oxidase.

ACKNOWLEDGMENTS

The portion of this research carried out at SSRL was approved under Project 105. The portion of this investigation carried out at the University of Pennsylvania was supported in part by grants from NIH HL-17826 and NIGMS 12202. The portion of this investigation carried out at Albert Einstein College of Medicine was supported in part by U.S. Public Health Service Research Grant HL-13399 from the Heart and Lung Institute. This is Communication No. 392 from the Joan and Lester Avnet Institute of Molecular Biology. Many thanks are due for preparations received by R. Capaldi, T.E. King and J. Harmon.

REFERENCES

1. Nichols, P. and Chance, B. (1974) in Molecular Mechanism of Oxygen Activation (O. Hayashi, ed.), Academic Press, New York.
2. Yong, E.C. and King, T.E. (1972) J. Biol. Chem. 247, 6348-6388.
3. Babcock, G.T., Vickery, L.E. and Palmer, G., (1976) J. Biol. Chem. 251, 7907-7919.

4. Palmer, G., Antolis, T., Babcock, G.T., Garcia-Iniguez, L., Tweedle, M., Wilson, L.J. and Vickery, L.E. (1978) in <u>Mechanism of Oxidizing Enzymes</u> (Singer, T.P. and Ondarza, R.N., eds.), Elsevier, New York, pp. 221-227.

5. Malkin, R. and Malmstrom, B.G. (1970 Acv. Enzymology <u>33</u>, 177-244.

6. Shulman, B., Yafet, Y., Eisenberger, P. and Blumberg, W. (1976) Proc. Nat'l. Acad. Sci. USA <u>73</u>, 1384-1388.

7. Powers, L., Blumberg, W., Chance, B., Barlow, C., Leigh, Jr., J.S., Smith, J., Yonetani, T., Vik, S. and Peisach, J., "The Nature of the Copper Atoms of Cytochrome Oxidase as Studied by Optical and X-ray Absorption Edge Spectroscopy", Biochim. Biophys. Acta (in press).

8. Peisach, J. and Blumberg, W., (1974) Arch Biochem. Biophys. <u>165</u>, 691-708.

9. Hu, V., Chan, S. and Brown, G. (1977) Proc. Nat'l. Acad. Sci. USA <u>74</u>: 3821-3825.

EFFECTS OF LIPIDS

Cytochrome Oxidase, T.E. King et al. eds.
© 1979 Elsevier/North-Holland Biomedical Press

CYTOCHROME c OXIDASE IN DETERGENT AND PHOSPHOLIPID VESICLES.
AN EPR AND NMR STUDY

TORE VÄNNGÅRD and KARL-ERIK FALK
Department of Biochemistry and Biophysics, University of Göteborg
and Chalmers Institute of Technology. Fack, S-402 20   Göteborg
(Sweden)

ABSTRACT

EPR of cytochrome c oxidase has been studied in reductive titrations with ferrocytochrome c in the pH range 6.4 - 8.4. It is
suggested that at least part of the g 6 and 2.6 signals appearing
on partial reduction are due to cytochrome a_3 undergoing a high-
to low-spin transition at higher pH.

For oxidase incorporated in phosphatidylcholine vesicles the
pH-dependence of the EPR signals was shifted, and the protein be-
haved as if it experienced a pH, one unit lower than that of the
surrounding medium.

The vesicle preparation was characterized by proton NMR of the
lipid. The protein molecules appear to penetrate the bilayer,
forming an immobilized lipid envelope about three lipid molecules
thick.

INTRODUCTION

In earlier work on cytochrome c oxidase from this laboratory
the EPR signals were characterized[1] and the spin stated of the
hemes were determined for the oxidized and reduced protein[2].
These investigations have now been extended to EPR reductive titrations with ferrocytochrome c performed at various pH, hopefully
contributing to the solution of the assignment problem in EPR.
Also, the preparation of oxidase in phosphatidylcholine vesicles
described earlier[3] has been further characterized through reductive titrations and a proton NMR study of the phospholipid. Part
of this work will appear elsewhere[4] (K.-E. Falk and B. Karlsson,
to be published).

MATERIALS AND METHODS

Cytochrome c oxidase was prepared from beef heart by the methods of van Buuren[5] or Rosén[6], slightly modified. The oxidase was studied in detergent solutions (0.5 % Tween 80) or incorporated into vesicles (L-α-phosphatidylcholine from egg yolk) as described earlier[3]. The buffer was 0.1 M phosphate unless noted otherwise. The activity at pH 7.4 was 25 - 50 s^{-1} in detergent. Vesicles were prepared for NMR in 0.05 M phosphate, 0.1 M KCl in D_2O, pH 7.4. For EPR work at pH 8.4 the phosphate was 0.08 M, and at pH 6.4 the ionic strength was adjusted with 0.05 M KCl.

NMR and EPR spectra were recorded on Bruker 270 MHz and Varian E-9 spectrometers, respectively.

The high-spin signals (g 6) were integrated according to the procedure given in reference 1. The low-spin signals of the oxidized and partially reduced protein with peaks at g 3, 2.2, 1.45 and g 2.6, 2.2, 1.86 are in this paper according to common practise named g 3 and g 2.6 although their total intensities for reasons of overlap with the cytochrome c signal were calculated[7] from the g 1.45 and 1.86 peaks, respectively.

RESULTS

^{1}H NMR of vesicles

Vesicles without protein sonicated for 15-30 min showed well resolved phosphatidylcholine NMR spectra as indicated by the splitting seen in the hydrocarbon methyl peak (Figure 1 A). The introduction of oxidase did not only cause a broadening of the spectrum (Figure 1 B) but also a decrease in the integrated intensity of all the peaks in the spectrum. This is illustrated in Figure 2, which gives the sum of the hydrocarbon methyl and methylene peaks as a function of the amount of oxidase incorporated. The line in Figure 2 extrapolates to a protein/lipid ratio of 1.4 w/w.

EPR of oxidase in detergent solutions at various pH

The signal around g 6 that appears on partial reduction arises from several species[1]. The rhombic component was virtually absent at lower pH (Figure 3 A). The g 2.6 signal, which is observed only at partial reduction and neutral or basic pH, was split into two

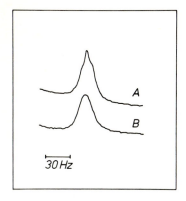

Fig. 1. The hydrocarbon methyl resonance of the 270 MHz ^{1}H NMR spectrum of phosphatidylcholine vesicles (13 g/l) at 14 C in (A) with no protein and in (B) with 5.8 µM cytochrome oxidase.

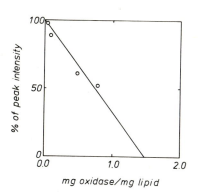

Fig. 2. Integrated NMR intensity of the hydrocarbon methyl plus methylene peaks of phosphatidylcholine vesicles.

peaks as reduction proceeded (not shown, cf reference 8). The total contributions of the species at g 6 and 2.6 in reductive titrations with cytochrome c^{2+} are given in Figure 4. The maximal amounts of the g 6 and 2.6 species at different pH are given in Table 1.

TABLE 1

MAXIMAL EPR SIGNAL INTENSITY IN TERMS OF aa_3 IN REDUCTIVE TITRATIONS WITH c^{2+}

pH	6.4		7.4		8.4	
Signal g-value	6	2.6	6	2.6	6	2.6
Preparation						
"van Buuren"	0.46	0	0.23	0.15	0.18	0.25
"Rosén"	0.58	0	0.13	0.22	0.12	0.32
"vesicle"	0.7	0	0.30	0	0.15	0.22

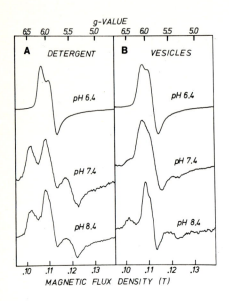

Fig. 3. EPR g 6 signals obtained at 9.12 GHz and 12 K of cytochrome oxidase reduced anaerobically with about two equivalents of ferrocytochrome c (incubation time 20 min at 22 C). The protein concentration was 100 µM in detergent and 30 µM in vesicles.

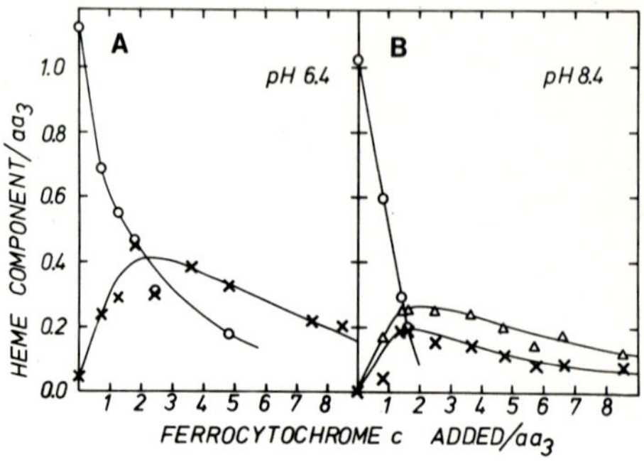

Fig. 4. Reductive titrations of 100 µM cytochrome oxidase followed by EPR. The symbols are: (o), g 3; (×), total g 6 signal; (Δ), g 2.6.

Fig. 5. Comparison of vesicle (pH 7.4) and detergent (pH 6.4) oxidase g 6 signals at varying degrees of reduction. The numbers within parantheses indicate the amount of of c^{2+} added per aa_3. Protein concentrations as in Figure 3.

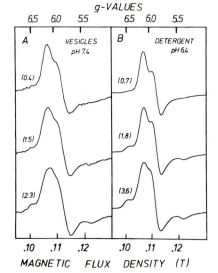

EPR of oxidase in vesicles

The shape of the g 6 signal differed from that of the oxidase in detergent at the same pH, see Figure 3 B. The changes in the shape as reduction proceeds at pH 7.4 are compared in Figure 5 to those occurring in detergent at pH 6.4. Attainment of equilibrium in titrations was slower than for detergent solutions. Table 1 gives the maximal amount of the g 6 and 2.6 signals.

DISCUSSION

Characterization of the vesicles

Vesicles exhibit a well resolved proton NMR spectrum, if they are small enough so that the lateral diffusion can contribute to the averaging motion. The vesicles used in this work fulfill this criterion, since even the triplet structure of the hydrocarbon methyl group was resolved (Figure 1). The introduction of oxidase caused a non-specific broadening of the whole phospholipid spectrum (for the methyl peak, see Figure 1 B), which, in conjunction with results from earlier electron microscopy and kinetic investigations[3], suggests that the protein molecules penetrate the lipid bilayer. Some of the molecules gave resonances that were broadened beyond detection (Figure 2). This indicates that there is an immobilized layer around the oxidase molecules of 0.7 mg lipid per mg protein, corresponding to an envelope roughly three lipid molecules thick. Spin label investigations by Jost et al.[9] indicated a layer only one molecule thick. The difference is expected, however, since the requirements on molecular motion for averaging in this case are higher for proton NMR than for spin label EPR.

Comparison of the vesicle and the detergent oxidase

Although the EPR spectra of the detergent and vesicle oxidases are not identical for the oxidized protein[4], the major differences appear on partial reduction. From a comparison between the g 6 lines in Figures 1 A and B and the fact that a g 2.6 line was not detected at pH 7.4 in the vesicles (Table 1) it appears, that the oxidase in vesicles behaves, as if the pH was about one unit lower than in the surrounding medium. Furthermore, the changes in the g 6 shape at increasing reduction in the vesicle oxidase at pH 7.4 were similar to those in the detergent oxidase at pH 6.4 (Figure 5). Recent studies by Maurel et al.[10] on the pH profile of the activity also indicate that the oxidase behaves as if being at a lower pH when anionic liposomes are added. However, their suggestion that the electrostatic field of the charged heads causes this change is not applicable here, since we observe effects with the zero-charge phosphatidylcholine at rather high ionic strength. An alternative explanation given by the same group is that the hydrocarbon chains prevent autoassociation of the protein. Although such effects cannot be excluded in our case, the simplest interpretation is that the apparent pK of a protein group has been changed through the introduction of the protein into the bilayer. It should be noticed that the preparation procedures of the lipid-oxidase complex are quite different in this work and in reference 10.

The assignment of intermediately appearing heme EPR signals

The assignment of all the EPR signals has not yet been made unambiguously. However, it is generally assumed that the g 3 signal represents cytochrome a. The work by Beinert and Shaw[11] then proves, that under certain conditions at least some of the g 6 signal arises from cytochrome a_3. The present results show that the g 6 and 2.6 signals follow each other closely in reductive titrations at various pH (Figure 4). Thus, the conversion from g 6 to 2.6 signals as the pH is increased is most easily interpreted as a high- to low-spin transition induced by an exchange of a heme ligand. In fact, the three g-values of the g 2.6 species closely resemble those of the hydroxide derivative of ferrimyoglobin[12]. In oxidase the heme available to external ligands is cytochrome a_3, which accordingly would be responsible for at least part of the g 6 and 2.6 signals. The decrease in midpoint potential expected on OH^- binding would also explain the apparent increase in the potential difference between the g 3 and g 6-2.6 species observed at increasing pH (see Figure 4). One difficulty with the present interpretation is that it also requires the midpoint potential of the undetectable copper, Cu_B, to be lower at higher pH. Otherwise, the sum of the g 6 and 2.6 signals, obtained only from molecules with Cu_B reduced, would probably be less than observed at high pH (Table 1). Evidently, the pH increase must induce several changes in the protein. This process may take some time, explaining why the g 2.6 signal does not appear immediately on a pH jump as described by Shaw et al.[13].

Babcock et al.[14] argue from quantitative work involving both MCD and EPR that a substantial fraction of the g 6 signal arises from cytochrome a undergoing a spin state change as cytochrome a_3 gets reduced. The authors do not report the intensity of the g 2.6 signal, but if it were of the same size as in our studies, it would make the argument less convincing. Furthermore, they have to postulate a strong positive redox interaction between Cu_B and cytochrome a_3 to explain the absence of signals from both of them. Simulations made as described earlier[15] indicate, however, that with such a coupling it is difficult to obtain the linear titration curves for both cytochrome a and a_3 reported from the MCD measurements[14] even of one introduces the suggested[14] negative coupling between the cytochromes.

ACKNOWLEDGEMENTS

The close co-operation with professor Bo G. Malmström and Dr. Boel Lanne on the EPR part and with Mr. Bo Karlsson on the NMR part of this work is most gratefully acknowledged. This study was supported by grants from the Swedish Natural Science Research Council, the Tercentenary Fund of the Bank of Sweden, Axel and Margaret Ax:son Johnson Foundation, Torsten and Ragnar Söderberg Foundation and Knut and Alice Wallenberg Foundation.

REFERENCES

1. Aasa, R., Albracht, S.P.J., Falk K.-E., Lanne, B. and Vänngård, T. (1976) Biochim. Biophys. Acta, 422, 260-272.

2. Falk, K.-E., Vänngård, T. and Ångström, J. (1977) FEBS Lett., 75, 23-27.

3. Karlsson, B., Lanne, B., Malmström, B.G., Berg, G. and Ekholm, R. (1977) FEBS Lett., 84, 291-295.

4. Lanne, B. Malmström, B.G. and Vänngård, T. (1978) Biochim. Biophys. Acta, in press.

5. van Buuren, K.J.H. (1972) Binding of Cyanide to Cytochrome aa_3, Ph. D. thesis, University of Amsterdam, Gerja, Waarland.

6. Rosén, S. (1978) Biochim. Biophys. Acta, 523, 314-320.

7. Aasa, R. and Vänngård, T. (1975) J. Mag. Res., 19, 308-315.

8. Hartzell, C.R. and Beinert, H. (1974) Biochim. Biophys. Acta, 368, 318-338.

9. Jost, P.C. Nadakavukaren, K.K. and Griffith, O.H. (1977) Biochemistry 16, 3110-3114.

10. Maurel, P. Douzou, P., Waldmann, J. and Yonetani, T. (1978) Biochim. Biophys. Acta, 525, 314-324.

11. Beinert, H. and Shaw, R.W. (1977) Biochim. Biophys. Acta, 462, 121-130.

12. Gurd, F.R.N., Falk, K.-E., Malmström, B.G. and Vänngård, T. (1967) J. Biol. Chem., 242, 5724-5730.

13. Shaw, R.W., Hansen, R.E. and Beinert, H. (1978) Biochim. Biophys. Acta, in press.

14. Babcock, G.T., Vickery, L.E. and Palmer, G. (1978) J. Biol. Chem., 253, 2400-2411.

15. Lanne, B. and Vänngård, T. (1978) Biochim. Biophys. Acta, 501, 449-457.

Cytochrome Oxidase, T.E. King et al. eds.
© *1979 Elsevier/North-Holland Biomedical Press*

THE LIPID-PROTEIN INTERFACE IN CYTOCHROME OXIDASE

O. HAYES GRIFFITH & PATRICIA C. JOST

Institute of Molecular Biology, University of Oregon, Eugene, Oregon 97403, USA

ABSTRACT

   Membrane proteins that penetrate through the phospholipid bilayer create a
protein-lipid interface. Electron spin resonance data on cytochrome oxidase
indicates that at this interface there is a boundary layer of lipids with
greatly reduced molecular motion. The lipid-protein association is dynamic and
exchange occurs between the protein sites and the fluid bilayer. A photo-spin
label has been prepared and preliminary data suggest that it can be covalently
linked to the protein, providing additional evidence of the lipid-protein
association. A multiple equilibria binding site model is developed which
accounts for the composite nature of the electron spin resonance spectra as a
function of the protein content of the detergent-free samples. The relative
binding constant of the immobilized lipid is low. The selection of lipid
species by the protein is, however, not completely random, and there appears
to be some preference for negatively charged lipid.

INTRODUCTION

   The multipeptide complex of cytochrome oxidase spans the inner mitochondrial
membrane and forms the terminal member of the electron transport chain. The
transmembranous[1,2] disposition of this biologically important and complex
protein necessarily involves close interaction with some of the membrane lipids.
This protein complex can be isolated from the membrane with its associated
lipids in a reasonably homogeneous preparation. This makes it possible to
examine the active complex by physico-chemical methods, and the protein retains
enzymic activity when lipid substitution is carried out. The successive steps
of removing most of the lipid, co-solubilizing with an excess of defined lipid
and a mild detergent (cholate or octyl glucoside) and centrifuging through a
linear sucrose gradient yields detergent-free membranous vesicles in which the
protein is active (Fig. 1). The lipid content is controllable, except for the
lipid held captive in or on the complex. This captive lipid is operationally
defined as not removable from the beef heart enzyme by neutral chloroform-
methanol mixtures[3] or mild detergents[4]. Using the protein associated with its
natural lipids, and more recently with defined lipids, we have been examining
the properties of the lipid-protein interactions. In addition to the captive

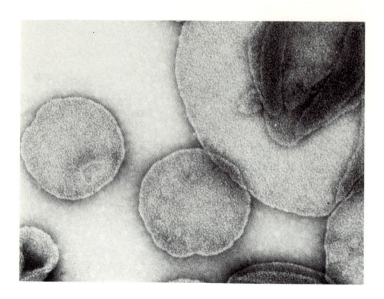

Fig. 1. Electron micrograph of negatively stained beef heart cytochrome oxidase preparation. The vesicles were formed by recombining lipids (mitochondrial) with lipid-depleted cytochrome oxidase solubilized in cholate and subsequent centrifugation into a linear (20-50%) sucrose gradient. All of the phospholipid polar head group classes (phosphatidyl-choline, phosphatidylethanolamine, phosphatidylglycerol and diphosphatidyl-glycerol) used to reconstitute the membrane environment of cytochrome oxidase yield the same closed vesicular structures. If any residual cholate is not removed, a few open holes or sheets may be seen, regardless of the type of lipid used. Sample contains 0.34 mg phospholipid/mg protein. Sodium phosphotungstate, pH 7.0; magnification 93,000X.

lipid there are two major lipid environments as detected by electron spin resonance spectroscopy: one is the ubiquitous bilayer seen in almost all membranes[5] and the second is the lipid in contact with the hydrophobic surface of the protein[6]. This protein-associated lipid is by no means fully characterized. However, sufficient data are available to provide broad general outlines. A diagrammatic sketch is shown in Fig. 2. The lipids at the protein boundary (boundary lipids) can be distinguished on the ESR time scale by their greatly reduced motion (immobilization). The number of lipids in contact with the protein (40-50) appears to correspond roughly to a layer or annulus of lipid surrounding the protein complex. A similar model has been proposed for the sarcoplasmic reticulum Ca^{2+}-ATPase to interpret activity data[7]. Boundary lipids are shown as the shaded lipid in Fig. 2, where the lipid tails are pictured as disordered in the potential wells of the protein surface, as deduced

Fig. 2. Sketch of cytochrome oxidase interacting with membrane lipids. This diagrammatically summarizes the main ideas about boundary lipids (shaded region). These lipids are in intimate contact with the hydrophobic surface but are in equilibrium (arrows) with the adjacent fluid bilayer regions.

from data on oriented samples[5]. This is a working interpretation of the data obtained so far, and not a final detailed model. The boundary lipids exchange with the lipids in the fluid phospholipid bilayer at a rate that must be slow on the ESR time scale ($\sim 10^7$ Hz)[8], but more rapid than the experimental times involved in sample preparation (hours). The major forces involved appear to be hydrophobic, but there is some evidence accumulating which indicates that electrostatic forces play some role.

In this paper we summarize some current experiments designed to examine lipid-protein associations, including the use of a photoreactive lipid label to constrain lipid at the interface, the development of a multiple equilibria approach to analyze lipid-protein binding, and preliminary data on the role of the polar head group.

THE USE OF PHOTO-SPIN-LABELING TO PROBE LIPID-PROTEIN INTERACTIONS

The concept of boundary lipid is consistent with current models of biological membranes. However, it still must be considered as a working hypothesis based largely on the equilibrium spin labeling data. Another way to test for the lipids in instantaneous direct contact with the hydrophobic surface of the protein is to introduce a functional group that can covalently link the lipid to its nearest neighbor. To form the link, a photoreactive phospholipid analog is useful. The labeled lipid can be present during reconstitution by standard methods. No reaction is allowed until after the detergent is removed and a natural configuration of the proteins and lipids is restored. Photolysis then activates the label and reaction occurs fairly indiscriminately with any nearest neighbor, which will include lipid and protein, as well as any other molecules present.

Photoreactive labels have been successfully used in a variety of enzymic and immunological studies[9]. Recently Chakrabarti and Khorana[10] have described the synthesis of a number of lipids containing photosensitive groups and demonstrated lipid-lipid crosslinking for some that contained the photosensitive group in the acyl chain. Although we are unaware of any reports of successful

protein crosslinking with lipid using this type of label, it appears to be a
promising approach to the study of lipid-protein interactions.

   In collaboration with J.F.W. Keana we have begun experiments aimed at
covalently linking the phospholipid head group to the protein and using a
spin-labeled side chain to monitor the behavior of the fatty acid tails. The
photo-spin label prepared for this purpose is the aryl azide (I) shown below.
This molecule is unreactive until it is photolyzed to generate the highly
reactive aryl nitrene (II) and N_2. The nitrene then unselectively inserts

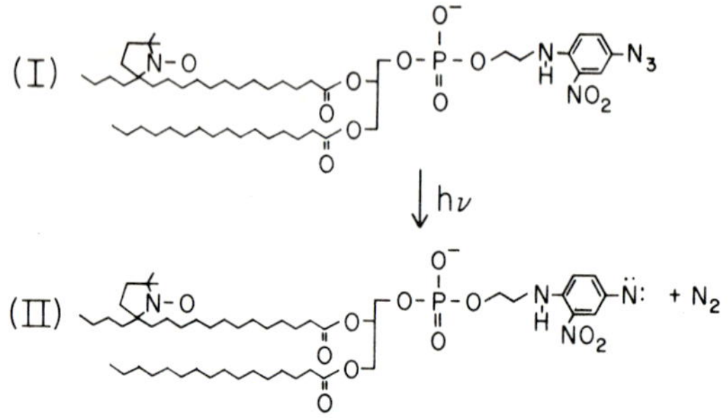

into nearest neighbor molecules. This new double label was synthesized by a
combination of the modified literature procedure[10] for the photolabile head
group and the synthesis of phosphatidylethanolamine containing 14-proxylstearic
acid at the C2 position[11]. The stable proxyl ring is not affected by the
synthetic steps leading to the aryl azide or by photolysis.

   Lipid-depleted cytochrome oxidase[4] was co-solubilized in the dark with
cholate and a 1:1 mixture of I and phosphatidylcholine (PC) containing ^{14}C-PC
as an unreactive marker. The cholate and excess lipid were removed by
dialysis, and the resulting vesicular preparation was diluted and photolyzed at
0° for 10 minutes, resolubilized in cholate and centrifuged through a linear
sucrose gradient containing cholate and a large excess of unlabeled PC. After
repeating the cholate and lipid gradient three times, the protein and its
associated lipid was recovered and the detergent removed by exhaustive
dialysis. The top spectrum of Fig. 3 shows the line shape obtained from this
final sample. Since the repeated centrifugations removed all of the ^{14}C-PC
originally present in the bilayer of the photolyzed sample, the presence of
the strong ESR signal is evidence for spin label that cannot be easily removed
from the protein. The line shape is characteristic for immobilized spin labels,
indicating interaction with the hydrophobic protein surfaces. The line shape
is complicated by the presence of exchange interactions, which are a direct

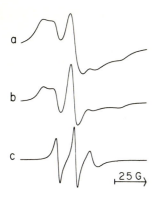

Fig. 3. ESR spectra of the photo-spin label after photolysis (all samples at 25°C). (a) Spectrum obtained from photo-spin-labeled cytochrome oxidase vesicles after reconstitution to replace all exchangeable lipid. The photoreactive lipid label represented 50 mole % of the total lipid present during photolysis. (b) Spectrum obtained after the same kind of reconstitution, photolysis, and removal of bilayer, except that the photoreactive lipid label represented 14 mole % of the total lipid during photolysis. (c) The photo-spin label in phosphatidylcholine bilayers at a labeling ratio of 1:100, after photolysis. The spectra are scaled to the same arbitrary center line height.

result of the close proximity of many of the spin labels to each other. This preparation in the presence of 2% cholate (before dialysis) exhibits the same general line shape. Under these conditions cholate solubilizes phospholipids, including the phospholipid spin label yielding a much narrower three-line spectrum (similar to the spectrum from the label in bilayers, Fig. 3c). Therefore, the observation that the line shape in Fig. 3a is the same in the presence and absence of cholate provides additional evidence that the photo-spin label is attached to the protein.

Since the labeling efficiency was higher than expected, as judged by the exchange interactions observed, the experiment was repeated, reducing the photo-spin label:PC molar ratio from 1:1 to 1:6. The corresponding line shape after repeated lipid dilutions and centrifugations to remove free label is shown in Fig. 3b. The exchange interactions are substantially reduced, but the immobilized line shape of the non-interacting lipid spin labels persists. In contrast the photolyzed spin label diluted 1:100 in unlabeled PC bilayers gives the narrow line shape shown in Fig. 3c. This line shape is characteristic of this spin label in bilayers with or without the photosensitive head group present. The samples containing protein (Fig. 3a and b) also have fairly high amounts of lipid (> 0.5 mg PC/mg protein). The absence of the characteristic bilayer line shape (Fig. 3c) in spectra a and b supports the interpretation that the lipid spin label is bound to the protein and is no longer able to exchange with the pool of unlabeled lipid in these vesicular samples. We tentatively conclude that the photo-spin label is covalently linked to the cytochrome oxidase protein complex. The nature of the bonds is unknown. They are stable to cholate solubilization, but appear to be labile to chloroform: methanol:ammonium hydroxide extraction.

These preliminary experiments show that a significant amount of lipid can be linked to the protein. Some cytochrome oxidase activity is still present in the labeled samples, but at a reduced level ($\sim$ 30 nmol O_2/sec/mg protein compared to the initial level of $\sim$ 150 nmol O_2/sec/mg protein in added PC). This reduced activity is consistent with the exposure to light and the long exposure to cholate at 4° during the successive gradients. The remaining activity is the same for both labeling levels.

With the labeled phospholipid now tethered to the protein by attachment of the head group at the lamellar interface, the behavior of the acyl chains can be examined in the presence of adjacent (unlabeled) bilayers. The large splitting and general line shape closely resemble the spectral component seen with conventional fatty acid and phospholipid spin labels that were originally used to identify and characterize boundary lipid. It is interesting to note that, although the lipid is attached to the protein only at the head group, the motion near the hydrocarbon terminus detected by the spin label is severely restricted. This is true even in the presence of extensive bilayer. The lipid side chain is evidently immobilized in potential wells in the irregular protein surface.

ESTIMATING THE MAGNITUDE OF THE RELATIVE LIPID BINDING CONSTANT

Lipid associated with cytochrome oxidase is highly immobilized in the ESR time scale[6,8]. This does not imply a high binding constant, and in fact the ESR spectral line shape tells us little about the binding constant. In order to obtain a measure of the relative binding constant, it is convenient to treat the lipid-protein association as an exchange reaction at equilibrium. Consider, for example, an exchange reaction in a membrane between a lipid spin label (L^*) and a membrane lipid (L) occupying a binding site on the hydrophobic surface of an integral protein (P). The reaction is

$$L^* + LP \; \rightleftarrows \; L + L^*P \tag{1}$$

and the relative binding constant is defined by the equation

$$K = \frac{[L][L^*P]}{[L^*][LP]} \tag{2}$$

This problem differs from the usual protein binding problem in that the lipid is the two-dimensional solvent and K is a measure of the relative affinity of one lipid compared to another. To solve for K, it is convenient to define the following three experimentally measurable quantities

$$y = \text{moles spin label in bilayer/moles bound spin label} \tag{3}$$

$$x = \text{total moles of unlabeled lipid/total moles of protein} \quad (4)$$
$$x^* = \text{total moles of labeled lipid/total moles of protein} \quad (5)$$

The quantity y is obtained from spectral analysis of the ESR data[11], whereas x and x^* are obtained from phosphorus and protein analyses and the known quantity of spin label added in the experiment. There are three mass balance equations. Two of them describe the conservation of labeled and unlabeled lipid and the third equation states in effect that all sites are occupied by either L^* or L. Any unoccupied hydrophobic sites on the protein would represent an energetically highly unfavorable state of the system.

When the protein has more than one binding site there are a set of equilibrium equations of the type

$$L^* + L_n P \rightleftarrows L + L^* L_{n-1} P \quad (6)$$

For mathematical convenience we assume a model in which there are n equivalent and independent sites. Combining the equilibrium equation and the three conservation equations yields the following general expression for lipid protein associations.

$$\frac{x}{yK + 1} + \frac{x^*}{y + 1} = n \quad (7)$$

Eq. (7) can be simplified further in the case of the spin labeling experiment. Dividing Eq. (7) by n and rearranging gives

$$\frac{x/n}{yK + 1} = 1 - \frac{x^*/n}{y + 1} \quad (8)$$

Because of the low concentration of spin label used, $x^* \ll x$. Furthermore, n is usually greater than one, further reducing the ratio x^*/n. Thus, $(x^*/n)(y + 1)^{-1} \ll 1$. Neglecting this term, Eq. (7) or (8) reduces to

$$y = \frac{x}{nK} - \frac{1}{K} \quad (9)$$

According to Eq. (9) a plot of the lipid to protein ratio (x) versus the observed ratio of bilayer to bound spin label (y) will yield a straight line with y intercept of $-1/K$, x intercept of n and slope of $1/nK$.

Eq. (9) is reminiscent of the Hughes-Klotz equation[13] or the Scatchard equation[14] describing multiple-binding site equilibria of water soluble proteins[13]. However, the mass balance conditions and experimentally measured quantities are not the same. Eq. (9) can be viewed as a membrane biochemistry version of the classical multiple binding site equilibrium problem. Experimental values of y for three different lipid-protein ratios using the spin

214

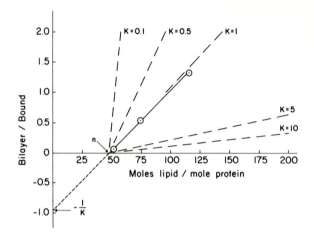

Fig. 4. Plot of Eq. (9). The three circles are experimental values of the ratio of bilayer to bound spin label (16-doxylstearic acid) in cytochrome oxidase vesicles at different lipid levels. The heavy dashed lines are derived from the multiple equilibrium site model for n = 48 independent equivalent sites with binding constants K ranging from 0.1 to 10.

label 16-doxylstearic acid are plotted in Fig. 4. From the y intercept it is clear that K is on the order of 1, and from the x intercept the number of binding sites is 45-50[*]. This type of plot is fairly sensitive to the value of the equilibrium constant, as shown by the theoretical plots for K from 0.1 to 10, calculated from Eq. (9) for n = 48.

Several interesting observations can be made from these preliminary binding data. A relative binding constant K of 1 indicates that the fatty acid spin label under the conditions used for the boundary lipid determinations (pH 7.4, mixed lipid bilayer) is mimicking the behavior of the natural unlabeled lipids, as assumed in the earlier work. The fact that the spin labeled fatty acid can displace phospholipids with a relative binding constant near unity suggests that many lipids in the boundary layer probably have relatively low specific affinity for this protein. Although the mixture of natural lipids present contains both neutral and acidic lipids and the fatty acid label is partially protonated, it is evident that any specificity of binding must be relatively low. This does not rule out some polar head group specificity. Sites with a high binding constant for a specific lipid would not be detected by this spin

---

[*]Of course, this interpretation may be model dependent, and we have not treated here models based on multiple equilibria with non-equivalent binding sites.

label. However, specificity can be detected by a modification of this procedure using spin-labeled phospholipids.

LIPID POLAR HEADGROUP SPECIFICITY

There are two solvent systems for integral membrane proteins, the aqueous phase and the lipid bilayer[15]. The bilayer solvent is a two-dimensional fluid composed of a number of lipids that differ in polar head groups and lipid side chains. In order to reduce the complexity of this solvent system, the lipid mixture can be replaced by a defined lipid. Some lipid, largely diphosphatidyl-glycerol, is not displaced from beef heart cytochrome oxidase by mild detergents or neutral chloroform-methanol extraction[3]. We have assumed that this lipid is captive[16] and does not exchange with the bilayer. The strategy we are using is to replace the remaining exchangeable lipid with a single lipid, and then introduce phospholipid spin labels with different polar head groups.

A recent experimental result using this approach is shown in Fig. 5. The two samples are parallel reconstitutions of lipid-depleted cytochrome oxidase in dioleoylphosphatidylcholine. The lipid to protein ratio is the same in the two samples. The sample of Fig. 5a contains the negatively charged spin label, phosphatidylglycerol. For Fig. 5b, the neutral spin label, phosphatidylcholine, is present.

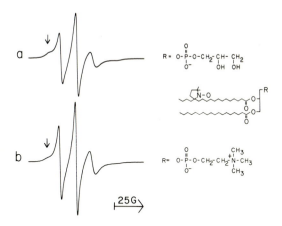

Fig. 5. ESR spectra of small quantities of spin labeled (14-proxyl) phosphatidylglycerol (a) and phosphatidylcholine (b) in cytochrome oxidase reconstituted in unlabeled dioleoylphosphatidylcholine. Both samples had the same lipid to protein ratio (0.34 mg PC/mg protein) and are suspended in 0.25 M sucrose, 1 mM EDTA, 20 mM Tris-HCL, pH 7.4 at 35°C. The spectra are scaled to reflect the same total concentrations of spin labels and the same spectrometer settings, and the arrows mark the positions of the low field immobilized spectral component.

These spectra are normalized to the integrated absorption, so that the gain is the same and the spectra can be directly compared. The line amplitudes are inversely related to the line widths. Both spectra indicate the presence of immobilized lipid and fluid bilayer components. The immobilized component is similar to that seen in Fig. 3b, but here the phospholipid spin label is free to exchange and equilibrate between the boundary and bilayer lipid regions. Although these preliminary experiments have not yet been analyzed quantitatively, the negatively charged phospholipid clearly shows a larger bound component (see arrows) and a correspondingly smaller bilayer component. In a series of three pairs of reconstitutions at different lipid to protein ratios this difference in behavior of the two spin labels was consistently observed. There are two possibilities that can account for these spectral differences. First, that the protein has some sites with a relatively higher affinity for the phosphatidylglycerol. Second, that the zwitterionic phosphatidylcholine bilayer tends to exclude the negatively charged label. A control experiment with the two phospholipid spin labels in dioleoylphosphatidylcholine bilayers in the absence of protein shows no evidence of clustering of the spin labels. We conclude that the protein has some sites that preferentially bind the negatively charged phospholipid, phosphatidylglycerol.

Using the equilibrium binding treatment described above, this approach can be extended to the determination of the relative binding constants and the number of binding sites for a variety of different phospholipids. The effects of any perturbation by the nitroxide group on the lipid tail are minimized because the final equilibrium constants are based on series of spin labels with different head groups but with the labeled hydrocarbon chain held constant[*].

The spin labeling data indicate a non-random distribution of lipids in the boundary layer, with the protein exhibiting some preference for the negatively charged lipid, as diagrammed in Fig. 6. This effect implies a non-random distribution of charged amino acid residues at the polar interface. We have recently seen a similar effect in the purple membrane of halophilic bacteria[17], the Na/K ATPase from the electric organ of electric eel[18] and the antenna protein of *Rhodopseudomonas sphaeroides* chromatophores[19]. The function of much of the lipid in contact with the protein is probably to provide a solvation layer, preventing indiscriminate aggregation in the plane of the

---

[*]The same procedure can be used to directly measure any perturbation of the lipid side chain by the label. In this case the spin label and the unlabeled lipid would be identical except for the presence of the label, and any deviation from K = 1 would measure the effect of perturbation.

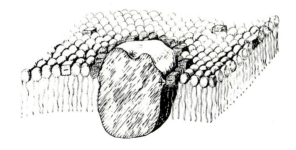

Fig. 6. Summary model of the cytochrome oxidase surrounded by boundary lipids incorporating the recent data on polar head group preference. Among the exchangeable lipids in contact with the protein there is some preference for negatively charged phospholipids, e.g., phosphatidylglycerol. This suggests a non-random distribution of polar amino acids on the protein with some positively charged residues at the lamellar interface, depicted here as a positive charge on the protein and a negative charge on a lipid.

membrane. The role of the charge preference is unknown, but indicates that the lipid matrix is not solely an inert solvent.

ACKNOWLEDGMENTS

We are greatly indebted to Tsoo E. King for many helpful discussions and for the cytochrome oxidase that made these experiments possible. Valuable contributions to the work described here were made by many of our colleagues, especially Debra A. McMillen (photo-spin labeling), Jaakko R. Brotherus (binding constant treatment), Pamela J. Bjorkman (polar head group specificity) and John F.W. Keana and Richard Roman (synthesis of new labels), and by the organizers of this symposium (for enforcing the deadline date that ensured completion of this manuscript). We acknowledge research support from Public Health Service grant GM-25698.

REFERENCES

1. Schneider, D.L., Kagawa, Y. and Racker, E. (1972) J. Biol. Chem., 247, 4074-4079.

2. Henderson, R., Leigh, J.S., and Capaldi, R.A. (1977) J. Mol. Biol., 112, 631-648.

3. Awasthi, Y.C., Chuang, T.F., Keenan, T.W. and Crane, F. (1971) Biochim. Biophys. Acta, 226, 42-52.

4. Yu, C., Yu, L. and King, T.E. (1975) J. Biol. Chem., 250, 1383-1392.

5. Jost, P.C., Griffith, O.H., Capaldi, R.A. and Vanderkooi, G. (1973) Biochim. Biophys. Acta, 311, 141-152.

6. Jost, P.C., Griffith, O.H., Capaldi, R.A. and Vanderkooi, G. (1973) Proc. Natl. Acad. Sci. USA, 70, 480-484.

7. Warren, G.B., Birdsall, N.J.M., Lee, A.G. and Metcalfe, J.C. (1974) in Membrane Proteins in Transport and Phosphorylation, Azzone, G.F. ed., North Holland Publishing Company, Amsterdam, The Netherlands, pp. 1-12.

8. Jost, P.C., Nadakavukaren, K.K. and Griffith, O.H. (1977) Biochemistry, 16, 3110-3114.

9. Bayley, H. and Knowles, J.R. (1977) in Methods of Enzymology, Vol. XLVI, Jakoby, W.B. and Wilchek, M. eds., Academic Press, New York, pp. 69-114.

10. Chakrabarti, P. and Khorana, H.G. (1975) Biochemistry, 14, 5021-5033.

11. Keana, J.F.W., Lee, T.D. and Bernard, E.M. (1976) J. Am. Chem. Soc., 98, 3052-3053.

12. Jost, P.C. and Griffith, O.H. (1978) in Methods of Enzymology, Hirs, C.H.W. and Timasheff, S.N. eds., Academic Press, New York, Vol. XLIX, pp. 369-418.

13. Klotz, I.M. (1953) in The Proteins, Neurath, H. and Bailey, K. eds., Academic Press, New York, Vol. 1B, pp. 727-806.

14. Scatchard, G. (1949) Ann. N.Y. Acad. Sci., 51, 660-672.

15. Singer, S.J. (1971) in The Structure and Function of Biological Membranes, Rothfield, L.I. ed., Academic Press, New York, pp. 145-222.

16. Griffith, O.H. and Jost, P.C. (1978) in Molecular Specialization and Symmetry in Membrane Function, Solomon, A.K. and Karnovsky, M. eds., Harvard University Press, Cambridge, Mass., pp. 31-60.

17. Jost, P.C., McMillen, D.A., Morgan, W.D. and Stoeckenius, W. (1978) in Light Transducing Membranes: Structure, Function and Evolution, Deamer, D. ed., Academic Press, New York, in press.

18. Brotherus, Jaakko, unpublished data.

19. Birrell, G.B., Sistrom, W.R. and Griffith, O.H. (1978) Biochemistry, in press.

Cytochrome Oxidase, T.E. King et al. eds.
© *1979 Elsevier/North-Holland Biomedical Press*

THE ROLE OF PHOSPHOLIPID IN CYTOCHROME OXIDASE*

CHANG-AN YU, LINDA YU, AND TSOO E. KING
Laboratory of Bioenergetics, State University of New York, Albany, N. Y. 12222,
U.S.A.

ABSTRACT

Isolated active cytochrome oxidase contains about 20% phospholipid (PL)* by
weight. The PL can be removed by several methods in the presence of detergents. The residual PL is mainly cardiolipin. The PL-depleted enzyme is inactive. The activity can be restored partially by PL or detergents, such as
Emasol-1130 or Tween-80. The efficiency of individual PL on restoring activity from depleted preparations depends greatly on what detergent the enzyme is
dispersed or dissolved in. Full activity of delipidated enzyme can be restored by azolectin in cholate. In Tween-80, all PL restore the same degree
of activity as Tween-80 alone does. The restoring efficiency of individual PL
is increased in the following order: cardiolipin > phosphatidyl choline >
phosphatidyl ethanolamine, when oxidase is in cholate, Emasol-1130 or octylglucoside. The difference in efficiency among PL decreases when the enzyme is
dissolved in low concentration of Triton X-100.

The requirement of PL on the electron transfer of oxidase is between cytochromes a and a_3. The oxidation of a_3 by oxygen or the reduction of a by
ferrocytochrome c is not greatly affected by the removal of PL. Depleted
enzyme also forms a complex with cytochrome c. The possible functional role of
PL is discussed.

INTRODUCTION

It is now generally agreed that cytochrome oxidase is a lipoprotein complex.
However, the essentiality of PL in activity has been suggested[1-6], questioned[7,8] and fully established[9-12]. The specificity of individual or mixed PL
in oxidase activity is still a matter of uncertainty. Some investigators have
claimed that cardiolipin[9,13] is indispensable and most effective in restoring
cytochrome oxidase activity from its PL depleted or insufficient preparations;
others have considered lysolecithin[3,14], phosphatidylethanolamine[1], phosphati-
dylinositol and a proper mixture of PL[10,4] essential. No requirement of PL but
just the presence of a dispersing detergent has also been propoposed[7]. In an

*Abbreviations used: PL, phospholipid(s); $AmSO_4$, ammonium sulfate.

attempt to clarify some of these conflicting points, we have re-examined the indispensibility of PL. The results indicate that azolectin can restore full activity of oxidase. Other PL tested, singly or in combination, show only partial effectiveness[10]. Recently the superiority of azolectin over other PL has been questioned[11]. No PL head groups preference in restoring oxidase activity but just the fluidity of fatty acid has been claimed[11]. In this communication we attempt to explain experimentally why the different results from different laboratories still exist and to discuss the role of PL in electron transfer within the oxidase complex.

PHOSPHOLIPID COMPOSITION OF ISOLATED ACTIVE AND DELIPIDATED CYTOCHROME OXIDASE

Phospholipid content of the isolated cytochrome oxidase varies greatly among the different preparations. It depends very much on how the preparation is made and what solubilizing agents are used. In general, the preparation made by cholate extraction suffers some degree of PL deficiency[16] unless great precaution is exercised[10]. Table I summarizes PL composition of some representative preparations.

TABLE I

PL COMPOSITION OF ISOLATED CYTOCHROME OXIDASES

PL Components[a]	Concentration of PL, µg/mg protein				
	Brierley & Merola[1]	Marinetti et al.[17]	Fleischer et al.[18]	Yu et al.[10][b]	Robinson & Capaldi[19]
Total	248	230	270	186	120
DPG	77	31	81	22	48
PC	68	77	71	91	28
PE	52	45	68	52	43
Other	51	77	48	35	–

[a] DPG, cardiolipin; PC, phosphatidylcholine; PE, phosphatidylethanolamine.
[b] The "PL-sufficient" preparation.

The maximal oxidase activity is observed when the PL content of the preparation is about 20%, *i.e.* 20 g PL per 100 g protein. Oxidase preparation with lower PL content usually shows stimulation by exogenous PL *except* if the PL-deficient enzyme is made in special detergents such as Tween-80[10,20]. As can be seen in Table I, the variation of individual PL is wide. The concentration of cardiolipin usually increases as the total PL decreases. This fact may indicate that cardiolipin is preferentially bound to cytochrome oxidase and it is important to the enzymic activity. Direct correlation between activity and cardiolipin in the mitochondria has been observed[13].

Phospholipids in isolated cytochrome oxidase can be removed by several methods including organic solvent extraction[1,19], phospholipase digestion[21,22] repeated AmSO$_4$ precipitation in the presence of cholate[10,4,23], Triton X-100[22], and gel column filtration in the presence of detergents[19]. None of the methods gives a complete removal of PL. Table II summarizes the composition of residual PL in various delipidated preparations.

TABLE II

PL COMPOSITION OF DELIPIDATED CYTOCHROME OXIDASE

					μg PL/mg protein		
						Gel filtration[19]	
				Triton			
Residual PL	2 CHCl$_3$ CH$_3$OH[19]	Acetone H$_2$O[1]	Cholate AmSO$_4$[10]	X-100 AmSO$_4$[22]	Triton X-100	DOC	Tween-80
Total	25	57	9	40	57	32	53
DPG	20	28	8	29	32	19	32
PC	1.5	8	0	0	6	3	9
PE	1.5	8	0	0	8	5	12
Others	1.5	16	1	11	8	4	-[a]

[a]Not given

Regardless of the method used, cardiolipin is the main constituent of the residual PL. The presence of other PL depends greatly on the degree of the removal of PL; the higher the residual PL the more the PC and PE. These results again suggest that cardiolipin is the most tightly bound PL in cytochrome oxidase. Among the methods reported, repeated cholate-ammonium sulfate precipitation[10] does, no doubt, remove more PL and yet yields an enzyme which can be fully re-activated by azolectin. The cholate-AmSO$_4$ delipidated oxidase has also been shown to be satisfactory in the reconstitution of oxidase vesicles used in the studies of respiratory controls and oxidative phosphorylation[24]. The other methods described give a slightly higher residual phospholipid and yield enzymes whose activity can be restored only partially. For example, only about 30% (∼6 mol O$_2$/s/mol heme a) of activity is restored by azolectin from the preparation made by Sephadex 4B gel filtration in the presence of 1 mM Triton X-100[19].

PROPERTIES OF DELIPIDATED CYTOCHROME OXIDASE

Removal of PL from oxidase, either by repeated cholate-AmSO$_4$ method or by gel filtrations, does not cause any loss of essential protein components, as judged by the protein pattern on SDS-polyacrylamide gel electrophoresis. No apparent difference in spectral properties between active and delipidated oxidase is observed.

Delipidated enzyme, although completely inactive, forms a complex with cytochrome c as the active oxidase does. A one to one stoichiometry between heme a and cytochrome c still exists. Delipidated oxidase, prepared by cholate-AmSO$_4$ method, is soluble in aqueous solution, apparently due to the residual detergent present in the preparation. When the detergent is removed by dialysis against water, the enzyme becomes insoluble and cannot be solubilized by cholate alone. The dialyzed delipidated enzyme is soluble in Triton X-100 (0.5%) in the absence of salt.

When delipidated oxidase is dispersed in 50 mM phosphate buffer, pH 7.4, containing 1% cholate, the enzymic activity is partially restored upon addition of various pure phospholipids. The maximal activity restored and the amount of PL used are summarized in Table III.

As can be seen from the table, azolectin restores delipidated oxidase activity completely, whereas other PL have only partial effectiveness, even the PL derived from mitochondria or cytochrome oxidase itself. Atempts to use a known combination PL mixture to restore the full activity has so far been unsuccessful.

The efficiency of PL in restoring oxidase activity depends greatly on the dispersing detergents. No difference in efficiency among PL can be detected if the delipidated enzyme is dissolved in dilute Tween-80 solution prior to the addition of PL.

Although data in the table clearly demonstrate the superiority of azolectin over the other individual PL, indicating that certain combination of several single PL is needed for maximal activity, the specificity of PL head groups has been, however, questioned[11]. The question is based on the fact given in the last two columns of Table III. The delipidated oxidase made by gel filtration method shows no preference on PL for activity restoration. There are several reasons to account for the observed discrepancy between the results from these two laboratories[10,11]; firstly, the presence of Tween-80 in the assay buffer may have masked the different efficiencies of PL on enzyme activity; secondly, the relatively high residual PL in delipidated preparation by the column method certainly can provide some essential yet minor PL, such as phosphatidylserine, phosphatidylinsonitol, etc. The minor, yet important, PL might be tightly

TABLE III

EFFICIENCY OF PL IN RESTORING ACTIVITY OF DELIPIDATED OXIDASE

	Specific activity[a] mol O_2/s/mol heme a	PL concentration[a] µmol Pi/mg protein	[b]Restored activity (%)[19]	
			5 mg PL/mg protein	
			by dilution	by dialysis
None	0.01	–	–	–
DPG	9.1	0.6	–	–
PC	9.0	0.1	90	100
PE	7.5	0.3	85	100
Azolectin	17.0	0.5	100	100
PL from mitochondria	11.0	0.75	90	120
"Native" oxidase	17.5	0.24	–	–

[a]Data from Ref. 10.
[b]Delipidated enzyme is made by gel filtration in the presence of 1 mM Triton X-100. The assay mixture might contain Tween-80. The activity is expressed by percentage and using azolectin as 100%. However, the 100% activity represents approximately 3 and 6 mol O_2/s/mol heme a at 25° for dialysis and dilution methods, respectively.

bound to the protein, and not easily removed, or it may become unstable after dissociating from the protein. In fact, the lack of some minor components in the PL prepared from mitochondria or cytochrome oxidase could be a reason for the lower efficiency in restoring enzyme activity. Thirdly, the low degree of activity restoration also would have decreased the difference in the relative efficiency of PL used. Finally, the ratio between the PL and dispersing detergents may also affect the efficiency in restoring the enzyme activity. Our recent unpublished results show that difference in efficiency among the different PL is enhanced as the concentration of Triton X-100 in the solution increases. At low concentration of Triton (0.5 mg/mg protein) all PL tested have similar activity. This phenomenon is not observed in other detergents such as octylglucoside, cholate, or deoxycholate. Figure 1 shows the titration of PL at different concentrations of detergents. It should be mentioned that among the common detergents used only Tween-80 and Emasol-1130 can restore a part of oxidase activity from the delipidated enzyme. The reason for their effectiveness has been claimed[11] to be the fluid environment for the enzyme complex. Whether or not this is the case and only reason remain to be experi-

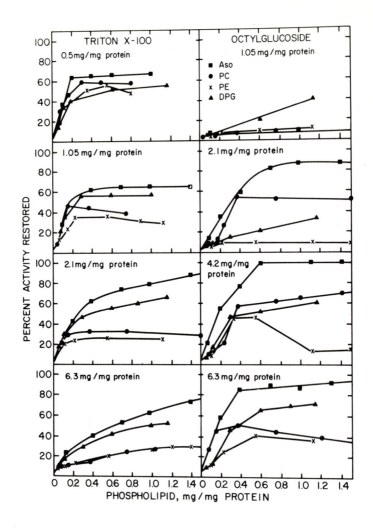

Fig. 1. Efficiency of phospholipids in restoring enzymic activity of delipidated cytochrome oxidase at various concentrations of detergents as a function of PL. Residual cholate in delipidated enzyme was removed by dialysis overnight against 50 mM phosphate buffer, pH 7.4, containing 0.25 M sucrose. The dialyzed enzyme was diluted to 5 mg/ml with 50 mM phosphate buffer before mixing with indicated amounts of Triton X-100 or octylglucoside in the presence of 3% saturation ammonium sulfate. The solubilized enzyme was then reacted with phospholipids micelle at given concentrations. 100% activity indicates 17 mol O_2/s/mg protein at 25°.

Abbreviations: Aso, PC, PE, and DPG represent azolectin, phosphatidyl choline, phosphatidyl ethanolamine and cardiolipin, respectively.

mentally substantiated. Table IV summarizes the affect of detergents on delipidated cytochrome oxidase activity.

TABLE IV

EFFICIENCY OF DETERGENTS ON RESTORING ENZYMIC ACTVITY FROM DELIPIDATED OXIDASE

Detergents	Concentration used	Specific activity[a]	
	mg/mg protein	detergents only	+ azolectin 0.5 mg/mg protein
Emasol-1130	5	6	13.5
Tween-80	5	5.8	5.9
Cholate or DOC[b]	5	0.1	17.0
Triton X-100	2	0.6	15.5
Octylglucoside	4	0.5	17.1

[a] mol O_2/sec-mol/heme a
[b] DOC, deoxycholate

THE POSSIBLE MECHANISM OF PL IN ELECTRON TRANSFER WITHIN THE OXIDASE COMPLEX

It has been shown from studies of EPR of spin labelled PL[25] that there are
two types (actually 3 types but the authors state only two*) of PL associated
with cytochrome oxidase; i.e. (i) fluid (environmental) bilayer lipid and (ii)
"boundary" lipid. The latter boundary lipid surrounds the oxidase complex, in
addition to the protein bound PL. The protein bound PL cannot be removed without denaturation of the protein[25]. Isolated, fully active oxidase possesses
only the boundary and protein-bound PL. Since the boundary PL is found to be
exchangeable with the bilayer fluid PL in oxidase[26] (but not exchangeable with
protein bound PL), it is reasonable to assume that such PL does not have strict
specificity toward the enzymic activity. Furthermore, it is known that, besides
enzymic activity there is no difference in physical, structural, and spectral
properties between active and delipidated oxidase.

The kinetic rate of the reduction of cytochrome a by ferrocytochrome c and
the oxidation of reduced cytochrome a_3 by oxygen show only slight difference
between the PL-sufficient and PL-depleted preparations[10]. The "active" site of
PL has been localized between a and a_3. Most investigators have found that the
tightly bound PL or the residual PL in the delipidated preparations are cardiolipin (see Table II). Cardiolipin is a negatively charged PL, usually with
relatively high content of unsaturated fatty acid constituents[27]. Its presence

*One of the authors (TEK) acknowledges the original authors in clarifying the
nomenclature.

may at least reinforce the binding of cytochrome c to the oxidase, although the formation of the complex is a pure protein-protein interaction (cf. Ref. 10).

The superiority of phospholipid with higher phase transition or more fluid fatty acid constituents[11] suggests that molecular movement within the cytochrome oxidase complex may be a requirement for enzymic activity.

Kinetic study of anaerobic oxidation of ferrocytochrome c by active oxidase has revealed that electron transfer between a and a_3 is too slow to account for the catalytic rate of the cytochrome oxidase reaction[28]. This indicates that the binding of oxygen to a_3 may have a profound effect on the electron transfer between a and a_3, and the binding of oxygen may be a prerequisite for enzymic activity. Spectral shift also has been observed in the dithionite reduced CO spectra of delipidated oxidase[29, 30]. All these results strongly indicate that PL modifies the environment of cytochrome a_3 to facilitate oxygen binding and then greatly accelerates electron transfer. In fact, change of secondary structure, by the addition of PL to delipidated oxidase, has been observed. Phospholipid has been found to increase the helix content of cytochrome oxidase by circular dichroic studies[21]. To correlate the specificity of individual PL on enzymic activity and the increment of helix content would provide useful information on the role of PL in cytochrome oxidase. Investigations on this aspect are currently going on in our laboratory.

ACKNOWLEDGEMENTS

This research was supported by NIH grant (U.S.) Nos. GM-16767 and HL-12576, and a grant from the American Heart Association, Northeastern New York Chapter, Inc.

REFERENCES

1. Brierley, G. P., and Merola, A. J. (1962) Biochim. Biophys. Acta, 64, 205-217.

2. Brierley, G. P., Merola, A. J., and Fleischer, S. (1962) Biochim. Biophys. Acta, 64, 218-228.

3. Igo, R. T., Mackler, B., Duncan, H., Ridyard, J. N. A., and Hanahan, J. (1960) Biochim. Biophys. Acta, 42, 55-60.

4. Tzagoloff, A., and MacLennan, D. H. (1965) Biochim. Biophys. Acta, 99, 476-485.

5. Wharton, D. C., and Griffiths, D. E. (1962) Arch. Biochem. Biophys., 96, 103-114.

6. Chuang, T. F., Sun, F. F., and Crane, F. L. (1970) Bioenergetics, 1, 227-235.

7. Morrison, M., Bright, J., and Rouser, G. (1966) Arch. Biochem. Biophys., 114, 50-55.

8.   Orii, Y., and Okunuki, K. (1965) J. Biochem. (Tokyo), 58, 561-568.

9.   Chuang, T. F., and Crane, F. L. (1973) Bioenergetics, 4, 563-378.

10. Yu, C. A., Yu, L., and King, T. E. (1975) J. Biol. Chem., 250, 1383-1392.

11. Vik, S. B., and Capaldi, R. A. (1977) Biochemistry, 16, 5755-5759.

12. Lemberg, M. R. (1969) Physiol. Rev., 49, 48-121.

13. Liskova, Z., Strunecka, A., and Drahota, Z. (1974) Physiol. bohemo., 23, 221.

14. Cohen, M., and Wainio, W. W. (1963) J. Biol. Chem., 238, 879-882.

15. Ambe, K. S., and Venkataraman (1959) Biochem. Biophys. Res. Commun., 1, 133-137.

16. Vanneste, W. H., Ysebaert-Vanneste, M., and Mason, H. S. (1974) J. Biol. Chem., 249, 7390-7401.

17. Marinetti, G. V., Erbland, J., Kochen, J., and Stotz, E. (1958) J. Biol. Chem., 233, 740-742.

18. Fleischer, S., Klouwen, H., and Brierley, G. P. (1961) J. Biol. Chem., 236, 2936-2941.

19. Robinson, N. C., and Capaldi, R. A. (1977) Biochemistry, 16, 375-381.

20. Horie, S., and Morrison, M. (1963) J. Biol. Chem., 238, 1855-1860.

21. Chuang, T. F., Awasthi, Y. C., and Crane, F. L. (1973) Bioenergetics, 5, 27-72.

22. Awasthi, Y. C., Chuang, T. F., Keenan, T. W., and Crane, F. L. (1971) Biochim. Biophys. Acta, 226, 42-52.

23. Greenlees, J., and Wainio, W. W. (1959) J. Biol. Chem., 234, 658-661.

24. Eytan, G. D., Matheson, M. J., and Racker, E. (1976) J. Biol. Chem., 251, 6831-6837.

25. Jost, P. C., Griffith, O. H., Capaldi, R. A., and Vanderkooi, G. (1973) Biochim. Biophys. Acta, 311, 141-152.

26. Jost, P., Nadakavukaren, K. K., and Griffith, O. H. (1977) Biochemistry, 16, 3110-3114.

27. Crane, F. L., and Sun, F. F. (1973) in Electron and Coupled Energy Transfer in Biological Systems, King, T. E., and Klingenberg, M. eds., Elsevier, Amsterdam, p. 477-587.

28. Andréasson, L. E., Malmström, B. G., Strömberg, C., and Vänngård, T (1972) FEBS Lett., 28, 297-301.

29. Martin, A. P., Doyle, G. E., and Stotz, E. (1965) Biochim. Biophys. Acta, 110, 290-300.

30. Orii, Y., and King, T. E. (1976) J. Biol. Chem., 251, 7487-7493.

EFFECTS OF TEMPERATURE

Cytochrome Oxidase, T.E. King et al. eds.
© *1979 Elsevier/North-Holland Biomedical Press*

EFFECTS OF TEMPERATURE ON CYTOCHROME OXIDASE -A NEW PROPOSAL FOR THE ARRHENIUS
BREAK MECHANISM OF MITOCHONDRIAL CYTOCHROME OXIDASE ACTIVITY-

SATOSHI YOSHIDA, YUTAKA ORII, SUGURU KAWATO AND AKIRA IKEGAMI
Department of Biophysical Engineering, Faculty of Engineering Science (S.Y.)
and Department of Biology, Faculty of Science (Y.O.), Osaka University, Toyonaka
Osaka 560 and Institute of Physical and Chemical Research, Hirosawa, Wako-shi,
Saitama 351 (S.K. and A.I.)

ABSTRACT

   Isolated and solubilized mammalian cytochrome oxidase exhibited an Arrhenius
plot with a break (T_b) around 20°C when its activity was assayed over a temperature range between 10 and 35°C. Isolated cytochrome oxidase was also incorporated into pure dipalmitoyl phosphatidylcholine (DPPC, the pase transition
temperature T_t = 40°C), dimyristoyl phosphatidylcholine (DMPC, T_t = 23°C) and
dioleoyl phosphatidylcholine (DOPC, T_t =-22°C). The DPPC system showed almost
the straight Arrhenius plot between 9 and 36°C. However, after the redissolution
of cytochrome oxidase from the DPPC vesicles, it yielded the bent plot when
assayed in solution. Cytochrome oxidase-DOPC was more active than the solubilized enzyme and exhibited a bent Arrhenius plot with T_b = 23°C. The plot of the
oxidase-DMPC was also bent (T_b = 26°C).

   These reults indicate clearly that the break in the Arrenius plot occurs
primarily reflecting the structural transition in the cytochrome oxidase molecule between "hot" and "cold" conformations as proposed previously (Orii *et al.*
(1977) J.Biochem., 81, 505). This transition as well as the molecular state
of cytochrome oxidase was demonstrated to be affected by the physical state
of the membrane lipids.

INTRODUCTION
   Using deoxycholate-treated heart muscle preparations, Smith and Newton[1]
determined the cytochrome oxidase activity over a temperature range between

---

Abbreviations used are: DPPC, DL-α-dipalmitoyl phosphatidylcholine; DMPC, L-α-
dimiristoyl phosphatidylcholine; DOPC, L-α-dioleoyl phosphatidylcholine; ECP,
50 mM Na-phosphate buffer (pH 7.4) containing 0.25 % (v/v) Emasol 1130 and
0.1 % (w/v) Na-cholate; pCMB, p-chrolo-mercuric benzoate; RCR, respiratory
control ratio; T_t, the phase transition (crystalline to liquid-crystalline)
temperature.

15 and 55°C and obtained an Arrhenius plot consisting of two straight slopes which intersected at 30°C. This result was tentatively related to the two-step electron transfer mechanism of the oxidase. On the other hand, Raison *et al.*[2] and Eresińsca and Chance[3] assayed the cytochrome oxidase activity of mitochondria from rat liver and pigeon heart, respectively, and obtained a bent Arrhenius plot with a break around 20°C. This phenomenon has been correlated with the membrane phase transition, although there are several pieces of evidence to indicate that the mitochondrial membrane is fluid at least above 0 - 10°C[4,5,6].

Through investigations of Shiff base formation upon alkalinization of solubilized cytochrome oxidase, Orii *et al.*[7] proposed that the enzyme assumed "hot" and "cold" types of conformations. This reaction proceeded in the rapid and slow steps and only the rapid phase yielded a bent Arrhenius plot. The break was at 24°C for the oxidized enzyme and between 19 and 16°C for the reduced form. Thus it is probable that the bent Arrhenius plot obtained in assays of the mitochondrial cytochrome oxidase activity does reflect the structural changes of the enzyme *per se* which occur independently of, but may be affected by, the membrane phase transition.

In the present study we demonstrate that even solubilized cytochrome oxidase exhibits in the activity assay an Arrhenius plot with a break, and investigate how the oxidase activity is affected by the physical states of artificial membranes which were prepared by incorporating the isolated enzyme into phospholipid liposomes of known chemical constitution.

MATERIALS AND METHODS

Preparation of Cytochrome Oxidase Vesicles. DPPC or DMPC in chloroform was dried *in vacuo* and the residue was suspended in 50 mM Na-phosphate buffer (pH 7.4) containing 2 mM $MgSO_4$, followed by the addition of Na-cholate (20 %, w/v) to a final concentration of 1 %. The lipid-cholate mixture was vortexed vigorously with occasional heating to 40-50°C until the solution became clear. After this solution had been cooled to 0°C cytochrome oxidase in ECP was added and the mixture in a dialysis tube was dialyzed against 1000 volumes of 50 mM Na-phosphate buffer (pH 7.4) - 2 mM $MgSO_4$ for 20 hours at 4-5°C with changes of the outer solution 2-3 times. DOPC in chloroform was dried either under a stream of dried N_2 gas or by evaporating the solvent *in vacuo* and otherwise was treated in the same way as described for DPPC or DMPC. Even after dissolution in 1 % cholate DOPC gave a slightly turbid solution.

Assay of Cytochrome Oxidase Activity. Cytochrome oxidase activity was determined by measuring the rate of oxygen uptake in the presence of ascobate and cytochrome c. A Beckman 39065 Polaragraphic Oxygen Sensor was used to monitor the oxygen concentration in the reaction system, the temperature of which was kept constant within the accuracy of $\pm$ 0.05°C.

After temperature equilibration of either ECP (for purified enzyme) or 50 mM Na-phosphate buffer (pH 7.4) plus 2 mM $MgSO_4$ (for cytochrome liposome system) in a reaction vessel (3.0 ml), a cytochrome oxidase solution (200 µM) was added. After another 5-10 minuites for temperature equilibration 0.1 ml of 1 M ascorbate solution (pH 6.5-7.0) was added and the rate of oxygen uptake was recorded as a control. Then cytochrome c was added to a final concentration of 14.7 µM. The net oxidase activity was obtained by subtracting the velocity of oxygen uptake in the absence of cytochrome c from that in its presence. The experimental activation energy, E, was calculated from the slope of the line obtained by plotting ln kr against 1/T, where kr is the rate constant. This was obtained from the initial velocity (v_0) in the unit of n mol O_2/(s x n mol heme a) and the oxygen concentration.

RESULTS

Temperature Dependence of the Activity of Soluble Cytochrome Oxidase. The activity of soluble cytochrome oxidase was determined in a temperature range between 10 and 35°C. The Arrhenius plot showed a break around 21°C (Fig.1). In these measurements a stock solution of cytochrome oxidase at 500-900 µM was diluted to around 16 µM and allowed to stand at 0°C until a series of assays were completed in 7 hours. During this period no activity change of the diluted oxidase solution was noticed when assayed intermittently at 25°C. On the other hand, after standing at 25°C for 24 hours, the diluted oxidase lost its enzymic activity by 30% when assayed at 25°C, and the Arrhenius plot did not give a break in the temperature range between 11 and 28°C.

The profiles of the Arrhenius plot changed when cytochrome oxidase was treated with pCMB. The enzyme (16.8 µM) was allowed to react at 25°C for 30 min with pCMB (46.7 µM), which was sufficient enough to block three sulfhydryl groups. With the pCMB-treated oxidase an inflection in the Arrhenius plot became less marked. (Fig.1.) The activation energy in a lower temperature range was almost the same for the both treated and untreated enzymes, whereas in a higher temperature range the activation energy for the treated was higher than that for the untreated as summarized in Table 1.

Temperature Dependence of Cytochrome Oxidase Activity in Phosphatidylcholine Vesicles. By the cholate-dialysis method cytochrome oxidase was incorporated efficiently into vesicles of DPPC, DMPC and DOPC, respectively. The protein to lipid ratio was 0.8 (mg/mg). The cytochrome oxidase-DPPC vesicles

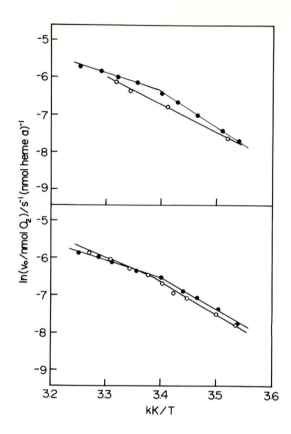

Fig.1. Arrhenius plots of the activity of cytochrome oxidase in ECP.
upper: ●— A stock solution of cytochrome oxidase was diluted to 16.7 μM
and used immediately. —O— , The diluted sample was stored at 0°C for 24
hours until used for the activity assay. lower: Effects of pCMB on the activity of cytochrome oxidase. —O— , Cytochorme oxidase was incubated with pCMB
(heme a : pCMB = 1 : 3) at 25°C for 30 minuites and then stored at 0°C until
used. —●— , Cytochrome oxidase was treated in the same way as above except
that pCMB was omitted .

were uniform in density (Fig.2.), and the formation of spherical vesicles
was also shown by electron microphotographs. Purified cytochrome oxidase in
ECP was sedimented at the bottom of a centrifugation tube (Fig.2.) and liposomes without the enzyme formed a single band at a density of 13 - 15 % (Fig.2.).
Cytochrome-DPPC vesicles were also formed even when the protein to lipid ratio
was increased to 3.2, and the lower the lipid content the denser the vesicles
(Fig.2.).

    A respiratory control ratio has been used by Racker[8] as a measure of the

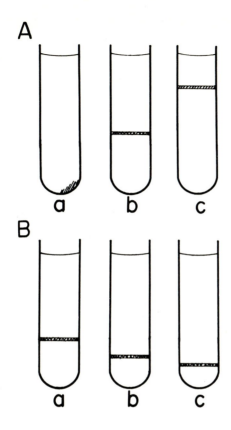

Fig.2. Sucrose density gradient ultracentrifugation patterns of cytochrome oxidase-DPPC vesicles. A: (a) Cytochrome oxidase in ECP was precipitated at the bottom of a tube. (b) Cytochrome oxidase-DPPC vesicles were prepared by the cholate dialysis method (DPPC, 2 mg/ml; cytochrome oxidase, 17 μM). (c) DPPC liposomes (2 mg/ml) without the protein were prepared by the dialysis method. B: Cytochrome oxidase-DPPC vesicles were prepared in different protein to lipid ratios. The concentration of cytochrome oxidase in each case was 17 μM and that of DPPC was 2 mg/ml (a), 1 mg/ml (b) and 0.5 mg/ml (c).

cytochrome oxidase vesicles formation. The addition of either dinitrophenol or pentachlorophenol to cytochrome oxidase-DPPC vesicles, however, did not increase the oxidase activity, thus giving RCR of 1. This result indicates that proton migrates freely through cytochrome oxidase phosphatidylcholine vesicles.

Cytochrome oxidase-DMPC vesicles gave an Arrhenius plot with a break at 26°C, which was higher than that for soluble cytochrome oxidase and than T_t of 23°C for DMPC.

Cytochrome oxidase in DOPC vesicles also showed the break at 24°C, the specific activity being higher than that of the soluble oxidase over a temperature range examined.

On the other hand, cytochrome oxidase in DPPC vesicles gave an almost straight Arrhenius plot irrespective of whether they were prepared by the dialysis method at 4°C or 25°C. The both samples were turbid and gave the almost identical slope for the plot. However, oxidase-DPPC vesicles prepared at 25°C exhibited a slight inflection between 22 and 25°C. The specific activity of these vesicles was the lowest among several types of cytochrome oxidase samples examined in the present study. In order to see if the decreased activity as well as the lack of an inflection point was due to irreversible alteration of cytochrome oxidase encountered during its incorporation into lipid vesicles, the attempt was made to redissolve the enzyme from its DPPC vesicles and to examine the activity.

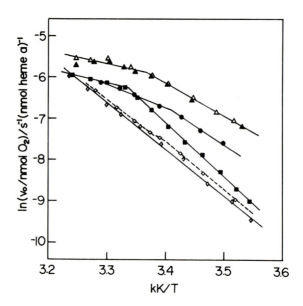

Fig.3. Arrhenius plots of the activity of cytochrome oxidase in various phosphatidylcholine vesicles. Cytochrome oxidase (23.4 μM) was incorporated into different phosphatidylcholine vesicles (3 mg/ml) as follows. ▲△, Incorporated into DOPC at 4°C. ■, Incorporated into DMPC at 4°C. Incorporated into DPPC at 4°C (◇) and 25°C (◇). ●, Cytochrome oxidase in DPPC vesicles (prepared at 4°C) was redissolved into ECP.

To 4.2 ml of a suspension of cytochrome oxidase-DPPC vesicles (5.5 mg/ml
of DPPC and 51.2 µM of cytochrome oxidase) was added 20 % sodium cholate to
a final concentration of 1 % at 4°C and the mixture was incubated at 41°C for
30 seconds to make the mixture completely clear. This solution was brought to
25 % saturation with ammonium sulfate and the precipitates were removed by
centrifugation at 24,000 x g for 30 minuites. The supernatant was passed through
a sheet of filter paper to remove floating white materials and the filtrate
was made to 35 % with ammonium sulfate. By centrifugation at 14,000 x g for
25 minuites the oxidase was sedimented as a green pellet and this was solubilized in ECP. The recovery was 66 % with respect to the protein content as
determined spectrophotometrically. When assayed in ECP the redissolved oxidase
gave the Arrhenius plot with an inflection at 21°C and the specific activity
was restored to 70 % of the original level (Fig.3.).

DISCUSSION

Solubilized cytochrome oxidase is monodisperse in solution and different
from cytochrome oxidase *in situ* in that the former is not affected by the
physical states of mitochondrial inner membrane any more. Accordingly, the
inflected Arrhenius plots observed with both solubilized and mitochondrial
enzymes indicate that this phenomenon is not necessarily related to the phase
change of the inner membrane. The phospholipid content of the purified prepa-

TABLE 1

ACTIVATION ENERGIES UNDER VARIOUS CONDITIONS

Cytochrome oxidase in		E Kcal mol^{-1}	
		H^a	L^a
ECP	fresh	9.3	18.9
	aged	14.7[b]	
ECP	+pCMB	12.4	16.8
	-pCMB	9.2	17.1
DMPC	incorporated	22.8[b]	
	solubilized	10.5	18.8
DOPC		6.6	15.8

[a]H: Higher temperature region. L: Lower temperature region.

[b]No break of the Arrhenius plot was observed in these cases.

ration was determined as 1.5 mol of phospholipid per mole of heme $\underline{a}$ based on
the value of 0.43 - 0.59 μg P/mg protein (Orii and Iba, unpublished result).
This is too low to account for the formation of membrane structure which undergoes phase transition. Also failure of the aged or pCMB-treated oxidase to
yield the bent Arrhenius plot must be ascribed to a structural alteration of
the soluble enzyme itself, since it is less likely that the physical states of
the phospholipids, Emasol or the both, which may be associated closely with
the enzyme, are altered by these treatments. Therefore, only the phase transition mechanism is difficult to explain the bent plot found with cytochrome
oxidase in mitochondrial inner membrane.

On the other hand, the shift of the rate limiting step in the cytochrome
oxidase reaction may explain a bent Arrhenius plot. The rate limiting step is
probably one of the electron transfer processes among multi-centers in this
enzyme (Orii, to be published) and heme $\underline{a}$ and copper are reported to be distributed among different constituent subunits[9,10]. Therefore, as the mutual
arrangement of the subunits is altered with temperature, as proposed by Orii
$et\ al.$[7], the shift of the rate determining step would ensue eventually. Thus
the primary cause of the bent Arrhenius plot for the oxidase reaction is ascribed to the structural transition of the enzyme even though phenomenally it
is correlated with the change in the kinetic behaviour.

Blazyk and Steim showed using a differential scanning calorimetric technique
that a reversible thermotropic phase transition of rat liver mitochondria
occurred centering at 0°C. Gulik-Krzywiki $et\ al.$ indicated from the X-ray
diffraction studies that the lipids of bovine heart mitochondria underwent a
broad transition between 10 and -10°C, and Shinitzky and Inbar demonstrated
that the temperature dependence of microviscosity of bovine heart mitochondrial
membrane showed no critical change between 0 and 40°C. All of these results
indicate that the mitochondrial membrane around 20°C is in the liquid-crystalline state which does not interfere with the intrinsic structural transition
of cytochrome oxidase $in\ situ$.

DOPC liposomes are in the liquid-crystalline state at room temperatures
and apparently an inflected Arrhenius plot reflects the structural transition
of cytochrome oxidase molecules themselves in those liposomes. As cytochrome
oxidase is transfered from ECP to DOPC liposomes, however, the activation
energies above and below the break point decreased by 3 Kcal/mol and this
point rose by 3°C. This result indicates that the molecular states of cytochrome oxidase are affected by the surrounding lipids. It is noteworthy that
in DPPC vesicles, which were crystalline over a temperature range examined,
there was no break in the plot whether cytochrome oxidase was incorporated

into the liposomes either at 4°C or 25°C. Judging from the activation energy of 22.8 Kcal/mol for the both cases it is likely that cytochrome oxidase in the "cold" conformation is trapped. These phenomena are a clear sign that the physical state of cytochrome oxidase is affected by the lipids in an artificial membrane.

The present investigations clearly indicate that the structural transition of cytochrome oxidase, which occurs unless there is no physical perturbance, is the cause for a break. Thus deviation of the plot from the linearity as observed with mitochondrial cytochrome oxidase primarily is the outcome of such mechanism, even though the structural transition of this enzyme is to be affected by the physical state of the lipids in the membrane as proposed by Raison *et al.*[11].

It is difficult, however, to understand at present what physiological significance this structural transition imply, because the body temperature of mammalians must be much higher than 20°C. It may be important to investigate the correlation between the temperature-adaptation of mammalians and the mechanism of Arrhenius break around 20°C of mitochondrial cytochrome oxidase activity.

ACKNOWLEDGMENTS

We wish to express our thanks to Dr. H.Nakamura and Dr.H.Morimoto for useful discussions, and to Mr.T.Takebayasi for technical assistance in taking the electron microphotographs.

This work was supported in part by grants from the Ministry of Education, Science and Culture of Japan, No. 248132.

REFERENCES

1. Smith, L. and Newton, N. (1968) in Structure and Function of Cytochrome Oxidase, Okunuki *et al.* eds., University Tokyo Press, Tokyo, pp.153-163

2. Raison, J.K., Lyons, J.M. and Thomson, W.W. (1971) Arch. Biochem. Biophys., 142, 83.

3. Erecińska, M. and Chance, B. (1972) Arch. Biochem. Biophys., 1515 304.

4. Blazyk, J.F. and Steim, J.M. (1972) Biochim. Biophys. Acta, 260, 737.

5. Glik-Krzywiki, T., Rivas, E. and Luzzati, V. (1967) J. Mol. Biol., 27, 303.

6. Shinitzky, M. and Inbar, M. (1976) Biochim. Biophys. Acta, 433, 133.

7. Orii, Y., Manabe, M., and Yoneda, M. (1977) J. Biochem., 81, 505.

8. Racker, E. (1972) J. Membrane Biol., 10, 221.

9. Tanaka, M., Haniu, M., Yasunobu, K.T., Yu, C.A., Yu, L., Wel, Y.H. and King, T.E. (1977) Biochem. Biophys. Res. Commun., 76, 1014.

10. Gutteridge, S., Winter, D.B., Bruynincky, W.J. and Mason, H.S. (1977)
    Biochem. Biophys. Res. Commun., 78, 945.

11. Raison, J.K., Lyons, J.M., Mehlhorn, R.J. and Keith, A.D. (1971) J. Biol.

    Chem., 246, 4036.

Cytochrome Oxidase, T.E. King et al. eds.
© *1979 Elsevier/North-Holland Biomedical Press*

FLUORESCENT PROBE STUDY OF TEMPERATURE-INDUCED CONFORMATIONAL CHANGES IN CYTOCHROME OXIDASE IN LECITHIN VESICLES AND SOLUBILIZED SYSTEMS

SUGURU KAWATO, AKIRA IKEGAMI, SATOSHI YOSHIDA AND YUTAKA ORII
The Institute of Physical and Chemical Research, Hirosawa, Wako-shi, Saitama 351 (S.K. and A.I.), the Department of Biophysical Engineering, Faculty of Engineering Science (S.Y.) and the Department of Biology, Faculty of Science (Y.O.), Osaka University, Toyonaka, Osaka 560, Japan

ABSTRACT

A protein-bound label N-(1-anilinonaphthyl-4)maleimide (ANM) was used to investigate conformational changes in cytochrome oxidase. The fluidity of cytochrome oxidase vesicles was monitored by a lipophilic probe 1,6-diphenyl-1,3,5-hexatriene (DPH). The fluorescence intensity and the emission anisotropy of these probes were examined between 4 and 60 $^{\circ}$C in the enzyme-dipalmitoyllecithin vesicles, in the enzyme-dioleoyllecithin vesilces, in the enzyme-dimyristoyllecithin vesicles and in the solubilized enzyme. The temperature dependent changes in these parameters indicated that there were two types of conformational changes in oxidized cytochrome oxidase: one was attributed to a enzyme intrinsic conformational change which occurred around 20 $^{\circ}$C, and the other was attributed to a conformational change induced by the lipid phase transition. These results were discussed in terms of the energy transfer between these fluorescent probes and heme a.

INTRODUCTION

Cytochrome c oxidase is a terminal enzyme of the respiratory chain in the mitochondrial inner membrane. Three-dimensional analyses of the structure of cytochrome oxidase in vesicle crystals have shown that the enzyme molecule sticked out on both side surfaces of the vesicle[1].

Raison et al.[2] and Erecińska and Chance[3] observed the break at around 20°C in the Arrehenius plot of cytochrome oxidase activity in rat liver and pigeon heart mitochondria, respectively. They attributed these breaks to the membrane phase transition. In solubilized and vesicle systems below 30 $^{\circ}$C, we[4] have shown a break around 20 $^{\circ}$C in the Arrehenius plot of cytochrome oxidase activity.

It is well known that the fluorescent probe technique is useful to monitor the structural changes in proteins[5,6]. By using this technique, we have investigated the temperature-induced conformational changes in cytochrome oxidase in three kinds of lecithin vesicles whose transition temperatures are 40 °C (dipalmitoyllecithin), -20 °C (dioleoyllecithin) and 23 °C (dimyristoyl-lecithin), and in phosphate buffer containing Emasol.

MATERIALS AND METHODS

Materials. Cytochrome oxidase [EC 1.9.3.1.] was prepared from beef heart muscle by the method of Okunuki et al.[7] with some modifications. The final ammonium sulfate precipitate was dissolved in 50 mM Na-phosphate buffer (pH 7.4) containing 0.25% (v/v) Emasol 1130 and 0.1% (w/v) Na-cholate (ECP) and dialyzed against a sufficient volume of the same buffer in the cold. In the present experiment, we used three lots of cytochrome oxidase preparations of different purity: A_{280nm}/A_{420nm} (oxidized) = 2.35, 2.55 and 2.57. The concentration of cytochrome oxidase was determined spectrophotometrically by using a millimolar extinction coefficient difference of 16.5 ($\Delta\varepsilon_{605-630nm}$, reduced). The dialysate was stored frozen in liquid nitrogen, and was used for experiments after thawing.

L-α-dipalmitoyllecithin (DPL), L-α-dioleoyllecithin (DOL) and L-α-dimyristoyllecithin (DML) were purchased from Sigma and used without further purification. ANM, DPH and p-chloromercuricbenzoate (pCMB) were purchased from Teika Seiyaku, Aldrich and Sigma, respectively.

Preparation of cytochrome oxidase vesicles. DPL or DML dissolved in chloroform was dried in vacuo and DOL dissolved in chloroform was dried under a stream of N_2 gas. Each of these residues was dispersed in 50 mM Na-phosphate buffer (pH 7.2) containing 2 mM $MgSO_4$ and a 20% (w/v) solution of Na-cholate was added to a final concentration of 1%. The lecithin-cholate mixture was vortexed vigorously with occasional heating to about 50 °C. After a few minutes of this treatment, DPL- and DML-cholate mixture became clear but DOL-cholate mixture was slightly turbid. The solution was cooled to 0 °C and cytochrome oxidase in ECP was added. This mixture in a dialysis tube was dialyzed against 1000 volumes of 50 mM Na-phosphate buffer (pH 7.2) containing 2 mM $MgSO_4$ for 20 h at 4 °C with 4 changes of the outer solution.

Fluorescence labeling. ANM was conjugated with SH groups of cytochrome oxidase. To 6 ml of the suspension of cytochrome oxidase vesicles (0.5 mg/ml in lecithin and 4.2 μM in the enzyme), 3 to 12.5 μl of 1 mM ANM in acetone was added and incubated at 18-19 °C for 2-3 h. To 3 ml of the solution of cytochrome oxidase in ECP containing 12 or 29 μM enzyme, 18 or 9 μl of 1 mM ANM in acetone was added respectively, and samples were incubated at about 17 °C for 2 h. The molar ratio of added ANM to the enzyme was in the range 1/10 to 1/2 in the present experiments. The free ANM is nonfluorescent and the reaction product of ANM with SH groups are strongly fluorescent[8]. The suspension of pure DPL vesicles (1 mg/ml) without the enzyme were prepared by sonication and 1 mM ANM in acetone was added to final concentration of 1 µM. After incubation at 43 $^{\circ}$C for 2 h, no fluorescence signal was observed. Therefore, ANM appears to be non-fluorescent in lipid vesicles without enzyme.

DPH was embedded in the hydrocarbon region of cytochrome oxidase vesicles. To 6 ml of the suspension of the vesicles (0.4 mg/ml in lecithin and 3.4 µM in the enzyme), 15.8 µl of 190 µM DPH in tetrahydrofuran was added, and the sample was incubated at 17-19 $^{\circ}$C for 2-3 h.

The reaction of ANM with cytochrome oxidase and the incorporation of DPH into vesicles were followed as an increase in the fluorescence intensity. After the incubation, when the fluorescence intensity reached a plateau, the fluorescence measurements were performed.

Fluorescence measurements. The steady-state fluorescence intensity and anisotropy were measured with a self-constructed instrument which was described previously[9]. The exciting light was polarized vertically and the fluorescence from the sample was automatically analyzed into vertically and horizontally polarized components, I_V and I_H, by an analyzer. These intensities were measured as numbers of photons per unit time interval. From these quantities, the total fluorescence intensity, I_T, and the emission anisotropy, r, were obtained as

$$I_T = I_V + 2I_H \qquad (1)$$
$$r = (I_V - I_H)/(I_V + 2I_H) \qquad (2)$$

The excitation wavelength was selected at 360 nm for ANM and DPH by a monochromator, all emission above 390 nm for ANM or above 420 nm for DPH was collected through cut-off filters (Toshiba UV-39 for ANM and Toshiba UV-39 and Hoya L-42 for DPH). Error in the emission anisotropy values was estimated to be within 1% in all experiments. The depolarization due to the scattering of exciting and emitted lights was negligible in the present vesicle suspensions of low lipid concentration (0.5 or 0.4 mg/ml). Temperature dependence of the fluorescence intensity and the emission anisotropy of probes was usually measured as temperature was raised.

Fluorescence spectra of samples were measured with a Hitachi MPF-3 fluorescence spectrophotometer.

Decays of the fluorescence polarization and the fluorescence lifetime were measured by a method similar to that described by Kawato et al.[10]. The excitation wavelength and the filters for emission were the same as those of the steady-state measurements.

RESULTS

   Characterization of ANM-labeled cytochrome oxidase. ANM reacts specifically
with SH groups in preference to other nucleophilic residues in proteins.
Cytochrome oxidase has seven SH groups per heme a, and three of them have been
allowed to react with pCMB in solubilized cytochrome oxidase[11]. After equimolar
ANM to the enzyme was incubated in ECP at 17 $^{\circ}$C for 2-3 h, 2.1 SH groups per
heme a were titrated by pCMB. After 2 molar equivalents of ANM for the enzyme
were incubated in ECP at 17 $^{\circ}$C for 2 h, the number of titrated SH groups by pCMB
were not less than 2.0 per heme a. Therefore, within experimental error, ANM
appears to react with the enzyme up to 1 mol/mol of heme a. Judging from the
increase in the fluorescence intensity of ANM, the reaction of ANM with
cytochrome oxidase finished for about 2 h at 18-19 $^{\circ}$C both in ECP and in
lecithin vesicles. The fluorescence intensity of ANM reached a plateau for about
2 h at 18-19 $^{\circ}$C and remained almost constant for more than 5 h.

   The excitation and the emission spectra of ANM reacted with oxidized
cytochrome oxidase in DML vesicles are shown in Figure 1.

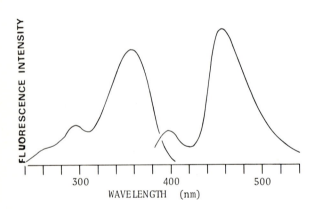

Fig. 1. Uncorrected excitation and emission spectra of ANM-cytochrome oxidase
in DML vesicles at 20 $^{\circ}$C. The concentrations of ANM, enzyme and DML are 63μM,
125 μM and 4 mg/ml respectively. Excitation and emission wavelength are 360 nm
and 436 nm respectively.

Because of the sharp absorption of heme _a_ prosthetic group, the emission spectrum had a depression around 420 nm. Similar spectra were obtained in other lecithin-enzyme vesicles and in ECP.

The activity of ANM-labeled cytochrome oxidase in ECP at 25 $^\circ$C was obtained by measuring the rate of oxygen uptake by a method similar to that described by Yoshida et al.[3]. The relative activity of ANM-labeled enzyme decreased as the molar ratio of ANM(added)/enzyme was raised up to 1, but when the molar ratio was raised further, the activity remained unchanged. The relative activity of ANM-labeled enzyme at the molar ratio of ANM(added)/enzyme above 1 was about three-fifth of the non labeled enzyme.

Cytochrome oxidase in DPL vesicles (Figure 2A). The steady-state fluorescence intensity, I_T, and the emission ansiotropy, r, of ANM and DPH in the enzyme-DPL vesicles are plotted against temperature in the figure. In I_T curve of ANM, two sharp peaks were observed in the range between 15 and 25 $^\circ$C and a great and broad peak was observed in the range between 33 and 60 $^\circ$C. A small sharp peak at 18 $^\circ$C and a depression around 38 $^\circ$C were observed in r curve of ANM. The I_T curve of DPH has two temperature inflections at about 20 and 38 $^\circ$C. The value of r of DPH decreased sharply around 40 $^\circ$C but its value decreased monotonously around 20 $^\circ$C.

Cytochrome oxidase in DOL vesicles (Figure 2B). A relatively small and a great peak were observed in I_T curve of ANM between 17 and 40 $^\circ$C. The r curves of ANM or DPH were relatively smooth.

Cytochrome oxidase in DML vesicles (Figure 2C). Four peaks at 16, 18, 22.5 and 27 $^\circ$C were observed in I_T curve of ANM. In r_1 curve of ANM, the zig-zag part was observed between 17 and 20 $^\circ$C. In r curve of ANM, the greatest difference of the value before and after the great decrease around 24 $^\circ$C was 0.7 (not shown in the figure) in six experiments. After the great decrease of the emission anisotropy finished, the value of r of ANM did not reverse and remained the similar value (r_2 curve) when temperature was decreased from 28.5 to 14.3 $^\circ$C. On the other hand, value of r of ANM reversed almost perfectly when temperature was decreased from 10-20 $^\circ$C to lower temperature (not shown in the figure). The I_T curve of DPH had a temperature inflection around 20 $^\circ$C and the value of r of DPH decreased sharply at about 24 $^\circ$C.

Cytochrome oxidase in ECP (Figure 2D). A peak was observed in I_T curve of ANM between 17 and 28 $^\circ$C. The rapid increase above 30 $^\circ$C in r curve of ANM seems to be caused by the denaturation of the enzyme.

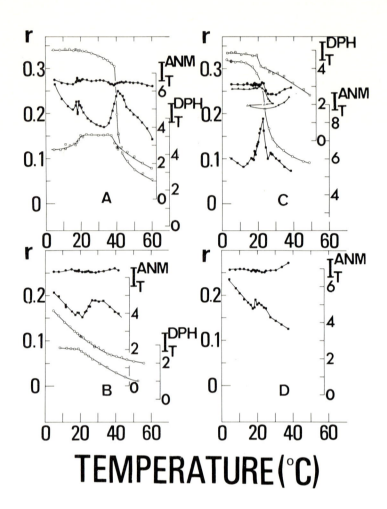

Fig. 2. Temperature dependence of the fluorescence intensity, I_T, [(■) ANM; (□) DPH] and the emission anisotropy, r, [(●) ANM; (○) DPH]. (A) DPL-enzyme; (B) DOL-enzyme; (C) DML-enzyme [(●) r_1 curve of ANM; (▲,△) r_2 curve of ANM]; (D) ECP-enzyme.

Time-dependent emission anisotropy and fluorescence lifetime of ANM. Decays of the emission anisotropy and the fluorescence intensity of ANM-labeled cytochrome oxidase were measured in ECP at several temperatures. Time courses of the emission anisotropy were biphasic, including a fast phase which corresponded to the rapid restricted rotation of ANM, followed by a slower phase which corresponded to the rotation of cytochrome oxidase molecule. The rotational correlation time of the enzyme in ECP was about 100 nsec. According to a double

exponential approximation in total fluorescence decay curves, the longer
component of the fluorescence lifetime was 6-7 nsec in all experiments.

DISCUSSION

   Rotation of cytochrome oxidase. Cytochrome oxidase does not rotate
appreciably in ECP within ANM fluorescence lifetime (see RESULTS). In lecithin
vesicles, the rotational correlation time of cytochrome oxidase would be greater
than 1 μsec considering recent results of protein rotation in membranes[12,13,14].
Since the fluorescence lifetime of ANM attached to the enzyme in lecithin
vesicles would be in the same order as that in ECP (about 10 nsec), within ANM
lifetime cytochrome oxidase will be perfectly immobilized in lecithin vesicles.
The steady-state emission anisotropy reflects the motion of the probe averaged
within its fluorescence lifetime. Therefore, the temperature dependent changes
in ANM steady-state emission anisotropy, r, will reflect only the changes in the
rapid restricted rotation of ANM molecule: a high value of r indicates a highly
immobilized state of the probe and a lower value of r a less immobilized state.
   Peaks in the temperature dependence of the fluorescence intensity of ANM.
The fluorescence intensity is very sensitive to the change in surroundings of a
fluorophore. One of most impotant factors of the quenching in the present
experiments will be the thermal quenching which will be caused by the collision
of the probe with solvent molecules or surrounding amino acid residues. Since
the collision will be more frequent at higher temperature, the fluorescence
intensity will monotonously decrease when there are no other quenching factors.
The present peaks in I_T curves of ANM indicate that there will be mechanisms of
increasing or decreasing the ANM fluorescence intensity at paticular
temperatures. One of most probable mechanisms of such changes is the change in
the realtive arrangement between ANM molecule and heme a. Since the fluorescence
spectra of ANM attached to SH groups of cytochrome oxidase partially overlaps
with the Soret band of the oxidized heme a (Figure 1), there will be the
resonance energy transfer between these two molecules[15,16,17]. Judging from the
slight decay of the emission anisotropy in the fast decreasing phase or from
rather high values of the steady-state emission anisotropy (above 0.25 except
the enzyme-DML vesicles in the liquid-crystalline phase) of ANM, ANM will be
fairly immobilized in the enzyme and the orientation between the emission moment
of ANM and the absorption moment of heme a will be restricted. Therefore, if the
conformational change in cytochrome oxidase involving the change in the relative
arrangement between ANM and heme a occurs, one would expect the change in the
efficiency of resonance energy transfer. This change would then be reflected in
the fluorescence intensity. The observed I_T curves of ANM may have peaks around

the structural changes in cytochrome oxidase under the above mechanism. There remain other possibilities for the change in the fluorescence intensity. However , the possibility that peaks in I_T curves may be caused by the change in the energy transfer between ANM and other fluorescent amino acid residues such as tryptophane, tyrosine and phenylalanine in the enzyme can be eliminated, because the excitation wavelength for ANM (360 nm) is longer than the absorption band of these fluorescent residues.

Temperature inflection in I_T curve of DPH. For similar reason in the case of ANM, temperature inflection in I_T curve of DPH would reflect the change in the arrangement between DPH molecule and heme a as a result of the conformational change in the enzyme. However, the inflection in I_T curve was not dramatic but smaller one. The result may be caused by following conditions: (1) Relatively small fraction of DPH would surround cytochrome oxidase compared with the fraction of DPH in bilayer region. Therefore, the conformational change in the enzyme would affect the small population of surrounging DPH. (2) DPH fluorescence was collected above 420 nm through cut-off filters. Therefore, the overlap between the DPH fluorescence spectrum through these filters and the Soret absorption band of oxidized heme a was less than that in the case of ANM. This suggests the lower efficiency of resonance energy transfer between DPH and heme a. Then, if the change in this efficiency of energy transfer is caused by the structural change in the enzyme, this change would be smaller than that in the case of ANM.

Temperature dependence of the emission ansiotropy of DPH. The r curves of DPH were smooth except the region of the lipid phase transition. The structural change in cytochrome oxidase appears to be unobservable in these curves.

Intrinsic conformational change in cytochrome oxidase. The peaks in I_T curve and the small peak in r curve of ANM around 18 $^\circ$C in the enzyme-DPL vesicles, the peaks in I_T curve of ANM between 17 and 40 $^\circ$C in the enzyme-DOL vesicles, the peak in I_T curve of ANM around 20 $^\circ$C of the enzyme-in-ECP, and peaks in I_T curve at lower temperature region and the zig-zag part of r curve of ANM in the enzyme-DML vesicles have not connections to lipid phase transition, because DPL and DML are in the gel phase in the above temperature region and DOL is in the liquid-crystalline phase above 0 $^\circ$C. Therefore, these peaks in I_T and r curves of ANM will be attributed to the intrinsic conformational change in cytochrome oxidase.

It is necessary to consider whether the conformational change that is independent of the lecithin phase transition is caused by the phase transition of remaining phospholipids originated from mitochondria. The phospholipid content of the purified preparation was determined as 1.5 mole of phospholipid per mol of heme $\underline{a}$[18]. Therefore, it is not plausible that the slight amount of phospholipids attached to the enzyme molecule could not cause the structural change in this bulky enzyme (molecular weight of about 200,000 daltons in ECP[19]).

<u>Conformational change induced by lipid phase transition</u>. The peak in I_T curve of ANM between 33 and 60 $^{\circ}$C and the depression in r curve of ANM around 38 $^{\circ}$C in the enzyme-DPL vesicles, the peaks in I_T curve of ANM in higher temperature region and the great decrease of r of ANM around 24 $^{\circ}$C in the enzyme -DML vesicles would have close connections to DPL or DML phase transition which was observed in r curve of DPH embedded in these vesicles. It will be likely that the lecithin phase transition changes the conformation of cytochrome oxidase . There is a question whether the above changes in I_T and r curves of ANM reflect only the lecithin phase transition but the structural change in the enzyme. In cytochrome oxidase vesicles, the lipid phase transition observed in the steady-state emission anisotropy of DPH was almost perfectly reversible (not shown in figures). Even after cytochrome oxidase-DML vesicles were incubated above 25 $^{\circ}$C for about 1 h, the emission anisotropy of DPH reversed to the values before this incubation when temperature was decreased below the phase transition . However, after incubation above 25 $^{\circ}$C for about 1 h, r of ANM did not reverse at all and remained at the value at incubation temperature (Figure 2C).

Purified cutochrome oxidase from beef heart[20] and yeast[21] contain a highly reactive SH group on subunit II which is exposed on the outer side of the mitochondrial inner membrane. Dockter et al.[16] have shown that the fluorescent SH reagent N-(iodoacetamidoethyl)-1-aminonaphthalene-5-sulfonic acid selectively reacts with the SH group of subunit II in yeast cytochrome oxidase. Therefore, the possibility that ANM also reacts with the SH group of subunit II which would be exposed on outer side of lecithin vesicles would be high. Thus, the reacted site of ANM would be exposed into water and ANM would not directly interact with lecithin molecules.

From above discussions, changes in I_T and r of ANM around the temperature of lecithin phase transition will not directly reflect the lipid phase transition and these will reflect the structural change in cytochrome oxidase.

ACKNOWLEDGEMENTS

The authors thank Professor S. Ebashi for kindly making his apparatus available for the experiment.

REFERENCES

1. Henderson, R., Capaldi, R.A. and Leigh, J.S. (1977) J. Mol. Biol., 112, 631-648.

2. Raison, J.K., Lyons, J.M. and Thomson, W.W. (1971) Arch. Biochem. Biophys., 142, 83.

3. Erecińska, M. and Chance, B. (1972) Arch. Biochem. Biophys., 151, 304-315.

4. Yoshida, S., Orii, Y., Kawato, S. and Ikegami, A., Biochemistry, submitted.

5. Mihashi, K. (1972) J. Biochem., 71, 607-614.

6. Ohyashiki, T., Sekine, T. and Kanaoka, Y. (1974) Biochim. Biophys. Acta, 351, 214-223.

7. Okunuki, K., Sekuzu, I., Takemori, S. and Yonetani, T. (1958) J. Biochem., 45, 847.

8. Kanaoka, Y., Machida, M., Machida, M. and Sekine, T. (1973) Biochim. Biophys. Acta, 317, 563-568.

9. Kinosita, K., Jr., Mitaku, S., Ikegami, A., Ohbo, N. and Kunii, T.L. (1976) Jpn. J. Appl. Phys., 15, 2433-2440.

10. Kawato, S., Kinosita, K., Jr. and Ikegami, A. (1977) Biochemistry, 16, 2319-2324.

11. Yoshida, S. and Orii, Y., to be published.

12. Cone, R.A. (1972) Nature New Biol., 236, 39-43.

13. Cherry, R.J., Bürkli, A., Busslinger, M., Schneider, G. and Parish, G.R. (1976) Nature, 263, 389-393.

14. Cherry, R.J., Heyn, M.P. and Oesterhelt, D. (1977) FEBS Lett., 78, 25-30.

15. Förster, T. (1965) in Modern Quantum Chemistry, Sinanoglu, O., ed., New York, N.Y., Academic Press, pp. 93-137.

16. Dockter, M.E., Steinemann, A. and Schatz, G. (1978) J. Biol. Chem., 253, 311-317.

17. Dale, R.E. and Eisinger, J. (1976) Proc. Natl. Acad. Sci. U.S.A., 73, 271-273.

18. Orii and Iba, unpublished.

19. Orii, Y., Matsumura, Y. and Okunuki, K. (1973) in Oxidases and Related Redox Systems, King, T.E., Manson, H.S. and Morrison, M., eds., pp. 666-672.

20. Kornblatt, J.A., Chen, W.L., Hsia, J.C. and Williams, G.R. (1975) Can. J. Biochem., 53, 364.

21. Birchmeier, W., Kohler, C.E. and Schatz, G. (1976) Proc. Natl. Acad. Sci. U.S.A., 73, 4334.

Cytochrome Oxidase, T.E. King et al. eds.
© *1979 Elsevier/North-Holland Biomedical Press*

TEMPERATURE-DEPENDENT STRUCTURAL CHANGES OF CYTOCHROME OXIDASE

YUTAKA ORII and TOSHIAKI MIKI
Department of Biology, Faculty of Science, Osaka University, Toyonaka, Osaka 560
(Japan)

ABSTRACT

Solubilized cytochrome oxidase has been proposed to assume hot and cold conformations depending on temperature[1]. As temperature was raised from 4° to 34°C
an absorption spectrum of the oxidized enzyme shifted toward longer wavelengths
in the visible and Soret regions and the change was reversible. A plot of
absorbance increment (ΔA) at a peak *versus* $1/T$ was discrete with a break around
20°C. On the contrary, spectral behavior of the reduced enzyme was complicated.
A temperature-difference spectrum (33°)-(4°) had a broad peak from 564 to 595 nm
and a sharp one at 614 nm. When temperature was lowered the sharp peak disappeared reversibly whereas the broad one remained. A ΔA-$1/T$ plot at 575 nm that
was obtained during the increase of temperature was sigmoidal, a midpoint being
at 19°C. The main peak shifted toward red upon raising the temperature. The
temperature-dependent conformational change thus occurs reversibly with the
oxidized enzyme whereas the transition from the hot to cold type of the reduced
oxidase takes a long time.

INTRODUCTION

In the preceeding two papers it was shown that solubilized cytochrome oxidase
yielded a bent Arrhenius plot when the activity was assayed at different temperatures. This phenomenon was correlated with the conformational transition
between the hot and cold types as proposed previously[1]. In the present investigation it is examined whether the spectral properties of this enzyme will be
affected by changing temperature in such a manner as to reflect the structural
transition.

MATERIALS AND METHODS

Bovine heart cytochrome oxidase was prepared as described previously[1] and
dissolved in 0.05 M sodium phosphate buffer, pH 7.4, containing 0.25% (v/v)
Emasol 1130 and 0.1% (w/v) sodium cholate. This medium was used throughout the
present study.

Absorption spectra were recorded on a computer-controlled single beam

spectrophotometer constructed of Union Giken spectrophotometer modules with
a microcomputer, system 71. For recording of a difference spectrum a reference
spectrum was stored in the memory of the computer. Temperature of a sample
solution was controlled by circulating water of desired temperature through a
cell holder and the temperature was monitored with a thermister, which was
calibrated with a standard thermometer, dipped into the solution. Absorbance
increments were corrected for volume change of a solution due to temperature
change.

Cytochrome oxidase activity was determined spectrophotometrically as described previously[2] to examine intactness of the enzyme.

RESULTS AND DISCUSSION

Temperature-difference spectrum of oxidized cytochrome oxidase. In a temperature range between 1.5° and 35°C cytochrome oxidase was stable and no
turbidity due to heat denaturation developed. Therefore, all the spectra were
recorded in this range. As shown in Fig. 1 an α-peak of the oxidized enzyme
shifted toward longer wavelengths upon raising temperature. A temperature-
difference spectrum (33.4°)-(1.5°) exhibited a peak at around 622 and a trough
at 596 nm. A plot of absorbance change at 596 nm *versus* 1/T as shown in Fig.
2-I was biphasic with a break at 20.5°C. This change occurred reversibly on
lowering temperature with a break at 20.8°C. These breaks are in the same
temperature range that was observed in bent Arrhenius plots for the oxidase
activity. Therefore, the same mechanism is suggested to be functioning in
yielding the break and this will be ascribed to a temperature-dependent structure change of cytochrome oxidase. If so, the oxidized enzyme changes the
conformation reversibly.

Spectral characteristics of reduced cytochrome oxidase. Figure 3 illustrates
the reduction process of the oxidized oxidase with dithionite. It is noticeable
that the spectrum at 9°C showed a small but distinct peak at around 562 nm
whereas at 34°C a trough between this and the main α-peak became obscure.

Spectral changes of reduced cytochrome oxidase induced by changing temperature were composed of reversible and irreversible portions. Cytochrome
oxidase was reduced with dithionite at around 4°C taking 30 min, and temperature
was raised slowly with occasional recording of the spectrum. As shown in Fig. 4
a difference spectrum (33.4°)-(3.7°) had a broad peak extending from 565 to
595 nm and a prominent peak at 614 nm. When the sample temperature was lowered again to the starting temperature the broad peak still persisted, indicating that a certain kind of irreversible change had occurred during the initial

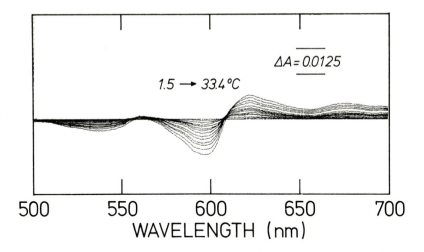

Fig. 1. Temperature-difference spectra of oxidized cytochrome oxidase. The concentration of the enzyme was 47.8 μM. A reference spectrum at 1.5°C was stored in the memory of a computer and the temperature was increased slowly. The spectrum was recorded every *ca.* 2.5°C interval.

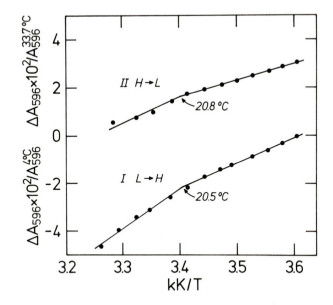

Fig. 2. Plots of absorbance change *versus* 1/T for the oxidized oxidase. The data were taken from Fig. 1.

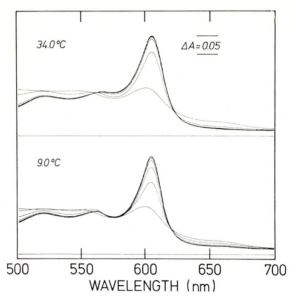

Fig. 3. Reduction of cytochrome oxidase with dithionite at 34° and 9°C. Cytochrome oxidase (15 µM) was reduced with a definite amount of dithionite and the spectra were recorded at 10 sec, 1, 2, 3, 4 and 5 min after addition of the reductant.

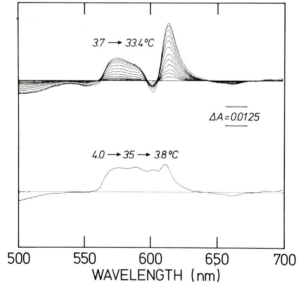

Fig. 4. Temperature-difference spectra of reduced cytochrome oxidase. Cytochrome oxidase (18.6 µM) was reduced with dithionite at 3.7°C for 30 min and the spectrum was recorded and stored in the memory. The temperature was increased slowly and the spectrum was recorded every *ca*. 3°C interval (top). Bottom: The sample temperature was changed as indicated.

temperature increase. A plot of the irreversible increase in absorbance at
575 nm *versus* 1/T was sigmoidal with a midpoint temperature of 18.6°C as if
the equilibrium was established between the two conformations at each temperature notwithstanding the fact that the reverse change did not occur practically. It is speculated, therefore, that the reduction of cytochrome oxidase at
a low temperature is effective only in reducing heme a and copper leaving the
protein moiety in the state of a stable oxidized oxidase. As the temperature
is increased this state is transformed into a hot type accompanying loosening
of the protein structure, or a large increase in entropy. This may explain
the apparently irreversible change from the hot to cold conformation of the
reduced oxidase. In this sense we can regard the absorption around 575 nm
of reduced cytochrome oxidase as that representing the conformational state.

The irreversible portion of spectral change can be removed from the overall
spectrum recorded. The enzyme was reduced with dithionite at around 4°C,
brought to 35° and again back to 4°C. The final spectrum was stored in the
memory of the computer and again temperature-difference spectra were recorded.
On raising temperature a red shift occurred as shown in Fig. 5 (top) and on

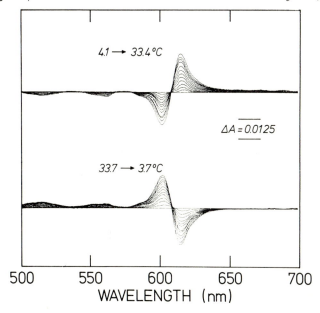

Fig. 5. Reversible portion of the spectral change of reduced cytochrome oxidase.
After complete reduction of cytochrome oxidase (18.6 μM) with dithionite at 4°C
the temperature was brought to 35° and back to 4°C. The final spectrum was
stored in the memory and the temperature was increased for recording of temperature-difference spectra (top). The spectrum at 33.7°C was in memory and
the temperature was lowered (bottom).

lowering difference spectra of mirror image were obtained (Fig. 5, bottom).
A plot of absorbance increment at a peak *versus* 1/T gave a straight line in
the both cases. This reversible change is correlated with a temperature-
dependent change in polarizability of the solvent, thus indicating its contact
with the chromophoric heme a in reduced cytochrome oxidase.

ACKNOWLEDGEMENT
   This work was supported in part by a grant from the Ministry of Education,
Science and Culture of Japan, No. 248132.

REFERENCES
1. Orii, Y., Manabe, M. and Yoneda, M. (1977) J. Biochem., 81, 505-517.
2. Orii, Y. and Okunuki, K. (1965) J. Biochem., 58, 561-568.

FUNCTIONS AND KINETICS

Cytochrome Oxidase, T.E. King et al. eds.
© *1979 Elsevier/North-Holland Biomedical Press*

STEADY STATE PROPERTIES OF CYTOCHROME OXIDASE

SHINYA YOSHIKAWA, KIMIYASU YAMAMOTO, KOICHI SUZUKI, TANHAKU SAI AND
HIROYUKI NISHI
Department of Biology, Faculty of Science, Konan University, Higashinada-Ku,
Kobe, Hyogo 658 (JAPAN).

SUMMARY

The spectral properties and enzymatic activity of purified beef heart cytochrome oxidase were investigated under turnover conditions in the presence of
ascorbate and cytochrome c. At low concentration of cytochrome c, the aerobic
steady state spectrum of the enzyme was different from that of fully oxidized
("resting") enzyme at ascorbate concentration <15 mM. As the ascorbate concentration was increased above 15 mM to 200 mM, a spectral change was observed
with a clear isosbestic point which the fully oxidized ("resting") enzyme did
not share at 620 nm. Further spectral change was not observed above 200 mM
ascorbate. This result suggests that under turnover conditions only two enzyme
species, both different from the resting enzyme, are present. Upon increasing
the ferrocytochrome c concentration at a fixed ascorbate concentration, the
activity of the enzyme increased and the absorbance of the 605 nm band also
increased proportionally. However, even at these higher cytochrome c concentrations the spectral differences between the high and low ascorbate forms of
the enzyme were maintained. The ascorbate induced spectral change stimulated
the enzymic activity. These results suggest that ferrocytochrome c reduces
specifically one heme a of the enzyme to form an enzyme species reactive with
O_2 and the other heme a, activated by ascorbate (and probably also by O_2),
stimulates the reaction on the subunit containing the ferrocytochrome c reactive heme a. Both of the enzyme species which appear at low cytochrome c
concentration show a similar reactivity to cyanide. This suggests that the
cyanide sensitive heme a which is probably the O_2 reactive site is the ferrocytochrome c reactive heme a.

INTRODUCTION

In 1960 Sekuzu et al.[1] reported that despite the fact that cytochrome oxidase was rapidly reduced by p-phenylene diamine, the component was not oxidized
by oxygen if cytochrome c was not present. On the basis of the result, they
proposed that the heme a protein is nonautoxidizable and cytochrome c is an
essential constituent of cytochrome oxidase. Yonetani[2] concluded, from a

similar observation, that cytochrome c mediated electron transfer between cytochromes c_1 and a as well as cytochrome a and a_3. On the basis of these studies
it seems possible that the aerobic steady state of cytochrome oxidase in the
presence of cytochrome c and ascorbate might provide a very convenient and
important means of examining the properties of cytochrome oxidase in the functioning state. However, it has received little attention since these studies.
In this paper we describe some results of an examination of the system under
turnover conditions which suggest a reaction mechanism involving four enzyme
species different from the static enzyme species (fully reduced and oxidized
forms and partially anaerobically reduced form).

MATERIALS AND METHODS

Cytochrome oxidase was purified from beef heart muscle, by a modified
Okunuki method[3]. The aerobic steady state was attained by introducing O_2 to
an anaerobic mixture of cytochrome oxidase, cytochrome c and ascorbate.
Absorbance of the aerobic steady state system containing high concentration
of cytochrome c was determined with a mixing apparatus (Union Giken Sample
mixing device MX-7). The cytochrome oxidase activity of the system was determined by the ratio of the initial concentration of dissolved oxygen to the
duration of the aerobic steady state after inroducing O_2 until the end of the
aerobic steady state which is accurately determined by a sudden spectral
change. The absorbance of cytochrome oxidase in the aerobic steady state in
the presence of high concentrations of cytochrome c up to 90 µM was determined
by subtracting the absorbance contribution of cytochrome c from the steady
state absorbance. The contribution of absorbance of cytochrome c was estimated
by the steady state absorbance at 550 nm. The reaction mixture contained
25~59 mM sodium phosphate buffer, pH 7.4 at 20°C. Enzyme concentration in
the reaction mixture was 13.5 µM in heme a, unless otherwise mentioned.

RESULTS

(1) Spectral change induced by ascorbate. Fig. 1 shows a series of difference spectra obtained from cytochrome oxidase (27 µM)-cytochrome c-ascorbate
mixture in which the concentration of cytochrome c was kept low (1 µM) and
constant but the ascorbate was varied from 15 mM to 200 mM; the reference was
fully oxidized oxidase. The spectra will be little influenced by the small
amount of cytochrome c present in the reaction mixture so the changes that are
observed must be due to the oxidase and are a reflection of the ability of
ascorbate to modify the oxidase under turnover conditions. This influence of
ascorbate upon the oxidase spectrum is exhibited only over a restricted

concentration range. Fig. 2 illustrates the invariance of the absorbance difference (turnover oxidase - resting oxidase) at ascorbate concentrations <15 mM. The absorbance difference was similarly independent of ascorbate when its concentration was >200 mM. However, within these concentration limits a clear isosbestic point was observed at 620 nm suggesting the presence of only two forms of the enzyme (E and E'). The form present at low ascorbate concentrations (E) appears to have a slightly different spectrum from the fully oxidized ("resting") enzyme. This is based upon the observation that the two cannot share the isosbestic point at 620 nm.

(2) <u>Spectral changes induced by ferrocytochrome c</u>. Increase in the ferrocytochrome c concentration at a fixed ascorbate concentration gave rise to additional spectral change in the visible and Soret region and to an increase in the enzymic activity. Absorbance changes from E and E' were examined at 0.015 M ascorbate and at 0.15 M ascorbate respectively, as shown in Fig. 3. The enzymic activity of the system is also shown in Fig. 3.

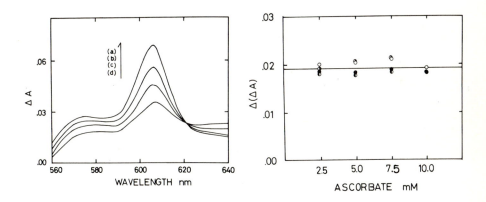

Fig. 1. (left) Difference spectra of cytochrome oxidase under turnover conditions against the fully oxidized enzyme. The reaction mixture contained cytochrome oxidase (27 μM), cytochrome c (1 μM) and ascorbate ((a): 200 mM, (b): 150 mM, (c): 50 mM, (d): 15 mM).

Fig. 2. (right) Absorbance difference between peak (599 nm) and trough (570 nm) of the aerobic steady state spectrum. The sample cuvette contained cytochrome oxidase (27 μM), ascorbate and cytochrome c (○: 1 μM; ●: 2 μM; ◐: 5 μM; ◑: 10 μM). Reference cuvette contained cytochrome oxidase and cytochrome c at the same concentration as in the sample cuvette.

Absorbance, A, and the enzymic activity, v, at various ferrocytochrome c concentrations, (C"), can be expressed by the following relations at each ascorbate concentration.

$$v = V_{max} \ (C")/(K_m^V + (C")) \tag{2}$$

$$A = A_0 + \Delta A_{max} \ (C")/(K_m^S + (C")) \tag{1}$$

The solid lines in Fig. 3 were obtained by introducing the experimentally observed values for A, v and C" into Equations 1 and 2 and adjusting the values of A_{max}, V_{max}, K_m^S and K_m^V to give the best fit. The best fit values of the parameters which are tabulated in Table I show that K_m^S (11.5 µM) at 605 nm is, within experimental accuracy, in agreement with that at 445 nm (13.3 µM),

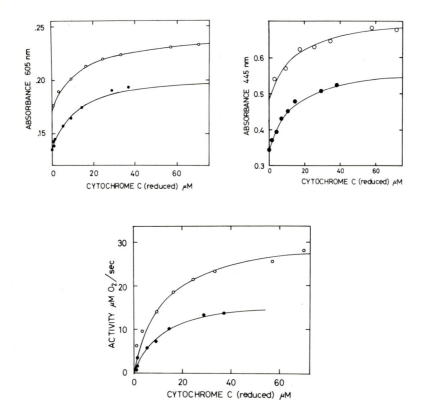

Fig. 3. Effect of ferrocytochrome c concentration on the aerobic steady state absorbance of cytochrome oxidase and the enzymic activity. Ascorbate concentration in the reaction mixture was 0.15 M (O) or 0.015 M (●). Solid lines are best fitted curves based on Eq. 1 or Eq. 2.

and that ΔA_{max} and K_m^S are independent of ascorbate concentration, that is, the spectral change induced by ferrocytochrome c is not affected by ascorbate concentration. Further examination of the parameters given in Table I indicates that K_m^V is also independent of ascorbate concentration, but that V_{max} depends on the concentration. The agreement between K_m^S and K_m^V suggests that the formation of enzyme-substrate complex is responsible for the spectral change induced by ferrocytochrome c. The enzyme substrate complex induced from E and E' are referred to as ES and ES', respectively.

Table I

Ascorbate (M)	V_{max} (μM O_2/sec)	K_m^V (μM)	ΔA_{max}		K_m^S (μM)	
			605nm	445nm	605nm	445nm
.15	31.8	11.5	.073	.231	11.5	13.3
.015	18.1	11.5	.073	.231	11.5	13.3

Change in enzyme concentration did not affect the mode of reaction, that is, K_m^V and K_m^S were independent of enzyme concentration and ΔA_{max}, V_{max} were proportional to enzyme concentration.

The aerobic steady state spectra of cytochrome oxidase in the presence of high concentrations of cytochrome c was determined by measuring the aerobic steady state absorbance at various wavelengths with the mixing apparatus and the data plotted to give the spectral curves shown in Fig. 4 for two cytochrome c concentrations. The aerobic steady state spectrum at 0 μM ferrocytochrome c and the fully reduced oxidase spectrum given in this figure were determined with a usual recording spectrophotometer. The spectra of ES and ES' were estimated by fitting Eq. 1 to the absorbance values at each wavelength. The best fitted curves for the extrapolated values at selected wavelengths are given in Fig. 4-a and 4-b respectively. These calculated spectral curves for ES and ES' are seen to resemble that for the fully reduced enzyme through the α band region. However, the fully reduced oxidase does not share the isosbestic point at 588 nm, and on this basis, is judged to be different from either ES or ES'. Fig. 4 also shows that the ferrocytochrome c induced absorbance is independent of ascorbate concentration at any wavelength examined (i.e., compare ΔA at any wavelength between Fig. 4-a and 4-b).

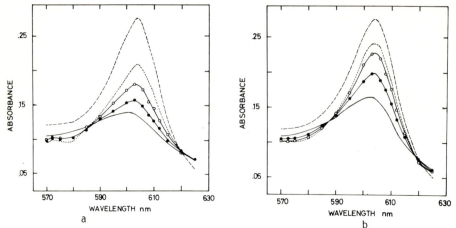

Fig. 4. Aerobic steady state spectrum of cytochrome oxidase. The reaction mixture contained ascorbate (a: 0.015 M; b: 0.15 M) and cytochrome c (—○—: 20 μM; —○—: 70 μM). The dotted lines (--------) show the calculated spectra of ES in a and ES' in b. The broken lines (------) and the solid lines (———) in a and b show spectrum of the fully reduced enzyme and the aerobic steady state spectrum at 0 μM, respectively.

(3) Reaction of cyanide with E. Despite the close similarity in the visible spectrum between E and fully oxidized oxidase as prepared, the reactivity of E to cyanide is much higher than that of the oxidized enzyme. The spectral change around the Soret band induced by cyanide in the presence of 5 mM ascorbate (Fig. 5) shows that at cyanide concentrations less than 1/2 heme a concentration, including no cyanide, the spectra show an isosbestic point at 422 nm, while at cyanide concentration much higher than the enzyme concentration, another isosbestic point appears at 435.5 nm. This reaction attains equilibrium within 20 minutes at 20°C. Reaction of cyanide to the fully oxidized enzyme as isolated is much slower than this reaction. For example, it has been reported that it took about one day to attain equilibrium for the cyanide binding reaction to the fully oxidized enzyme[4,5]. The spectral change at less than 10 μM cyanide, reported by them, is much smaller than that obtained here at the same concentration of cyanide.

The reaction observed in the presence of low concentration of cyanide was examined at 435.5 nm at the isosbestic point of the spectral change induced by higher concentration of cyanide. The spectral change with cyanide concentration (Fig. 7) indicates that the concentration of cyanide required for the maximal absorbance change is about 40% of heme a concentration in the system and the equilibrium constant calculated from the curve is 0.093 ± 0.059 μM. A rectangular hyperbolic relation (Fig. 6) is found between the initial velocity of the cyanide binding reaction and cyanide concentration. This

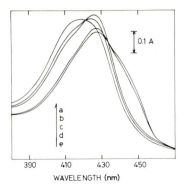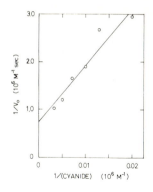

Fig. 5. (left) Effect of cyanide on the Soret band of E. The reaction mixture contained cytochrome oxidase (18 μM), cytochrome c (2 μM), ascorbate (5 mM) and cyanide ((a): 0 μM; (b): 2 μM; (c): 4 μM; (d): 50 μM: (e): 300 μM).

Fig. 6. (right) Initial velocity of the cyanide binding reaction. The reaction mixture contained cytochrome oxidase (10 μM), cytochrome c (2 μM), ascorbate (5 mM) and various concentrations of cyanide.

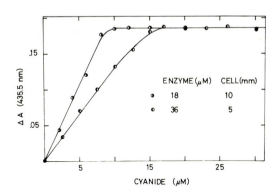

Fig. 7. Absorbance change of E at 435.5 nm induced by cyanide. Composition of the reaction mixture except for cyanide was the same as that in Fig. 5.

relation suggests the following scheme for cyanide binding,

$$E + CN \underset{k_{-1}}{\overset{k_1}{\rightleftarrows}} E^+CN$$

$$E^+CN \underset{k_{-o}}{\overset{k_o}{\rightleftarrows}} ECN$$

in which E^+CN denotes an intermediate form of cyanide complex which has the same spectrum as that of E. The numerical values of k_1 and k_o obtained from the results are listed in Table II. The reaction between cyanide and E' (Compound I) has a similar aspect[6]. Table II also contains the numerical values of parameters of the cyanide-Compound I reaction reported[6]. These parameters in Table II are in good agreement with each other, except for the k_o value.

Table II

	$Kd(\mu M)$	e_t/heme a	$k_1 (M^{-1}sec^{-1})$	$k_o(sec^{-1})$
E	.093 ± .059	.395 ± .082	1.7×10^3	.27
E'	.277 ± .074	.417 ± .017	2.5×10^3	.02

DISCUSSION

The absorbance change in the enzyme at the aerobic steady state can be taken as composed of two components, one which depends only on ferrocytochrome c concentration and the other, only on ascorbate concentration. The former is proportional to the enzymic activity at a fixed ascorbate concentration, while the latter increases the V_{max} value. Each spectral change has a set of isosbestic points. One possible interpretation of these observations is as follows: ferrocytochrome c reduces specifically one heme a of the functional unit of the enzyme to form an enzyme species which reacts with O_2, that is, enzyme-substrate complex, and the other heme a, activated by ascorbate, stimulates the reaction which takes place on the ferrocytochrome c reactive heme a (or the subunit containing the heme a). This interpretation does not require any effect of ferricytochrome c on the enzymic reaction. Further, the independency of the mode of the enzymic reaction on the enzyme concentration suggests that any kind of stable enzyme-cytochrome c complex is not involved in the enzymic reaction. Thus, the following scheme is possible for the enzyme reaction,

$$E + C'' \xrightarrow{k_1} ES + C''' \, , \quad ES \xrightarrow{k_2} E$$

$$E' + C'' \xrightarrow{k_1'} ES' + C''' \, , \quad ES' \xrightarrow{k_2'} E'$$

$$C''' + A \longrightarrow C''$$

$$E + A \rightleftharpoons E'$$

where E, ES, E' and ES' are defined as above, A, C'' and C''' denote ascorbate and ferrous- and ferric-cytochrome c, respectively. The relation, $k_1/k_1' = k_2/k_2'$, gives the apparent independency between the two spectral changes, that is, the fact that the ferrocytochrome c induced spectral change is not changed by ascorbate concentration and visa versa.

The distinction between E and the oxidized enzyme as prepared is clearly shown by the slight difference in their visible spectra (Fig. 1) and the clear difference in their reactivity with cyanide. The close similarity between their visible spectra suggests that E is a kind of conformational variant of the fully oxidized enzyme. Thus it could be referred to as another form of the oxidized enzyme according to Orii and Okunuki[7].

Spectrum of E' has characteristics of Compound I reported by Orii and King[8,9], that is, absorption peak at 603 nm and a shoulder around 580 nm in the difference spectrum versus the oxidized enzyme (Fig. 1).

The spectral change induced by ferrocytochrome c is clearly different from both redox difference spectra of cytochromes a and a_3[10]. For example, the ratio, ΔA_{max} at 605 nm/ΔA_{max} at 445 nm, for the ferrocytochrome c induced spectrum is 0.32 at both ascorbate concentrations, while this ratio for the redox difference spectra of cytochromes a and a_3 are 0.22 and 0.07, respectively[10]. This strongly suggests that the ferrocytochrome c induced spectral change is caused by neither reduction of cytochrome a nor that of cytochrome a_3.

The reaction of E with low concentration of cyanide has a close resemblance to that of E' (or Compound I), as shown in Table II, suggesting that the state of cyanide binding site in E is closely similar to that in E'. Therefore, the cyanide binding site (probably the same as that of O_2) is likely to be the ferrocytochrome c reactive heme a.

ACKNOWLEDGMENTS

The authors are grateful to Prof. H. Matsubara and his collaborators for their encouragement and valuable criticism during the course of this study. They wish to thank Prof. W. J. Wallace for valuable discussion and assistance in the correction of syntax and grammatical errors.

REFERENCES

1. Sekuzu, I., Takemori, S., Orii, Y. and Okunuki, K. (1960) Biochim. Biophys. Acta 37, 64-71.
2. Yonetani, T. (1960) J. Biol. Chem. 235, 3138-3143.
3. Yoshikawa, S., Choc, M.G., O'Teol, M.C. and Caughey, W.S. (1977) J. Biol. Chem. 252, 5498-5508.
4. Orii, Y. and Okunuki, K. (1964) J. Biochem. 55, 37-48.
5. Yoshikawa, S. and Orii, Y. (1973) J. Biochem. 73, 637-645.

6. Yoshikawa, S., Ueno, T. and Sai, T. (1977) J. Biochem. 82, 1361-1367.
7. Orii, Y. and Okunuki, K. (1963) J. Biochem. 53, 489-499.
8. Orii, Y. and King, T.E. (1972) FEBS Lett. 21, 199-202.
9. Orii, Y. and King, T.E. (1976) J. Biol. Chem. 251, 7487-7493.
10. Yonetani, T. (1961) J. Biol. Chem. 236, 1680-1688.

Cytochrome Oxidase, T.E. King et al. eds.
© *1979 Elsevier/North-Holland Biomedical Press*

MECHANISMS FOR THE STEADY STATE KINETICS OF REACTION OF CYTOCHROME c WITH
MITOCHONDRIAL CYTOCHROME OXIDASE

BEVERLY ERREDE AND MARTIN D. KAMEN
Department of Radiation Biology and Biophysics, University of Rochester School
of Medicine and Dentistry, Rochester, New York 14642, and the Section of
Molecular Biology, Department of Biological Sciences, University of Southern
California, Los Angeles, California 90007

ABSTRACT

The same general rate law accounting for the observed kinetic properties of
the reaction of ferrocytochrome c with mitochrondrial cytochrome oxidase (ferrocytochrome c:oxygen oxidoreductase E.C.1.9.3.1), can be derived from each of
four proposed mechanisms. The required assumptions as well as implications of
these mechanisms are discussed with respect to kinetic results obtained from
the reaction of variant cytochromes c with oxidase. These considerations
suggest either of two proposed mechanisms provide a reasonable basis for interpretation of kinetic results. Both involve extension of a Minnaert scheme
[Minnaert, K. (1961), Biochim. Biophys. Acta 50, 23] to include reaction of two
substrate molecules per oxidase molecule.

INTRODUCTION

The functional competence of variant cytochromes c[1-7] as well as chemically
modified cytochromes c[8,9,10] is commonly quantitated by application of steady
state kinetic analyses for reactions of cytochrome c with mitochrondrial cytochrome oxidase (ferrocytochrome c:oxygen oxidoreductase E.C.1.9.3.1). Reliable
estimation of kinetic constants on which to base mechanistic interpretations
requires observance of certain limiting conditions that have been described
previously[6]. Moreover, the bases for the interpretation of kinetic constants
need to be defined explicitly. The interpretation of kinetic results requires
choice of one among many mechanisms that may rationalize the data. We show that
at least four mechanistic schemes are in adequate accord with the data we have
obtained for the oxidase reaction, including two based upon diametrically
opposed assumptions — on the one hand, that productive complexes be formed in
a significant fraction of collisions, and on the other, that nonproductive complexes are usual and reaction proceeds through a rarer transition state. The
minimum assumptions and implications of each mechanism are described in order
to clearly define the premises upon which the various possible interpretations

are made. Results obtained from the kinetic comparison of variant cytochromes c reaction with oxidase are used to determine which of the proposed schemes most reasonably account for the available data.

RESULTS AND DISCUSSION

Definition of the reaction.

The oxidation of ferrocytochrome c catalyzed by cytochrome oxidase is summarized by equations 1a - 1c.

$$4 \text{ ferrocytochrome } c + \text{oxidase}_{ox} \longrightarrow \text{oxidase}_{red} + \text{ferricytochrome } c \quad (1a)$$

$$\text{oxidase}_{red} + 4H^+ + O_2 \longrightarrow \text{oxidase}_{ox} + 2 H_2O \quad (1b)$$

$$4 \text{ ferrocytochrome } c + 4H^+ + O_2 \longrightarrow \text{ferricytochrome } c + 2 H_2O \quad (1c)$$

Polarographic measurements of oxygen uptake catalyzed by this system demonstrate that rates of reaction are independent of oxygen concentration under the aerobic assay conditions generally employed for steady state kinetic determinations. It follows that the reaction represented by equation 1b is rapid and does not determine the measured rate of reaction. Therefore, the proposed mechanisms will consider the oxidation of cytochrome c catalyzed by cytochrome oxidase to be equivalent to a one-substrate, enzyme-catalyzed reaction.

Properties of the reaction.

Kinetic characterization of the oxidase-catalyzed oxidation of ferrocytochrome c has been presented elsewhere[6,11], and these observations have been reported previously by other investigators[12-16]. A summary of the pertinent results is given:

1) The oxidase reaction does not show saturation by ferrocytochrome c. All cytochromes that have been studied follow a first order time course at all concentrations employed.

2) k_{obs}, the pseudo-first order rate constant, is directly proportional to the oxidase concentration.

3) k_{obs} decreases with increasing cytochrome c concentration.

4) The value of k_{obs} is a function of total cytochrome c concentration and does not change when the initial ratio of ferrocytochrome c to ferricytochrome c is varied.

A number of mechanisms have been proposed that account for these observations[14,17-21]. However, these proposals result in rate laws that define k_{obs}

as a simple hyperbolic function of total cytochrome c concentration. As such, they predict linear Lineweaver-Burk and Eadie-Hofstee plots for rate dependence on total cytochrome c concentration. More recent evidence has demonstrated that these plots are nonlinear when the cytochrome c concentration range employed for analysis is more extensive than that used in the earlier studies[5,11]. A rate law of the form given by equation 2 has been shown to predict rate dependence on cytochrome c concentration over a very large concentration range and for several cytochromes c of greatly differing reactivities[6,11].

$$\text{velocity} = \frac{\alpha_1 + \alpha_2[C]}{1 + \beta_1[C] + \beta_2[C]^2} \text{ [oxidase][ferrocytochrome } c] \qquad (2)$$

α_1, α_2, β_1 and β_2 are constants and $[C]$ is total cytochrome c concentration. Equation 2 includes additional terms in total cytochrome c concentration in both denominator and numerator which are not present in the rate laws derived from previously proposed mechanisms. The denominator terms of the rate law represent the distribution of oxidase in its various forms at steady state. The quadratic term in total cytochrome c concentration indicates that under certain conditions two molecules of cytochrome c may be bound to the oxidase at the moment of electron transfer. Experiments that measure stoichiometry for cytochrome c binding to c depleted mitochondria or that directly measure cytochrome c binding to oxidase support the kinetic evidence for (cytochrome $c)_2$-oxidase complexes[5,22-24].

Proposal of mechanisms consistent with the rate law for cytochrome c reaction with oxidase.

Minnaert's "Mechanism IV"[14] is widely accepted and we have used it as the basis for deriving four mechanisms that predict the exact form of the rate law for this reaction by inclusion of complexes involving two molecules of cytochrome c per oxidase molecule. Each mechanism proposed represents schematically different ways in which two cytochrome c molecules can interact with the oxidase.

"Dependent Site" Mechanism. The mechanism presented in Table 1 is a direct extension of Minnaert's "Mechanism IV"[14] in that an electron transfer occurs between cytochrome c and an oxidase molecule that is already complexed with one molecule of cytochrome c. It is important to note that the site 2 electron transfer occurs only when a sufficient concentration of cytochrome c:oxidase complex has been formed and thus refers only to reaction of the second bound cytochrome c. The rate law derived from this mechanism, using the steady state approximation, is given in Table 1 and is of the form specified by equation 2. The assumptions required to obtain the expected form of the rate equation are:

1) reverse reaction (reduction of cytochrome c) is negligible [i.e., $k_4 = k_{10} = k_{16} = 0$]; 2) reaction of cytochrome c with ferricytochrome c:oxidase complex is equivalent to reaction with ferrocytochrome c:oxidase complex [i.e., $k_7 = k_{13}$, $k_8 = k_{14}$, $k_9 = k_{15}$, $k_{11} = k_{17}$, and $k_{12} = k_{18}$]; 3) the functions of rate constants that determine the distribution of oxidase complexes between ferrocytochrome c-bound and ferricytochrome c-bound forms are equal [i.e., $k_1(k_3 + k_5) = k_6(k_2 + k_3)$, $k_7(k_9 + k_{11}) = k_{12}(k_8 + k_9)$].

The rate equations derived from the various mechanisms define the constants α_1, α_2, β_1 and β_2 of equation 2 in terms of various functions of rate constants for the proposed reaction sequences. For convenience we introduce and define notations to represent these functions. The constants k_1° and k_2° represent limiting values of first order rates obtained at zero cytochrome c concentration for the site 1 and site 2 reactions, respectively. K_1 represents the high affinity binding constant and K_2 the low affinity binding constant.

Table 1. "Dependent Site" Mechanism*

$$S + E \xrightleftharpoons[k_2]{k_1} U \xrightarrow{k_3} V \xrightleftharpoons[k_6]{k_5} P + E$$

$$S + U \xrightleftharpoons[k_8]{k_7} W \xrightarrow{k_9} X \xrightleftharpoons[k_{12}]{k_{11}} P + U$$

$$S + V \xrightleftharpoons[k_{14}]{k_{13}} Y \xrightarrow{k_{15}} Z \xrightleftharpoons[k_{18}]{k_{17}} P + V$$

Rate Equation:

$$v = \frac{k_1^{\circ} + K_1 k_2^{\circ}[C]}{1 + K_1[C] + K_1 K_2 [C]^2} [E^{\circ}] [S]$$

Definition of Mechanistic Constants:

Limiting Rate of Reaction

$$k_1^{\circ} = \frac{k_1 k_3}{k_2 + k_3}$$

$$k_2^{\circ} = \frac{k_7 k_9}{k_8 + k_9}$$

Binding

$$K_1 = \frac{k_6}{k_5} = \frac{k_1}{k_2}$$

$$K_2 = \frac{k_{12}}{k_{11}} = \frac{k_7}{k_8}$$

*
Symbols: S = ferrocytochrome c; P = ferricytochrome c; E = free oxidase; U = ferrocytochrome c:oxidase complex; V = ferricytochrome c:oxidase complex; W = 2 ferrocytochrome c:oxidase complex; X=Y = ferrocytochrome c + ferricytochrome c:oxidase complex; Z = 2 ferricytochrome c:oxidase complex; E$^{\circ}$ = total oxidase; C = total cytochrome c [i.e., S + P].

The dependent site mechanism of Table 1 specifically defines k_1° as the product of the rate constant for electron transfer, k_3, and the ratio of rate constants that determine the steady state concentration of cytochrome c:oxidase complex, $k_1/(k_2 + k_3)$. k_2° is similarly defined for the site 2 reaction.

Within the framework established by the three assumptions specified for the derivation of the rate law using this mechanism, the constants K_1 and K_2 are defined as binding constants for ferricytochrome c (product) with oxidase and cytochrome c:oxidase complex, respectively. The additional assumption that the binding constants for product and reactant are equivalent is not required to obtain the expected form of the rate equation. However, the observation that k_{obs} is independent of the initial ratio of ferricytochrome c to ferrocytochrome c[11,12] does suggest that this additional assumption may be justified.

"Independent Site" Mechanism. The mechanism presented in Table 2 provides an alternative scheme that will accommodate the presence of two cytochrome c molecules bound to oxidase. This mechanism postulates the existence of two independent cytochrome c reaction sites on the oxidase. The rate law derived from this mechanism using the steady state approximation is also of the required form. The assumptions that were made for the dependent site mechanism are also required to obtain the rate equation given in Table 2 [i.e., $k_4 = k_{10} = 0$; $k_1(k_3 + k_5) = k_6(k_2 + k_3)$; $k_7(k_9 + k_{11}) = k_{12}(k_8 + k_9)$].

Table 2. "Independent Site" Mechanism*

$$S + E_1 \underset{k_2}{\overset{k_1}{\rightleftarrows}} U_1 \overset{k_3}{\longrightarrow} V_1 \underset{k_6}{\overset{k_5}{\rightleftarrows}} P + E_1$$

$$S + E_2 \underset{k_8}{\overset{k_7}{\rightleftarrows}} U_2 \overset{k_9}{\longrightarrow} V_2 \underset{k_{12}}{\overset{k_{11}}{\rightleftarrows}} P + E_2$$

Rate Equation:

$$v = \frac{(k_1^\circ + k_2^\circ) + (K_1 k_2^\circ + K_2 k_1^\circ)[C]}{1 + (K_1 + K_2)[C] + K_1 K_2 [C]^2} [E^\circ][S].$$

Definition of Mechanistic Constants:

Limiting Rate of Reaction

$$k_1^\circ = \frac{k_1 k_3}{k_2 + k_3}$$

$$k_2^\circ = \frac{k_7 k_9}{k_8 + k_9}$$

Binding

$$K_1 = \frac{k_6}{k_5} = \frac{k_1}{k_2}$$

$$K_2 = \frac{k_{12}}{k_{11}} = \frac{k_7}{k_8}$$

* Symbols are as defined for Table 1. The subscripts 1 and 2 used for E, U and V refer to oxidase site 1 and site 2, respectively.

k_1°, k_2°, K_1 and K_2 are defined by the same functions of reaction pathway rate constants for both the dependent site and independent site mechanisms. The important distinction between the rate equations derived from the two mechanisms is that each defines the constants of equation 2 as different functions of $k_1°$, $k_2°$, K_1 and K_2. (Compare the rate equations given in Tables 1 and 2.)

"Dead End Complex" Mechanism. Stable enzyme-substrate complexes have been postulated as intermediates on the pathway of electron transfer in both of the above mechanisms. However, one may postulate that stable complexes of cytochrome c with oxidase are not on the pathway of electron transfer, but rather, that complex formation decreases the equilibrium concentration of oxidase available for reaction[12,18-20]. Such complexes of cytochrome c and oxidase are "dead end" rather than "active". For any fixed concentration of cytochrome c there will be a constant amount of oxidase available for reaction. The resulting pseudo-first order kinetics imply that electron transfer proceeds through a transition state type of interaction between cytochrome c and oxidase.

The mechanism presented in Table 3 parallels the ordered binding mechanism of Table 1. However, "dead end" complexes are proposed and electron transfer does not proceed through intermediate complexes of any finite lifetime. The rate law derived from this mechanism using the rapid equilibrium approximation also is of the form required by equation 2. As before, it is assumed that the reactions of cytochrome c with ferrocytochrome c:oxidase complex and ferricytochrome c:oxidase complex are equivalent [i.e., $k_7 = k_{13}$, $k_8 = k_{14}$, $k_9 = k_{15}$, $k_{11} = k_{17}$ and $k_{12} = k_{18}$] and that reverse reactions are negligible [i.e., $k_4 = k_{10} = k_{16} = 0$]. It is also necessary to assume that the equilibrium binding constants for complex formation with product and reactant are equivalent [i.e., $k_1/k_2 = k_6/k_5$ and $k_7/k_8 = k_{12}/k_{11}$].

Table 3. "Dead End Complex" Mechanism[*]

$$U \xrightleftharpoons[k_2]{k_1} S + E \xrightarrow{k_3} P + E \xrightleftharpoons[k_6]{k_5} V$$

$$W \xrightleftharpoons[k_8]{k_7} S + U \xrightarrow{k_9} P + U \xrightleftharpoons[k_{12}]{k_{11}} X$$

$$Y \xrightleftharpoons[k_{14}]{k_{13}} S + V \xrightarrow{k_{15}} P + V \xrightleftharpoons[k_{18}]{k_{17}} Z$$

Rate Equation:

$$v = \frac{k_1° + K_1 k_2°[C]}{1 + K_1[C] + K_1 K_2[C]^2} [E°][S]$$

Definition of Mechanistic Constants:

Limiting Rate of Reaction

$$k_1° = k_3$$

$$k_2° = k_9$$

Binding

$$K_1 = \frac{k_1}{k_2} = \frac{k_6}{k_5}$$

$$K_2 = \frac{k_7}{k_8} = \frac{k_{12}}{k_{11}}$$

[*] Symbols are as defined for Table 1.

The characteristic constants of equation 2, α_1, α_2, β_1 and β_2 are the same functions of the mechanistic constants $k_1{}^\circ$, $k_2{}^\circ$, K_1 and K_2 for this mechanism and the dependent site mechanism of Table 1. However, it should be noticed that the constants $k_1{}^\circ$ and $k_2{}^\circ$ are different functions of the reaction pathway rate constants.

"Exchange" Mechanism. In each of the mechanisms already presented it has been assumed that the second molecule of cytochrome c participates in the transition state through some interaction with oxidase. Because the *in vitro* assay system requires use of concentrations of cytochrome c in large excess of oxidase, it is possible that the apparent "second site" reaction is a consequence of a non-catalytic side reaction. We have considered one such possibility — that free cytochrome c undergoes electron exchange with cytochrome c bound to oxidase and have found that a rate law of the required form results. This proposal is formalized by the mechanism presented in Table 4. The assumptions required in the derivation of the rate equation using the steady state approximation are those used by Minnaert[14] for "Mechanism IV" [i.e. $k_4 = 0$; $k_1 = k_6$; $k_2 = k_5$] with the additional assumption that the rate of electron exchange between ferrocytochrome c:oxidase complex and free ferricytochrome c is equivalent to that for exchange between ferricytochrome c:oxidase complex and free ferrocytochrome c [i.e., $k_7 = k_8$].

$k_1{}^\circ$ and K_1 are the same functions of the reaction pathway rate constants as specified for the dependent site mechanism. However, "site 2" constants are now related to reactions involving electron exchange between free cytochrome c and oxidase-bound cytochrome c.

Table 4. "Exchange" Mechanism*

$$S + E \underset{k_2}{\overset{k_1}{\rightleftharpoons}} U \xrightarrow{k_3} V \underset{k_6}{\overset{k_5}{\rightleftharpoons}} P + E$$

$$S + V \underset{k_8}{\overset{k_7}{\rightleftharpoons}} U + P$$

Rate Equation:

$$v = \frac{k_1{}^\circ + K_1 k_3 k_e}{1 + (K_1 + k_e)[C] + K_1 k_e[C]^2} [E^\circ][S]$$

Definition of Mechanistic Constants:

Limiting Rate of Reaction: $k_1{}^\circ = k_1 k_3/(k_2 + k_3)$

Binding: $K_1 = k_1/k_2$

Exchange Reaction: $k_e = k_7/(k_2 + k_3)$

$K_1 k_e = k_1{}^\circ(k_7/k_2)$

*Symbols are as defined for Table 1.

Comparison of proposed mechanisms.

Four mechanisms for the reaction of ferrocytochrome c with cytochrome oxidase have been proposed. Each mechanism predicts the necessary form of the rate law when certain assumptions (as specified in the preceding section) are imposed. These schemes will be compared on the basis of their ability to account for the results obtained from kinetic analyses of three structurally distinct cytochromes c.

"Exchange" Mechanism. The exchange mechanism (Table 4) has been proposed to consider the possibility that the second site reaction may be a consequence of a non-catalytic side reaction unique to the *in vitro* assay system employed. The values for the constants defined by the exchange mechanism (Table 4) for horse c and *Rhodospirillum rubrum* c_2 are given in Table 5,C. Also included in Table 5

Table 5. Values for the Kinetic Constants Defined by the "Dependent Site" (A), "Dead End Complex" (A), "Independent Site" (B) and "Exchange" (C) Mechanisms.*

Cytochrome	I (mM)		$k_1°$ $(M^{-1}s^{-1})$	K_1 (M^{-1})	$k_2°(k_7/k_2)^†$ $(M^{-1}s^{-1})$	$K_2(k_e)^†$ (M^{-1})	$k_7^§$ $(M^{-1}s^{-1})$
horse c	44	A	2.50×10^8	6.21×10^6	4.01×10^6	3.84×10^4	
		B	2.47×10^8	6.17×10^6	2.48×10^6	3.87×10^4	
	4	A	1.90×10^8	1.20×10^7	4.20×10^6	9.20×10^4	
		B	1.87×10^8	1.19×10^7	2.78×10^6	9.27×10^4	
		C	1.90×10^8	1.19×10^7	2.65×10^5	9.27×10^4	4.17×10^2
Rhodospirillum rubrum c_2	4	A	8.65×10^5	3.56×10^5	2.90×10^5	6.37×10^4	
		B	7.02×10^5	2.72×10^5	1.64×10^5	8.32×10^4	
		C	8.65×10^5	2.72×10^5	1.19×10^5	8.32×10^4	1.02×10^3
Paracoccus denitrificans c_{550}	44	A	6.05×10^5	1.29×10^5	1.54×10^5	4.36×10^3	
		B	4.62×10^5	1.24×10^5	1.43×10^5	4.51×10^3	

* Assay conditions: 25°C; I=44 mM, 0.10 M 2-(N-morpholino)ethanesulfonic acid, pH 6.0; I = 4 mM, 0.01 M 2-(N-morpholino)ethanesulfonic acid, pH 6.0. Details are given in reference 6.

† $k_2°$ and K_2 apply to the values given in lines A and B. The constants in parentheses are for the "exchange" mechanism and are associated with the values given in line C. The units for k_7/k_2 are M^{-1}.

§ Dr. G. Smith, personal communication of unpublished results: the electron exchange rate constants, k_7, were determined with a 220 megahertz NMR spectrometer using the technique of monitoring the decrease in longitudinal relaxation times of the methionine methyl protons of ferrocytochrome c under conditions of low ionic strength (I≈0) at room temperature.

are the values of the electron exchange rate constant, k_7, determined by Dr. G. Smith (personal communication of unpublished results) under similar conditions. The value for k_7 obtained by this independent procedure allows calculation of a value for the dissociation rate constant, k_2, based on the estimate made for k_7/k_2 from steady state kinetic analysis. The values calculated for k_2 in this manner are 1.57×10^{-3} s^{-1} and 8.57×10^{-3} s^{-1} for horse c and *rubrum* c_2, respectively. However, van Gelder et al.[25] have reported that the value for the dissociation constant for horse c is 3×10^2 s^{-1} (0°C, I=0). To obtain agreement between this value and the value calculated from the data presented in Table 5, it would be necessary to make the improbable assumption that exhange between oxidase-bound cytochrome c and free cytochrome c is five orders of magnitude greater than exchange between free cytochromes c. This discrepancy contraindicates the use of the exchange mechanism for the steady state kinetics of cytochrome c reaction with oxidase.

"Dependent Site" Mechanism vs. "Dead End Complex" Mechanism. The characteristic constants of equation 2 are the same functions of the mechanistic constants, $k_1°$, $k_2°$, K_1 and K_2, for the mechanisms presented in Tables 1 and 3, however, the interpretation of these constants is totally different. Even though K_1 and K_2 are defined as equilibrium binding constants in both mechanisms, their physical meaning differs. In the dependent site mechanism (Table 1) electron transfer occurs only when cytochrome c is in a "stable" complex with oxidase. Therefore, binding, which controls the rate of complex formation, is an important functional parameter. In the mechanism of Table 3, binding limits the amount of free oxidase available for reaction and is not a parameter related directly to function. Within the framework established by the dead end complex mechanism, only the values for $k_1°$ and $k_2°$ provide information that reflects the functional competence of cytochrome c. The two mechanisms also differ with respect to the relationship between limiting rates of reaction and binding constants. If the values of the rate constants k_3 and k_9 remain unchanged, it is expected that decreases in K_1 and K_2 (reflecting lowered complex formation) will be accompanied by decreases in $k_1°$ and $k_2°$ for the dependent site mechanism contrasted to increases in $k_1°$ and $k_2°$ for the dead end complex mechanism.

It is apparent from the comparison of the limiting rates of reaction and the binding constants obtained for the cytochromes c represented in Table 5,A that large relative decreases in the values for K_1 and K_2 are paralleled by decreases in $k_1°$ and $k_2°$. A similar correlation has been found for comparisons of native and chemically modified cytochromes c[8,9]. Although "dead end" complex formation is not ruled out by these considerations, it does appear that available data are consistent with the interpretation provided by mechanisms postulating "active" complex formation.

"Dependent Site" Mechanism vs. "Independent Site" Mechanism. The mechanisms presented in Tables 1 and 2 both require active complex formation and both provide for the binding of two cytochrome c molecules to the oxidase. In addition, both require identical assumptions in order to obtain the established form of the rate law. However, the two mechanisms differ in fundamental character. The dependent site mechanism (Table 1) postulates that the values of the kinetic constants are determined by the "binding state" of the oxidase. The constants $k_1°$ and K_1 define reaction of cytochrome c with free oxidase and the constants $k_2°$ and K_2 define the reaction of cytochrome c with cytochrome c:oxidase complex. Sites 1 and 2 refer only to the reaction of unbound and singly bound oxidase and not necessarily to the existence of distinct sites of reaction on the oxidase molecule. The mechanism of Table 2 does postulate the existence of two sites on the oxidase. Therefore, the constants $k_1°$ and K_1 define reaction at site 1 while the constants $k_2°$ and K_2 define the reaction at site 2. The kinetic properties reflected by these constants for either site are independent of the binding state of the other. As a consequence of this distinction the dependent site mechanism accounts for one monocytochrome c:oxidase complex in which cytochrome c is bound only at "site 1", while the independent site mechanism accommodates two kinetically distinct monocytochrome c:oxidase complexes where one complex has cytochrome c bound at site 2 and the other has cytochrome c bound at site 1.

The values for K_1 and K_2 (Table 5,B) that have been determined for the various cytochromes c can be used to approximate the equilibrium distribution of total oxidase among the four forms predicted by the independent site mechanism. Such approximations indicate that for horse c and *Paracoccus denitrificans* c_{550}, formation of only the high affinity species of monocytochrome c:oxidase complex is significant so that the ordered binding postulated by the dependent site mechanism is realized. However, for reaction with *Rhodospirillum rubrum* c_2 the site 2 monocytochrome c:oxidase complex can account for as much as 23% of the concentration of all cytochrome c-bound forms of oxidase possible. Inasmuch as two forms of monocytochrome c:oxidase complex are consistent with the values obtained for K_1 and K_2 for *rubrum* c_2, the independent site mechanism provides a reasonable basis for interpretation of the kinetics of oxidase reaction with this cytochrome c.

SUMMARY

Four mechanisms for the reaction of ferrocytochrome c with cytochrome oxidase have been proposed. Each mechanism predicts the form of the rate law deduced from experiment. Three of the proposed schemes assume productive complex formation and are extensions of the generally accepted "Mechanism IV" of Minnaert[14] in which we have taken into account interactions involving two cytochrome c molecules per oxidase. One mechanism, designated "dependent site", proposes productive complex formation between cytochrome c and free oxidase in addition to further complex binding of a second cytochrome c to the initially formed oxidase complex. An alternative scheme represented by the "independent site" mechanism assumes the existence of two independent cytochrome c binding sites on the oxidase and allows for random binding at either site in contrast to the ordered binding of the "dependent site" mechanism. The third variation is given by the "exchange" mechanism and accommodates the second molecule of cytochrome c by assuming free cytochrome c participates in electron exchange with a molecule of cytochrome c bound to the oxidase. The fourth mechanism proposed — termed "dead end complex" — assumes that nonproductive complex formation is usual and electron transfer occurs through a rarer transition state. At present it is not possible to reject definitively any of these mechanisms. However, we believe either the "independent" or "dependent" site mechanisms can most easily account for the available data from comparative kinetic studies.

ACKNOWLEDGEMENTS

We are indebted to Professor G. P. Haight for important discussions in the early phases of this research and to L. Galbaiti for helpful criticism and discussion. Financial support has been provided by grants from the National Institutes of Health (GM-18528) and the National Science Foundation (BMS 75-3708 and BMS 75-13608).

REFERENCES

1. Davis, K.A., Hatefi, Y., Salemme, F.R. and Kamen, M.D. (1972) Biochem. Biophys. Res. Commun., 49, 1329-1335.

2. Yamanaka, T. (1972) Adv. in Biophys., 3, 227-276.

3. Smith, L., Nava, M.E. and Margoliash, E. (1973) in Oxidases and Related Redox Systems, Vol. 2, King, T.E., Mason, H.S. and Morrison, M. eds., University Park Press, Baltimore, MD, pp. 629-638.

4. Smith, L., Davies, H.C. and Nava, M.E. (1976) Biochemistry, 15, 5827-5831.

5. Ferguson-Miller, S., Brautigan, D.L. and Margoliash, E. (1976) J. Biol. Chem., 251, 1104-1115.

6. Errede, B. and Kamen, M.D. (1978) Biochemistry, 17, 1015-1027.

7. Kamen, M.D. (1972) Proteins, Nucleic Acids, Enzymes, 18, 753-773.

8. Smith, H.T., Staudenmayer, N. and Millett, F. (1978) Biochemistry, 16, 4971-4974.

9. Ferguson-Miller, S., Brautigan, D.L. and Margoliash, E. (1978) J. Biol. Chem., 253, 149-159.

10. Ferguson-Miller, S., Brautigan, D.L. and Margoliash, E. (1977) in The Porphyrins, Dolphin, D., ed., Academic Press, in press.

11. Errede, B., Haight, G.P. and Kamen, M.D. (1976) Proc. Nat. Acad. Sci. USA, 73, 113-117.

12. Smith, L. and Conrad, H.E. (1956) Arch. Biochem. Biophys., 63, 403-413.

13. Yonetani, T. (1960) J. Biol. Chem., 235, 3138-3143.

14. Minnaert, K. (1961) Biochim. Biophys. Acta, 50, 23-34.

15. McGuinness, E.T. and Wainio, W.W. (1962) J. Biol. Chem., 237, 3273-3278.

16. Nicholls, P. (1974) Biochim. Biophys. Acta, 346, 261-310.

17. Cope, F.W. (1963) Arch. Biochem. Biophys., 103, 352-365.

18. Hollocher, T.C. (1962) Arch. Biochem. Biophys., 98, 12-16.

19. Hollocher, T.C. (1964) Nature, 202, 1006-1007.

20. Nicholls, P. (1964) Arch. Biochem. Biophys., 106, 25-48.

21. Yonetani, T. and Ray, G. (1965) J. Biol. Chem., 240, 3392-3398.

22. Nicholls, P. (1965) in Oxidases and Related Redox Systems, Vol. 2, King, T.E., Mason, H.S., and Morrison, M., eds., John Wiley and Sons, New York, pp. 764-783.

23. Vanderkooi, J., Erecinska, M. and Chance, B. (1973) Arch. Biochem. Biophys., 157, 531-540.

24. Erecinska, M. (1975) Arch. Biochem. Biophys., 169, 199-208.

25. van Gelder, B.F., van Buuren, K.J.H., Wilms, J. and Verboom, C.N. (1975) in Electron Transfer Chains and Oxidative Phosphorylation, Quagliariello, E., Papa, S., Palmieri, F., Slater, E.C. and Siliprandi, N., eds., North Holland Publishing Co., Amsterdam, Netherlands, pp. 63-68.

Cytochrome Oxidase, T.E. King et al. eds.
© 1979 Elsevier/North-Holland Biomedical Press

KINETICS AND BINDING OF CYTOCHROMES C WITH PURIFIED MITOCHONDRIAL OXIDASE AND
REDUCTASE

Shelagh Ferguson-Miller

  Dept. of Biochemistry, Michigan State University, East
  Lansing, Michigan

Hanns Weiss

  European Molecular Biology Laboratory, Heidelberg, W.
  Germany;

S. H. Speck, D.L. Brautigan, N. Osheroff, and E. Margoliash

  Dept. of Biochemistry and Molecular Biology, Northwestern
  University, Evanston, Illinois, USA

INTRODUCTION

   Despite extensive knowledge of the structure of cytochrome c, the mechanism
by which it transfers electrons in the respiratory chain is not understood.
There is still controversy concerning the requirement for a classic
"enzyme-substrate" type complex between cytochrome c and its mitochondrial redox
partners[1-9]. It also remains to be established whether cytochrome c
oxidase and cytochrome c reductase can react simultaneously with cytochrome c
(requiring two spacially separate interaction domains) or whether cytochrome c
must dissociate from one to react with the other (requiring mobility and
dissociation rates consistent with physiological rates of electron
transfer)[5,8,10].

   As an approach to answering these questions, a comparison was made between
the kinetics and the binding of cytochrome c with purified reductase and oxidase
from Neurospora[11] and beef[12,13]. This comparison was carried out
under a variety of conditions and with a variety of modified cytochromes c in
which different single lysines had been converted from positively to negatively
charged residues by reaction with carboxydinitrobenzene chloride[14,15].

   The results show that under most conditions the steady-state kinetics with
both enzymes are dominated by the binding parameters, indicating that a stable
complex is an intermediate in the electron transfer reaction. In addition, the
purified oxidase and reductase are found to have the same preferred interaction
domain on cytochrome c. The significance of these findings in terms of the
mechanics and kinetics of electron transfer in the mitochondrial respiratory
chain are discussed.

MATERIALS AND METHODS

   Neurospora mitochondrial cytochrome c oxidase and reductase were solubilized
with Triton X-100, purified by cytochrome c affinity chromatography, and
demonstrated to be in a mono-disperse state[16-18]. Beef heart mitochondrial
particles were prepared as previously described[5]. Beef heart cytochrome c
reductase (Complex III) was prepared according to Reiske et al[19].
Cytochrome c oxidase activity was measured using the ascorbate-TMPD-oxygen
uptake assay system previously described[5,8,20]. Cytochrome c reductase
activity was measured spectrophotometrically as described by Speck et
al.[12,13] using an analogue of coenzyme Q_2, 2,3-dimethoxy-5-methyl-6--
decylhydroquinone.

   Binding measurements were made using the gel-filtration technique of Hummel
and Dryer[20] as modified by Weiss and Juchs[18]. Preparation and
purification of the carboxydinitrophenyl-derivatives of cytochrome c
(CDNP-cytochrome c) are described elsewhere[14,15,22].

RESULTS

Cytochrome c binding and kinetics with oxidase.

   Native and CDNP-cytochromes c. In Figure I, the binding and the activity of
native and modified horse cytochrome c with purified Neurospora cytochrome
oxidase are compared. Under the conditions used (25 mM Tris Acetate, pH 7.5,
0.25% Tween 20), native horse cytochrome c binds with high affinity (apparent
$K_D < 10^{-8}M$) to an average of 1.6 sites and shows a lower affinity reaction
(apparent $K_D \sim 10^{-6}M$) at an additional site (Fig. IA). It is not clear
whether the ratio of high affinity sites greater than one is real or
artifactual. An overestimate of the number of binding sites may occur if true
equilibrium has not been achieved in the time period of the filtration when very
tight binding is involved[23]. Extra binding may also be related to the
detergent used in the preparation (Triton X-100), since neither beef[5,8] nor
yeast[24] oxidases prepared with deoxycholate and Tween 20 exhibited the
effect.

   The kinetics of the reaction of native horse cytochrome c with Neurospora
oxidase show a pattern very similar to that observed in the binding studies,
when measured under identical conditions using the ascorbate-TMPD-oxygen uptake
assay system[5,20] (Fig. IB). The apparent K_m values (5 x $10^{-8}M$ and
$10^{-6}M$) are comparable to the K_D values obtained by binding measurements.

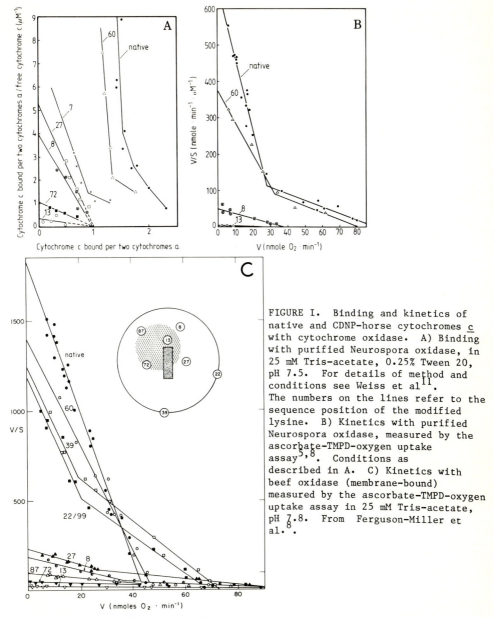

FIGURE I. Binding and kinetics of native and CDNP-horse cytochromes c with cytochrome oxidase. A) Binding with purified Neurospora oxidase, in 25 mM Tris-acetate, 0.25% Tween 20, pH 7.5. For details of method and conditions see Weiss et al[11]. The numbers on the lines refer to the sequence position of the modified lysine. B) Kinetics with purified Neurospora oxidase, measured by the ascorbate-TMPD-oxygen uptake assay[5,8]. Conditions as described in A. C) Kinetics with beef oxidase (membrane-bound) measured by the ascorbate-TMPD-oxygen uptake assay in 25 mM Tris-acetate, pH 7.8. From Ferguson-Miller et al.[8].

When CDNP-lysine derivatives of horse cytochrome c were compared to the native protein by activity and binding procedures, increases in apparent K_m were parallel with increases in apparent K_D, and the relative effects of derivatization at different positions were the same (Fig. IA,B). The pattern of activity and binding changes is very similar to that seen in the reaction with beef oxidase (Fig. IC).

These data define an interaction domain on cytochrome $\underline{c}$ for Neurospora cytochrome oxidase that is centered at the same position as that for beef cytochrome oxidase, shown in the inset diagram in Figure IC.

Ionic Strength Effects. The changes in K_D and K_m values caused by converting single lysines to negatively charged carboxydinitrophenyl-derivatives can also be produced by increasing the ionic strength. Figure II shows the changes in apparent K_D (panels A and B) and apparent K_m (panel C) of native horse cytochrome $\underline{c}$ reacting with Neurospora cytochrome oxidase, at ionic strengths from 0.025 to 0.050. The binding of both reduced and oxidized cytochrome $\underline{c}$ was measured[11]. Reduction of cytochrome $\underline{c}$ and oxidase was maintained by the addition of 0.5mM KCN and 0.5mM ascorbate. As previously demonstrated kinetically[3], the reduced and oxidized forms bind with very similar affinities to the oxidase over the ionic strength range tested, though oxidized cytochrome $\underline{c}$ consistently has slightly lower K_D values. The kinetic measurements (Fig. IIC) show increasing apparent K_m values with increasing ionic strength, that are in good agreement with the changes in K_D.

The comparative binding and kinetic constants are summarized in TABLE I. The values given are for the initial phases and are estimates from the data shown in Figures I and II, with no corrections applied for contribution of a second interaction. Such corrections would tend to decrease both the apparent K_m and K_D values to a similar extent. Notwithstanding this qualification, it is clear that good qualitative and quantitative agreement exists between the binding and kinetic measurements, indicating that derivatization and ionic strength effects on the steady-state kinetics are predominantly the result of changes in the binding of cytochrome $\underline{c}$ to cytochrome oxidase.

Cytochrome c Binding and Kinetics with Reductase.

Native and CDNP-cytochrome c. Two different mono-disperse preparations of Neurospora reductase, each containing two cytochromes $\underline{b}$ per one cytochrome $\underline{c_1}$[18], were tested for binding and electron transfer activity with cytochrome $\underline{c}$. Both showed similar binding behavior with cytochrome $\underline{c}$, but only the preparation to which an equivalent of purified non-heme-iron protein had been added showed electron transfer activity[11,18].

As shown in Figure IIIA, native horse cytochrome $\underline{c}$ binds to Neurospora reductase with high affinity ($K_D < 10^{-8}$ M) at an average of two sites per two cytochromes $\underline{b}$. As discussed for the oxidase binding studies, this high ratio (>1:1) may reflect incomplete equilibrium under tight binding conditions[23], and may also include nonspecific binding dependent on the detergent.

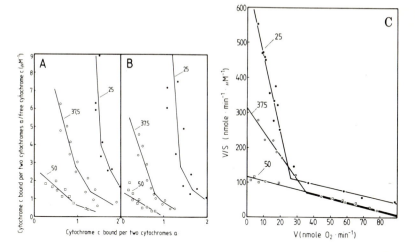

FIGURE II. Effects of ionic strength on binding and kinetics of native horse cytochrome c with purified Neurospora oxidase. A) Binding of ferricytochrome c. Conditions as described by Weiss et al.[11]. The numbers on the lines refer to the millimolar concentration of acetate in the buffer. B) Binding of ferrocytochrome c. Conditions as in A. C) Kinetics of reactions of cytochrome c with Neurospora oxidase measured by the ascorbate-TMPD-oxygen uptake assay[5,8]. Conditions as in A.

TABLE I

Comparison of K_D and K_m values for the reaction of Cytochrome c with Neurospora Cytochrome Oxidase.

Cytochrome c	Concentration of Tris-Acetate pH 7.5	K_M	K_D (Ferricytochrome c)
	mM	M	M
Native horse	25	5×10^{-8}	$\leq 10^{-8}$
CDNP-60 [a]	25	9×10^{-8}	3×10^{-8}
CDNP-8	25	7×10^{-7}	3×10^{-7}
CDNP-13	25	8×10^{-6}	4×10^{-6}
Native horse	37.5	2×10^{-7}	1×10^{-7}
Native horse	50	1×10^{-6}	4×10^{-7}

[a]The number indicates the sequence position of the lysine residue that was modified.

In Figure IIIB, the kinetics of reaction of native horse cytochrome $\underline{c}$ with reconstituted Neurospora reductase are shown[11]. The apparent K_m value (5×10^{-7} M) obtained by the spectral assay procedure[12,13] is considerably lower than the apparent K_D (Figure IIIA). The same K_m was found for horse cytochrome $\underline{c}$ reacting with purified beef reductase under identical assay conditions[13]. The discrepency between the K_m and K_D values of two orders of magnitude is unlikely to be accounted for by error in the binding measurements and therefore probably reflects the involvement of other rate constants in the K_m, in addition to those determining the K_D (See Speck et al.[13] for further discussion). This interpretation is

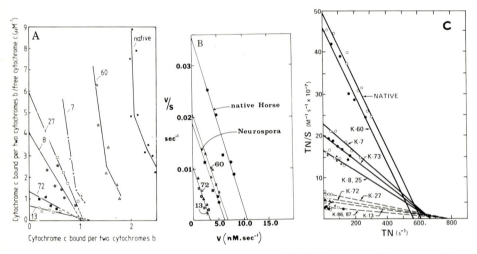

Figure III. Binding and kinetics of native and CDNP-horse cytochromes $\underline{c}$ with cytochrome $\underline{c}$ reductase. A) Binding with purified Neurospora reductase. Conditions as described for Figure IA. B) Kinetics with purified Neurospora reductase, measured by the spectral assay system of Speck et al.[12,13], in 50 mM Tris-acetate, 0.25% Tween, 0.5 mM EDTA, 0.01% BSA, pH 7.5. C) Kinetics with purified beef reductase (Complex III) measured by the spectral assay of Speck et al.[12,13], in 100 mM Tris-acetate, 0.5 mM EDTA, 0.01% BSA. For details see Speck et al.[13].

substantiated by the fact that derivatization of cytochrome $\underline{c}$ causes changes in the maximal velocity rather than the K_m with the Neurospora reductase, even when the binding has been lowered substantially such that no equilibration artifact can exist (Figure IIIA and B). Nevertheless, the relative effects of derivatization at different positions on cytochrome $\underline{c}$ are consistent with the binding behavior, in that CDNP-lysyl-13 and -72 cytochromes $\underline{c}$ are the most inhibited[11]. A complete analysis of the kinetics of the reconstituted Neurospora reductase has not yet been done, since the enzyme was somewhat unstable under the conditions used.

With the beef reductase, (Figure IIIC) a better correlation between K_m and K_D values is found[10] at higher ionic strengths, suggesting that under some conditions the K_m values measured by this spectral assay procedure may become strongly reflective of the true K_D.

It can be concluded from the differential inhibitory effects of the various CDNP-lysyl cytochromes c on both binding and kinetics with the Neurospora[11] and beef[13] enzymes that the interaction domain on cytochrome c for cytochrome c reductase is similar to that defined for cytochrome c oxidase[8].

Ionic Strength Effects. Increasing ionic strength decreases the binding and activity of cytochrome c with Neurospora and beef reductases in a manner similar to the effects of derivatization[11,12,13] (data not shown), demonstrating the generality of the influence of the binding behavior on the kinetic behavior.

DISCUSSION

The kinetics of reaction of cytochrome c with purified Neurospora cytochrome oxidase, measured by the ascorbate-TMPD-oxygen uptake assay procedure, yield apparent Michaelis constants that are in good quantitative agreement with the apparent dissociation constants measured by gel filtration. Both the kinetics and the binding show a biphasic pattern similar to that found with membrane-bound and purified beef oxidase[5,8,6,25]. The results are consistent with the interpretation that TMPD is acting preferentially as a reductant of cytochrome c bound to the oxidase and thereby giving a measure of the amount of complex present[5,8].

It was postulated[8] that this high affinity complex was not usually observed in the spectral assay system[1] because cytochrome c in the concentration range required to define this phase (0.01 to 1 µM) was not readily measurable by conventional spectrophotometry. In fact, when the spectral kinetics are determined at the low concentrations of cytochrome c and the low ionic strength conditions used in the TMPD assay, a high affinity phase is observed with an apparent $K_m = 6 \times 10^{-8}$ M (Ferguson-Miller - unpublished observations) consistent with the kinetic and binding parameters reported above.

The correspondence between the kinetic and binding data demonstrates that electron transfer between cytochrome c and cytochrome oxidase is controlled by rates of association and dissociation of cytochrome c. Furthermore, the same kinetic and binding parameters are observed for the cytochrome c-dependent restoration of electron transfer to the complete mitochondrial respiratory chain[5,26-28]. Therefore, it can be concluded that the tightly binding form of cytochrome c is a competent link in the electron transfer chain. A further

corollary of the steady-state kinetic and binding data is that there is more
than one way for cytochrome $\underline{c}$ to transfer electrons to oxidase. Which pathway
is operative under particular experimental and physiological conditions becomes
an important question.

The scheme depicted in Figure IV is an attempt to clarify the various
interactions between cytochrome $\underline{c}$ and cytochrome oxidase and how they are
related.

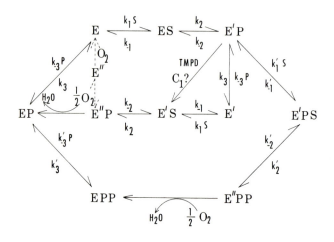

FIGURE IV. Kinetic scheme for the reactions of cytochrome $\underline{c}$ with cytochrome
oxidase. S=reduced cytochrome $\underline{c}$, P=oxidized cytochrome $\underline{c}$, E=fully oxidized
oxidase, E'=one-quarter reduced oxidase (one electron added), E"=one-half
reduced oxidase (two electrons added). Dotted lines indicate a possible
alternative pathway that requires cytochrome $\underline{c}$ to dissociate before electrons
can be transferred to oxygen.

After the initial interaction and electron transfer event has occurred, there
are several possible routes for donation of a second electron. In the presence
of a reducing agent such as TMPD, the product, ferricytochrome $\underline{c}$, is able to
accept an electron without dissociating from the oxidase[5,8]. The
possibility is also suggested that this pathway could occur physiologically if
cytochrome c_1 were able to perform the same function as TMPD. However, in the
absence of a reducing agent with access to the complex, there are two other
possibilities: the product of the first interaction may dissociate and allow a
second substrate molecule to occupy the same site, or the product may remain
bound while a second cytochrome $\underline{c}$ binds simultaneously and donates its electron
to oxidase. That this latter alternative can and does occur is demonstrated by
steady-state kinetic analysis[5,8,6,25]. The second addition appears to be

dependent on the first[6,25], and negative cooperativity and a different orientation of the second cytochrome $\underline{c}$ molecule has been suggested[5,8].

The major factors dictating the pathway that will be followed are the concentration of substrate, the rate constants k'_1 and k_3, and the differential influence of pH and ionic strength on these rate constants. Estimates of k'_1 (10^7 to 10^8 $M^{-1}s^{-1}$) and k_3 (10 to 50 s^{-1}) can be made from the combined data of the steady-state kinetics[5,6,8,25], pre-steady state kinetics[29], and binding measurements[11,26-28]. Assuming values within this range, the influence of cytochrome $\underline{c}$ concentration on the pathway of electron transfer can be illustrated. If $k_3 = 20s^{-1}$, $k_1 = 5 \times 10^7 M^{-1}s^{-1}$, and cytochrome $\underline{c}$ = 10 μM, the predominant pathway will be a second addition of cytochrome $\underline{c}$ with a first order rate of 500 s^{-1}, since the first molecule will turnover at only $20s^{-1}$. Conversely, at 0.1 μM cytochrome $\underline{c}$ the fastest turnover will be at the first site, since the first order rate of association of the second molecule will be only 5 s^{-1}.

These considerations may be important in the interpretation of pre-steady state kinetics where the rapid addition of one or two electrons is followed by a slow step[30-33]. One interpretation originally suggested by Gibson et al.[28] was that dissociation of product was rate limiting. It is apparent that under some conditions the dissociation of the first cytochrome $\underline{c}$ may be an obligatory and slow step, but under others, turnover of cytochrome $\underline{c}$ at the second site may be able to deliver electrons rapidly and bypass the slow dissociation. Whether presence of one or two molecules of cytochrome $\underline{c}$ bound to the oxidase has any influence on rates of electron transfer within the enzyme is an interesting possibility for control that has not been thoroughly investigated.

The questions originally posed in this paper concerned whether cytochrome $\underline{c}$ formed a stable complex as an intermediate in electron transfer, and if so, whether it could do so simultaneously with its two redox partners or must alternate between them. The results obtained with the oxidase demonstrate the existence and competency of a complex with cytochrome $\underline{c}$. However, with the purified reductase the binding and kinetic parameters were not consistent under all conditions, suggesting that the rate controlling step is not always the association or dissociation of cytochrome $\underline{c}$. Nevertheless, the differential inhibitory effects of the cytochrome $\underline{c}$ derivatives allow the definition of the preferred interaction domain on cytochrome $\underline{c}$ for the reductase, and show it to be essentially the same as for the oxidase. These data imply that simultaneous interaction of oxidase and reductase with cytochrome c is unlikely. However, the requirements for the alternative possibility, that cytochrome c functions as a mobile carrier, also do not appear to be satisfied, since the estimated dissociation rates for cytochrome c from the oxidase under a variety of conditions (10 to $50s^{-1}$) are somewhat lower than physiological rates of electron transfer (70 to 100 s^{-1})[35,36].

This apparent conflict can be resolved by suggesting that cytochrome c does not truly dissociate from the oxidase, but only rotates[10] slightly within a binding site formed by a close association of cytochrome c_1 and oxidase[5,8,34]. An analysis of the kinetics and binding of cytochrome c with the complete mitochondrial membrane, and of the kinetic effects of cross-linking cytochrome c to its redox partners, should provide conclusive evidence to resolve the question of the mechanics of cytochrome c function.

REFERENCES

1. Smith, L. and Conrad H. (1956) Arch. Biochem. Biophys. 63, 403-413.

2. Orii, Y., Sekuzu, I. and Okunuki, K. (1962) J. Biochem. (Tokyo) 51, 204.

3. Yonetani, T. and Ray, G.S. (1965) J. Biol. Chem. 240, 3392.

4. Nicholls, P. (1974) Biochim. Biophys. Acta 346, 261.

5. Ferguson-Miller, S., Brautigan, D.L. and Margoliash, E. (1976) J. Biol. Chem. 251, 1104-1115.

6. Errede, B., Haight, G.P., and Kamen, H. (1976) Proc. Natl. Acad. Sci. USA 73, 113-117.

7. Chance, B., Saronio, C., Waring, A. and Leigh, J.S. (1978) Biochem. Biophys. Acta. 503, 37-55.

8. Ferguson-Miller, S. Brautigan, D.L. and Margoliash, E. (1978) J. Biol. Chem. 253, 149-159.

9. Roberts, H. and Hess, B. (1977) Biochim. Biophys. Acta 462, 215-234.

10. Chance, B. (1967) Biochem. J. 103, 1.

11. Weiss, H., Ferguson-Miller, S., Speck, S., Brautigan, D.L., Osheroff, N. and Margoliash, E. in preparation.

12. Speck, S., Ferguson-Miller, S., Brautigan, D.L., Osheroff, N., and Margoliash, E. (1978) Fed. Proc. 37 Abstract.

13. Speck, S., Ferguson-Miller, S., Osheroff, N. and Margoliash, E. (1978) Proc. Nat. Acad. Sci. USA (in press).

14. Brautigan, D.L., Ferguson-Miller, S. and Margoliash, E. (1978) J. Biol. Chem. 253, 130-139.

15. Brautigan, D.L., Ferguson-Miller, S., Tarr, G.E. and Margoliash, E. (1978) J. Biol. Chem. 253, 140-148.

16. Weiss, H. and Sebald, W., Meth. Enzymol. (in press).

17. Weiss, H. and Ziganke, B., Meth. Enzymol. (in press).

18. Weiss, H. and Juchs, B. (1978) Eur. J. Biochem. 88, 17-28.

19. Reiske, J.S., Meth. Enzymol. 10, 239-245.

20. Brautigan, D.L., Ferguson-Miller, S. and Margoliash, E., Meth. Enzymol. (in press).

21. Hummel, J.P. and Dryer, W.J. (1962) Biochim. Biophys. Acta 63, 530-532.

22. Osheroff, N., Brautigan, D.L. and Margoliash, E., in preparation.

23. Dixon, H.B.F. (1976) Biochem. J. 159, 161-162.

24. Dethmers, J., Ferguson-Miller, S. and Margoliash, E., in preparation.

25. Errede, B. and Kamen, H. (1978) Biochemistry 17, 1015-1027.

26. Vanderkooi, J., Erecinska, M. and Chance, B. (1973) Arch. Biochem. Biophys. 157, 531-54.

27. Cooper, S.C., Lambek, J., and Erecinska, M. (1975) FEBS Letters 59, 241-244.

28. Erecinska, M. (1975) Arch. Biochem. Biophys. 169, 199-208.

29. Van Gelder, B.F., Van Buuren, K.J.H., Wilms, J., and Verboom, C.M. (1975) in "Electron Transfer Chains and Oxidative Phosphorylation" (Quagliariello, E. et al. Eds.) pp. 63-68. North Holland Publishing Co., Amsterdam.

30. Antonini, E., Brunori, M., Greenwood, C., and Malmstrom, B.G. (1970) Nature Lond. 228, 936-937.

31. Antonini, E., Brunori, M., Colosimo, A., Greenwood, C. and Wilson, M.T. (1977) Proc. Natl. Acad. Sci. USA, 74, 3128-3132.

32. Andreasson, L.-E., Malmstrom, B.G., Stromberg, C. and Vanngard, T. (1972) FEBS Letters 28, 297.

33. Van Buuren, K.J.H., Van Gelder, B.F., Wilting, J. and Braams, R. (1974) Biochim. Biophys. Acta 333, 421.

34. Orii, Y. and King, T.E. (1978) in "Frontiers of Biological Energetics: from Electrons to Tissues" (Eds. B. Chance, P.L. Dutton, J.S. Leigh, and A. Scarpa) in press.

35. Wilson, D.F., Owen, C.S. and Holian, A. (1977) Arch. Biochem. Biophys.
    182, 749-762.

36. Nicholls, P. and Kimelberg, H.K. (1972) in "Biochemistry and Biophysics of
    Mitochondrial Membranes" (Eds. Azzone, G.F., Carafoli, E., Lehninger, A.
    Quagliariello, E. and Siliprandi, N.) pp. 17-32, Academic Press, New
    York.

Cytochrome Oxidase, T.E. King et al. eds.
© 1979 Elsevier/North-Holland Biomedical Press

KINETICS OF REACTION OF CYTOCHROME C WITH CYTOCHROME C OXIDASE

LUCILE SMITH, HELEN C. DAVIES AND MARÍA ELENA NAVA

Department of Biochemistry, Dartmouth Medical School, Hanover, New Hampshire, 03755 and the Department of Microbiology, University of Pennsylvania Medical School, Philadelphia, Pennsylvania, 19104, U.S.A.

ABSTRACT

The spectrophotometric and polarographic methods for measuring the kinetics of oxidation of cytochrome c by cytochrome c oxidase show very different responses to changes of experimental conditions. Under some conditions the O_2 uptake rates measured by the polarographic method in the presence of TMPD plus ascorbate are the same as the rates calculated from the spectrophotometric measurements of the oxidation of cytochrome c; under other conditions the measured O_2 uptake is much greater. "Apparent K_M" values calculated from plots of v/S against v derived from the two methods are different. In Tris-cacodylate buffer, pH 7.8, the rates of O_2 uptake in the presence of TMPD plus ascorbate with increasing concentrations of cytochrome c correlate with simultaneously measured concentrations of cytochrome c which remain *oxidized* in the aerobic state. The data suggest that some conditions promote the formation of a cytochrome c-cytochrome oxidase complex favorable to increased electron transfer between the two.

INTRODUCTION

Recently a number of studies of the kinetics of oxidation of cytochrome c by cytochrome c oxidase have been made using different species or derivatives of cytochrome c as a means of gaining insight into its mechanism of action[1,2,3,4,5]. Two different kinds of methodology have been utilized, yielding somewhat different conclusions.

Spectrophotometric measurements of the oxidation of pure cytochrome c by O_2 catalysed by cytochrome c oxidase (on mitochondrial membrane fragments or purified by the use of detergents) showed the reaction to be first order in ferrocytochrome c with concentrations of cytochrome c from 0.04 to 100 μM in buffers such as 0.05 M Tris-acetate, Tris-maleate or phosphate or 0.1 M MES or phosphate at pH values between 6 and 7.4[3,6,7]. However, the first order rate

ABBREVIATIONS: SMP, submitochondrial particles; DOC, sodium deoxycholate; TMPD, N,N,N',N' tetramethyl-p-phenylenediamine; MES, 2, (N-morpholino) ethanesulfonic acid

constant decreased with increasing concentration of cytochrome c in the reaction mixture, but not in a linear fashion. These unusual kinetics have been interpreted as reflecting rates of association and dissociation of cyto-chrome c and the oxidase[3,8,9]. Most recently Errede and Kamen[3], with their extended data, could explain the kinetics by assuming the binding of two molecules of cytochrome c with different affinities for the oxidase.

The oxidation of cytochrome c by the oxidase is also assayed by measuring O_2 uptake in the presence of a reductant of cytochrome c. Ascorbate has often been utilized, and the rate of O_2 uptake divided by 4 found to equal the rate of cytochrome c oxidation measured spectrophotometrically when the cytochrome c was nearly completely reduced during the reaction[10]; quite high concentra-tions of ascorbate are required. Ferguson-Miller et $al.$[1] employed a combina-tion of TMPD plus a lower concentration of ascorbate and reported that, with a concentration of TMPD around 0.7 mM, the cytochrome c remained more than 98% reduced under their experimental conditions. (TMPD can reduce the endogenous cytochrome c of mitochondrial membranes, while ascorbate does not[11]). Eadie-Hofstee-type plots (v/S against v) of O_2 uptake with increasing concen-trations of added cytochrome c gave two linear parts. Calculations of "apparent K_M" from these two lines were interpreted by Ferguson-Miller et $al.$[1,4] as evidence of binding constants of two molecules of cytochrome c to the oxidase, one with "high affinity", the other with "low affinity". These binding constants were different from those calculated by Errede and Kamen[3], but the experimental conditions were quite different in the experiments from the two laboratories.

We have compared the O_2 uptake rates of the oxidase with cytochrome c in the presence of TMPD plus ascorbate with the spectrophotometrically measured rates of oxidation of cytochrome c with cytochrome c concentrations between 0.05 and 5 μM, sometimes run simultaneously. Under some conditions the corresponding rates derived from the two methods were similar. Under other conditions the measured O_2 uptake rates were as much as 30-fold greater than the rates calculated from the observed rate of oxidation of cytochrome c. Changes of pH or of ionic strength or even of buffer type had very different effects on the rates measured by the two methods.

We would like to report some of our experience[12] of the last several years in working with the two methods and to discuss the implication such observa-tions may have for the mechanism of the reaction of cytochrome c with cytochrome oxidase.

MATERIALS AND METHODS

Preparations. Most of the experiments reported here were made with SMP prepared from beef heart mitochondria by the method of Lee and Ernster[13]. Usually the cytochrome c was first removed from the frozen and thawed mitochondria by stirring them with cold 0.075 M phosphate buffer, pH 7.5, using 1 ml of buffer for each 30 mg of protein, then collecting the washed mitochondria by centrifugation. The suspension was acidified to pH 5 with acetic acid, stirred for 15 minutes, centrifuged, and the pellet suspended in cold 0.5 M sucrose containing 0.01 M phosphate buffer, pH 7.0. The pH was brought to near 7 by addition of 0.2 M Na_2HPO_4, then the treated mitochondria were collected by centrifugation and suspended in cold 0.25 M sucrose containing 0.02 M phosphate buffer, pH 7.4. The SMP were stored at -20°.

When the SMP were treated with detergent, 10% sodium deoxycholate (DOC) or Triton X-100 was added to a concentrated suspension (around 14 mg protein per ml) to make the mixture 1 mg detergent per ml protein, then the mixture diluted the desired extent with cold distilled water[14].

The content of cytochrome aa_3 in the oxidase preparations was measured following the method of Vanneste[15] and that of cytochrome c by the method of Williams[16].

Cytochrome c oxidase was purified by the method of Hartzell and Beinert[17]. Cytochrome c was isolated from beef heart according to Margoliash and Walasek[18], and further purified by isoelectric focusing. For use in the spectrophotometric oxidase assays, the cytochrome c was reduced with $NaBH_4$[19]. The concentration of cytochrome c in solution was calculated from the absorbance at 550 nm of the reduced pigment, using 27.6 as the millimolar extinction coefficient[20].

Chemicals. Sodium ascorbate, obtained from Sigma Chemical Co., was recrystallized from hot water. TMPD was also obtained from Sigma.

Assay Methods. Cytochrome c oxidase was assayed polarographically with a Clark O_2 electrode at 26° in a water-jacketed chamber containing 2.6 ml[10]; rates are expressed as μMolar O_2 uptake per second. The contents of the reaction chamber are given with the figures and table. Oxidase activity was measured spectrophotometrically with the Aminco DW2a Dual Wavelength Spectrophotometer at 418 or 550 nm in the split beam mode following the method of Smith and Conrad[6]; first order rate constants were calculated[6]. Velocities of reaction were calculated from the product of the rate constants and the concentrations of ferrocytochrome c. These were divided by 4 to obtain the equivalent rates of O_2 uptake.

Steady state reduction of cytochrome c in the presence of the oxidase preparation, TMPD, ascorbate and O_2 was measured with the Aminco DW2a Spectrophotometer in the dual wavelength mode at 550 minus 535 nm. First the absorbance of oxidase plus cytochrome c was recorded (cytochrome c oxidized), then the increased absorbance on addition of TMPD plus ascorbate (aerobic steady state), then the further increase in absorbance (cytochrome c reduced) when the aerobic suspension became anaerobic as a result of the O_2 uptake. The experimental conditions were the same as those used in the corresponding O_2 uptake measurements. Any change in absorbance due to endogenous pigments of the oxidase preparation as the mixture became anaerobic was assessed by adding 0.75 mM KCN to a similar aerobic suspension. Such changes proved to be immeasurably small in most of the experiments.

RESULTS

Spectrophotometric Method. As previously observed in many laboratories, the oxidation of ferrocytochrome c followed a first order course throughout in different buffers at pH values between 6 and 7.4. We also found that the reaction was first order in ferrocytochrome c in Tris-cacodylate buffer at pH 7.8, but the rates were lower at this pH. The rate constants decreased with increase in the total concentration of cytochrome c between 0.05 and 5 µM, as at lower pH values. Plots of v/S against v derived from such data show two linear parts, as seen in Figure 1, from which "apparent K_M" and V_{max} can be calculated. These values are found in Table 1.

Polarographic Method. We found a number of difficulties in working with the polarographic method, which had to be eliminated or corrected for. For example, cytochrome c adsorbs to the electrode membrane and must be removed by a wash with buffer (we used 50 mM phosphate, pH 7) between each assay. Also some samples of cytochrome c and of TMPD and all of the samples of sucrose we tested contained substances which gave rise to KCN-insensitive O_2 uptake in the presence of ascorbate. Since we found that the inclusion of sucrose made no difference in the rates obtained in the buffers used, we omitted it from our assays. We also found that the use of syringes with metal tips and plungers in introducing ascorbate into the O_2 electrode chamber resulted in erratic effects on the O_2 uptake rates. With the purified cytochrome aa_3 preparation, the measured rates depended upon the extent of dilution of the preparation (up to about 20-fold) with 1% Tween 20 in 0.01 M phosphate buffer, pH 7, *before* addition to the reaction mixture. Such dilution had only small effects on rates obtained with SMP preparations. However, with either diluted or

TABLE 1

"APPARENT K_M" AND V_{MAX} VALUES DERIVED FROM POLAROGRAPHIC AND SPECTROPHOTOMETRIC ASSAYS

All assays were run in 25 mM Tris-cacodylate buffer (25 mM in cacodylate), pH 7.8. In the polarographic assays TMPD was 0.75 mM and ascorbate 10 mM. The "apparent K_M" and V_{max} values were derived from plots of v/S against v, as described in Methods.

	SMP treated with DOC		Purified cytochrome aa_3	
	polaro.	spectrophot.	polaro.	spectrophot.
"High affinity"				
apparent K_M, μM	0.055	0.133	0.036	0.22
turnover rate[a]	78.2	5.1	17.5	2.3
"Low affinity"				
apparent K_M, μM	1.65	4.17	0.314	3.85
turnover rate[a]	293	59.3	33.6	13.1

[a] μM cytochrome c sec.$^{-1}$ at V_{max} ÷ concentration of cytochrome aa_3

undiluted purified preparations or with any form of SMP a linear relationship was observed between O_2 uptake rates and quantity of preparation.

Measured rates of O_2 uptake increased sharply with increasing concentration of added cytochrome c up to 0.1 to 0.25 μM, then increased more gradually with further increase of cytochrome c concentration; the break in the curve at this low concentration of cytochrome c was seen with different kinds of preparations, irrespective of the content of cytochrome aa_3. Plots of the data in the form of v/S against v (Figure 2) gave smooth continuous curves with most preparations, without the sharp breaks observed by Ferguson-Miller et al.[1], although some plots of data derived from SMP not treated with detergent are nearly like theirs. In spite of the curvature, we calculated "apparent K_M" and V_{max} values from the initial and final parts of the traces as measures of "high affinity" and "low affinity" binding[1]. Because of the curvature the values are inaccurate. Some of these are listed in Table 1, along with the values obtained from the spectrophotometric assays, where similar plots showed two sharp lines. The values obtained with data from the two methods show considerable differences with any kind of preparation.

298

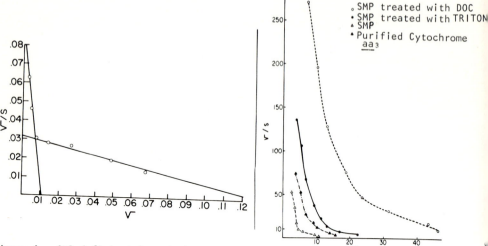

Figure 1. SMP deficient in cytochrome
c and treated with DOC containing
0.0026 mg protein and 0.0022 µM cytochrome aa₃ was used in each assay in
25 mM Tris-cacodylate buffer, pH 7.8

Figure 2. All SMP were deficient in
cytochrome c. △ Untreated SMP, 0.056
mg protein; O SMP treated with DOC,
0.028 mg protein; ● SMP treated with
Triton X-100, 0.056 mg protein; ▲
purified cytochrome aa₃. All assays
were run in 25 mM Tris-cacodylate
buffer, pH 7.8. v is O_2 sec.$^{-1}$ ÷
concentration of cytochrome aa₃.

Comparison of Polarographic and Spectrophotometric Methods. Figure 3 plots
results obtained from simultaneous assay of a cytochrome c-deficient SMP
treated with DOC by the two methods with increasing concentrations of cytochrome c in Tris-cacodylate buffer, pH 7.8. The only difference was the
presence of TMPD plus ascorbate in the polarographic method and the use of 5
times more preparation (but of the same initial dilution) in the polarographic
assays. The rates measured with the O_2 electrode were 27-fold greater at low
concentrations of cytochrome c than the corresponding rates calculated from
the spectrophotometric assays. At other pH values or in other kinds of buffer
there was a difference in the response of the two methods. For example at pH
6.0 in 25 mM phosphate buffer the measured and calculated rates were very
similar, particularly at low concentrations of cytochrome c (Figure 4). In
Tris-cacodylate buffer, pH 6.0, the rates were in agreement at 0.05 and 0.1 µM
cytochrome c, then at higher concentrations of cytochrome c, the measured
rates were greater (data not shown). The measured rates were considerably
greater than the calculated rates at higher concentrations of cytochrome c in

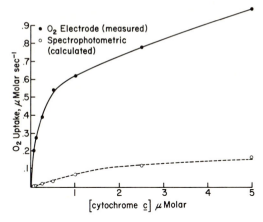

Figure 3. ● Measured O_2 uptake rates in polarographic assays with SMP (cytochrome c deficient) treated with DOC, with 0.022 μM cytochrome aa_3. ○ O_2 uptake rates corresponding to spectrophotometrically measured rates calculated for the same concentration of cytochrome aa_3. All assays were made in 25 mM Tris-cacodylate buffer, pH 7.8.

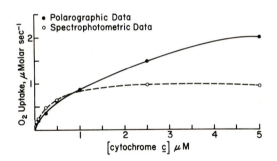

Figure 4. Assays were run in 25 mM phosphate buffer, pH 6.1, with SMP (cytochrome c-deficient) treated with DOC containing 0.045 μM cytochrome aa_3 for the polarographic measurements and 0.00045 and 0.0009 μM for the spectrophotometric measurements. All data were calculated for O_2 uptake with 0.045 μM aa_3. TMPD was 0.75 mM and ascorbate 10 mM in the polarographic experiments.

25 mM borate buffer, pH 7.8, but the difference was not as great at low concentrations of cytochrome c (0.05 - 0.1 μM) (data not shown).

Variations of pH or of ionic strength had quite different effects on the rates measured by the two methods. The pH optimum for the spectrophotometric method was around 6 at all concentrations of cytochrome c tested, while that for the polarographic method was 7.8 at low concentrations of cytochrome c and 7.3 with higher concentrations (5 μM). The O_2 uptake rates were relatively insensitive to changes of ionic strength, addition of 0.025 M NaCl to assays in 0.025 M Tris-cacodylate having no or a slightly inhibitory effect, while a similar addition stimulated the spectrophotometrically measured rates.

Steady-State Reduction of Cytochrome c in the Presence of TMPD, Ascorbate and O_2. The extent of reduction of added cytochrome c in the presence of TMPD, ascorbate and O_2 under the conditions of the polarographic assays was measured in the Aminco DW2a Spectrophotometer in the dual wavelength mode at 550 minus 535 nm. Since very small differences in absorbance were measured between the aerobic and the anaerobic states, the data are not very accurate, but are quite reproducible. Typical data are plotted in Figure 5 along with simultaneous measurements of O_2 uptake with identical mixtures. The rates of O_2 uptake follow the same pattern as the concentrations of cytochrome c which remain *oxidized* during the aerobic state, particularly at the lower concentrations of cytochrome c.

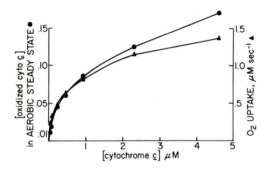

Figure 5. ▲ O_2 uptake rates measured polarographically with SMP (deficient in cytochrome c) treated with DOC containing 0.0226 µM cytochrome aa$_3$. ● Concentrations of oxidized cytochrome c in the aerobic steady state under the conditions of the polarographic assays. All measurements were made in 25 mM Tris-cacodylate buffer, pH 7.8.

Comparison of O_2 uptake with Endogenous and Exogenous Cytochrome c. O_2 uptake rates with SMP which were not made deficient in cytochrome c were measured in the presence of TMPD plus ascorbate with no added cytochrome c (endogenous cytochrome c only) and compared with the rates which followed on addition of exogenous cytochrome c. Figure 6 shows that the rates with endogenous cytochrome c and with added 0.1 µM exogenous cytochrome c (total minus endogenous) showed similar variations with changing pH in 25 mM Tris-cacodylate buffers (25 mM in cacodylate). In this experiment the concentration of endogenous cytochrome c was 0.043 µM. Similar experiments with several different preparations in different levels of endogenous cytochrome c (0.027 to 0.043)

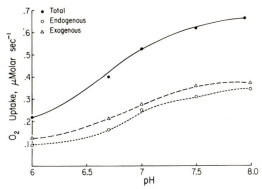

Figure 6. 0 O_2 uptake with TMPD plus ascorbate and SMP (not deficient in cytochrome c) containing 0.114 mg protein per ml and no added cytochrome c. ● O_2
uptake rates with addition of 0.1 µM cytochrome c. Δ Total minus endogenous.
Assays were run in 25 mM Tris-cacodylate buffers of varying pH.

all showed rates about the same as the endogenous rate in the presence of 0.1
µM exogenous cytochrome c.

DISCUSSION

The data presented show new aspects of the interaction of cytochrome c with
cytochrome c oxidase.

We found the oxidation of ferrocytochrome c, measured spectrophotometrically, to be first order in ferrocytochrome c in all buffers tested up to pH
7.8. Yonetani reported "slight" deviations from first order kinetics with
some, but not all, kinds of oxidase preparations at pH 7.4 and 8.0, but the
deviations only appear to occur at concentrations of cytochrome c above about
10 µM[9]. Thus our data are not in accord with the conclusion of Ferguson-Miller
et al.[21] that first order kinetics are seen only under unphysiological
conditions.

Plots of v/S against v with the data derived from the spectrophotometric
assays show two linear parts from which "apparent K_M" and V_{max} values can
easily be calculated according to Eadie-Hofstee (see below). In Tris-
cacodylate buffer, pH 7.8, the two linear parts broke sharply at the rate
obtained with 0.25 µM cytochrome c.

We encountered a number of difficulties in working with the polarographic
method, as described in the Methods section. Even when these were eliminated
or corrected for, plots of v/S against v gave continuous curves, except with
some SMP preparations not treated with detergent. Thus our estimation of

"apparent K_M" and V_{max} values from the initial and final slopes of the curves are relatively inaccurate. Even so, the data show clearly that these values are different when derived from the polarographic as compared with the spectrophotometric assays. As we have previously pointed out[2], in the spectrophotometric assays the cytochrome c must contact the oxidase, be oxidized, then dissociate so that another molecule can react. There is general agreement that binding of cytochrome c to and dissociation from the oxidase affects the kinetics seen in the spectrophotometric method[3,8,9]. In phosphate buffer, pH 6, the measured O_2 uptake rates and those calculated from the spectrophotometrically measured rates of oxidation of cytochrome c agree fairly well, so that the dissociation constant is being measured in both methods. Under other conditions, the "apparent K_M" values derived from the two methods are different, both at "low" and "high" concentrations of cytochrome c; thus the values derived from the O_2 uptake rates are not measures of the K_D. Also different conditions of pH and of buffer type affect the values derived from the two methods to different extents. The different effects of ionic strength on rates obtained with the two methods give additional evidence that different rate-limiting reactions are being measured for some experimental conditions.

The extra O_2 uptake observed under some conditions over that equivalent to the observed oxidation of soluble cytochrome c can be large. The simplest explanation would seem to be the formation of a combination between cytochrome c and cytochrome c oxidase which is reduced repeatedly by TMPD before it dissociates. Ferguson-Miller et al. have expressed a similar conclusion[4], although they did not measure rates of oxidation of cytochrome c at concentrations between 0.05 and 0.7 μM, that is, at the concentrations associated with the "high affinity" site evidenced in the O_2 uptake measurements.

In spite of the fact that the spectrophotometric and the polarographic assays can be measuring different rate-limiting reactions, both show the greatest activity as a function of cytochrome c in the presence of concentrations between 0.05 and 0.25 μM and show relatively less increase with increasing concentrations above this level. This effect was similar for all oxidase preparations, irrespective of the content of cytochrome aa_3; it depended only on the concentration of cytochrome c. In addition, this is the concentration of added cytochrome c which gave rates of O_2 uptake similar to those obtained with the endogenous cytochrome c of SMP. Our data are suggestive of a conformational form influenced by the concentration of cytochrome c.

We feel that our data are evidence for the formation of an especially reactive complex between cytochrome c and cytochrome c oxidase under some experimental conditions, which promotes increased interaction between the components, or the capacity for more rapid reduction or oxidation of the proper components. The availability of cytochrome c to form such a complex varies with the state of the oxidase preparation and the experimental conditions. The most favorable condition in our experiments was in 25 mM Tris-cacodylate buffer, pH 7.8, with SMP treated with DOC. Optimal activity of the O_2 uptake with endogenous cytochrome c of SMP was also seen at pH 7.8.

Observations of the state of the added cytochrome c during the aerobic steady state in the presence of TMPD plus ascorbate under the conditions of the polarographic assays showed a correlation between the simultaneously measured rates of O_2 uptake and the concentration of cytochrome c which remained *oxidized*. This also points to a form of cytochrome c which is very rapidly oxidized and remains oxidized until the O_2 in solution is exhausted and gives further evidence for the turnover of a highly reactive complex between cytochrome c and cytochrome c oxidase.

The turnover rates of the oxidase at V_{max} of the 'high affinity" reaction of SMP treated with DOC using the polarographic method in 25 mM Tris-cacodylate, pH 7.8, are the largest measured under the conditions tested. With a number of different SMP preparations treated with DOC, where the oxidase appears to be maximally exposed for reaction with cytochrome c[14], these rates were never larger than 100 sec.$^{-1}$ at 26°. This is barely rapid enough to account for oxidase rates observed with intact mitochondria[22], so perhaps we have not yet found the optimal conditions for the formation of the maximally-reactive cytochrome c-cytochrome c oxidase complex. The maximal rates obtained with the spectrophotometric method under the same conditions are lower, as expected if a complex formed must dissociate before another molecule of cytochrome c can be oxidized. This seems to be evidence, in agreement with some from other experimental approaches[23,24], that the cytochrome c does not dissociate from the membrane during normal electron transport. The turnover rates with SMP treated with DOC at higher concentrations of cytochrome c are as high as 300 sec.$^{-1}$. Thus if there are two sites on the oxidase where it can bind cytochrome c, electrons could apparently be introduced in tandem.

All turnover rates with the purified preparation are low compared to those seen with SMP treated with DOC.

ACKNOWLEDGEMENTS

This research was supported by a research grant from the U.S.P.H.S., No. GM06270. The assistance of George McLain in purifying cytochrome c is greatly appreciated.

REFERENCES

1. Ferguson-Miller, S., Brautigan, D. and Margoliash, E. (1976) J. Biol. Chem., 251, 1104-115.
2. Smith, L., Davies, H.C. and Nava, M.E. (1976) Biochemistry, 15, 5827-5831.
3. Errede, B. and Kamen, M.D. (1978) Biochemistry, 17, 1015-1027.
4. Ferguson-Miller, S., Brautigan, D. and Margoliash, E. (1978) J. Biol. Chem., 253, 149-159.
5. Staudenmayer, N., Smith, M.B., Smith, H.T., Spies, F.K. and Millett, F. (1976) Biochemistry, 15, 3198-3205.
6. Smith, L. and Conrad, H. (1956) Arch. Biochem. Biophys., 63, 403-413.
7. Smith, L., Nava, M.E. and Margoliash, E. (1973) in Oxidases and Related Redox Systems, King, T.E., Mason, H.S. and Morrison, M. eds., University Park Press, Baltimore, Vol. 2, pp. 629-638.
8. Minnaert, K. (1961) Biochim. Biophys. Acta, 50, 23-34.
9. Yonetani, T. and Ray, G.S. (1965) J. Biol. Chem., 240, 3392-3398.
10. Smith, L. and Camerino, P.W. (1963) Biochemistry, 2, 1428-1432.
11. Mochan, E. and Nicholls, P. (1972) Biochim. Biophys. Acta, 267, 309-319.
12. Smith, L., Davies, H.C. and Nava, M.E. (1978) Fed. Proc., 37, 1326.
13. Lee, C.P. and Ernster, L. (1967) Methods Enzymol., 10, 543-548.
14. Smith, L. and Camerino, P.W. (1963) Biochemistry, 2, 1432-1439.
15. Vanneste, W.H. (1966) Biochim. Biophys. Acta, 113, 175-178.
16. Williams, J.N. (1968) Biochim. Biophys. Acta, 162, 175-181.
17. Hartzell, C.R. and Beinert, H. (1974) Biochim. Biophys. Acta, 368, 318-338.
18. Margoliash, E. and Walasek, O. (1967) Methods Enzymol., 10, 339-348.
19. Smith, L., Davies, H.C. and Nava, M.E. (1974) J. Biol. Chem., 249, 2904-2910.
20. Margoliash, E. and Frohwirt, N. (1959) Biochem. J., 71, 570-572.
21. Ferguson-Miller, S., Brautigan, D.L. and Margoliash, E. (1978) in The Porphyrins, Dolphin, D. ed., (in press).
22. Nicholls, P. and Chance, B. (1974) in Molecular Mechanisms of Oxygen Activation, Hayashi, O. ed., Academic Press, New York, pp. 479-534.
23. Erecinska, M., Vanderkooi, J.M. and Wilson, D. (1975) Arch. Biochem. Biophys., 171, 108-116.
24. Davies, H.C., Pinder, P.B., Nava, M.E. and Smith, L. (1976) Fed. Proc., 35, 1598.

Cytochrome Oxidase, T.E. King et al. eds.
© *1979 Elsevier/North-Holland Biomedical Press*

THE ELECTRON-ACCEPTING SITE OF CYTOCHROME c OXIDASE

B.F. VAN GELDER*, E. VEERMAN**, J. WILMS* AND H.L. DEKKER*

*Laboratory of Biochemistry, B.C.P. Jansen Institute, University of Amsterdam,
Plantage Muidergracht 12, 1018 TV AMSTERDAM (The Netherlands)

**Physical Laboratory, Department of Molecular Biophysics, State University of
Utrecht, Sorbonnelaan 4, 3584 CA UTRECHT (The Netherlands)

SUMMARY

The reduction of heme a by reduced (porphyrin) cytochrome c has
been studied in cytochrome c oxidase and in its complexes with
(porphyrin) cytochrome c using the techniques of stopped flow and
pulse radiolysis. It was found that the value of the 'on' and 'off'
rate constants and the apparent dissociation constant is independent irrespective as to whether cytochrome c or its porphyrin derivative was bound to the high-affinity site on cytochrome c oxidase. These results suggest that the low-affinity site for cytochrome c is the primary electron-accepting site on cytochrome c
oxidase.

No difference in the value of the rate constants (k $2 \cdot 10^7$ $M^{-1} \cdot s^{-1}$ at
5 mM potassium phosphate, pH 7.0 and 21°C) was found when reduced cytochrome c or its porphyrin anion radical was used as substrate for the
reduction of the heme of cytochrome c oxidase, indicating that the electron transfer occurs via a mechanism involving a similar type of 'redox
orbitals' which might be the π^* orbitals of the porphyrin ring.

INTRODUCTION

Cytochrome c oxidase, the terminal component of the respiratory
chain, catalyses the oxidation of cytochrome c by dioxygen. The
enzyme is a lipoprotein complex of Mw 450 000, localized in the
inner membrane of mitochondria. The complex is a dimer[1] containing
at least 7 subunits and per protomer 2 hemes, designated as the
components: cytochrome a and cytochrome a_3, and two protein-bound
copper ions.

It has been demonstrated that cytochrome c oxidase possesses a
reductive side to which cytochrome c delivers its electron, and
an oxidative side where dioxygen and inhibitors of the enzymic
activity such as cyanide, azide, sulfide, CO and NO react with the
enzyme. Studies of the steady-state kinetics of cytochrome c oxidase with ascorbate, TMPD and cytochrome c as reducing substrates

have revealed two binding sites for cytochrome c on cytochrome c oxidase[2,3]. These binding sites are distinguished by a high-affinity site (K_m 50 nM) and a low-affinity site (K_m 1 µM). At low-ionic strength Orii et al.[4] isolated a complex of 1 cytochrome c per protomer of cytochrome c oxidase by column chromatography. This complex rapidly dissociated when brought at high salt concentrations. As shown in this paper, a 1:1 complex of porphyrin cytochrome c with cytochrome c oxidase can be isolated in a similar way. Porphyrin cytochrome c, demonstrated to be a competitive inhibitor of the enzymic activity of cytochrome c oxidase[5], has a much lower redox potential than cytochrome c. Thus it can hardly be reduced by reducing agents such as ascorbate and reduced cytochrome c, and therefore porphyrin cytochrome c can be used as a masking agent of the reductive sites of cytochrome c oxidase.

Hydrated electrons have been used as a reducing agent in studying the kinetics of reduction of the heme of cytochrome c oxidase by cytochrome c. It has been shown that hydrated electrons reduce cytochrome c fast ($k = 4.5 \cdot 10^{10} M^{-1} \cdot s^{-1}$) with a high yield[6] but that the reduction of the hemes of cytochrome c oxidase in the absence of cytochrome c occurs with a very low yield[7]. Since hydrated electrons are a potent reducing agent ($E_0 \simeq -3V$), they can reduce porphyrin cytochrome c, which subsequently transfers electron to cytochrome c, as has been shown by De Kok et al.[8]. The kinetics of the reaction between the porphyrin anion radical of cytochrome c and cytochrome c oxidase indicate that the heme edge of cytochrome c mediates in electron transfer.

The main objective of this study is to investigate via which of the reductive sites electrons are donated to cytochrome c oxidase. Using the isolated complexes of cytochrome c oxidase with native or iron-freed cytochrome c it could be demonstrated that electrons probably enter cytochrome c oxidase via its low-affinity site for cytochrome c.

METHODS

Cytochrome c oxidase was isolated from beef heart according to the methods of Fowler et al.[9] and MacLennan and Tzagoloff[10], as modified by Van Buuren[11]. Cytochrome c was prepared from horse heart as described by Margoliash and Walasek[12]. Porphyrin cytochrome c was obtained as reported by De Kok et al.[8]. Absorbance coefficients used for cytochrome c oxidase (reduced minus oxidized) were 24.0 mM$^{-1} \cdot$cm^{-1} at 605 nm[13], for cytochrome c (reduced minus oxidized) 21.1 mM$^{-1} \cdot$cm^{-1} at 550 nm[14], and for porphyrin cytochrome c 13.0 mM$^{-1} \cdot$cm^{-1} at 504 nm[8].

The complex of cytochrome c oxidase and (porphyrin) cytochrome were isolated essentially according to the method of Orii et al.[4] using ACA-54 in stead of Sephadex G-75. All handling of porphyrin cytochrome c was performed under conditions that minimize exposure to light.

Pulse radiolysis experiments were carried out as described previously[7]. For stopped flow, a Durrum-Gibson apparatus was used.

RESULTS AND DISCUSSION

The rate of reduction of cytochrome c oxidase and of its complex with cytochrome c by reduced cytochrome c has been studied at low-ionic strength (5 mM phosphate) in a stopped-flow apparatus The results are collected in a Guggenheim graph (Fig. 1), where the apparent first-order rate constant is plotted versus the cytochrome c concentration. It is obvious that the value of the second-order rate constant $(1.2 \cdot 10^7 \text{M}^{-1} \cdot \text{s}^{-1})$, that of the first-order 'off' constant (35 s^{-1}) and the value of the apparent dissociation

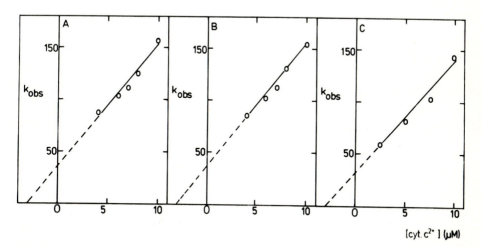

Fig. 1. Reaction of reduced cytochrome c with cytochrome c oxidase and with its complex with (porphyrin) cytochrome c. The first-order rate constant was calculated from absorbance changes at 444 nm, measured in a stopped-flow apparatus. Conditions: 5 mM potassium phosphate (pH 7.0), 1% Tween-20, temp., 10°C. A, cytochrome c oxidase, 1 µM; B, 1:1 complex of cytochrome c - cytochrome c oxidase, 1 µM; C, 1:1 complex of phorphyrin cytochrome c - cytochrome c oxidase, 1 µM.

constant (3 µM, calculated from the intersection with the cytochrome c axis) are the same irrespective as to whether cytochrome
c oxidase was complexed with oxidized cytochrome c or unbound. The
correspondence in kinetic behaviour of cytochrome c oxidase and
its complex with cytochrome c suggests that the electron-accepting
site on cytochrome c oxidase is the low-affinity site for cytochrome c. Support for this suggestion is the value of the kinetically measured apparent K_D of the interaction of both cytochromes
that corresponds with the K_m-value for the low-affinity site in
the catalytic oxidation of cytochrome c mediated by cytochrome c
oxidase[2]. Additional support is the observation that the tight
complex of cytochrome c with cytochrome c oxidase can not be isolated at high-ionic strength, whereas, as has been shown by Van
Gelder et al.[15], the 'off' constant in the reaction of reduced cytochrome c with cytochrome c oxidase decreases at higher ionic
strength without affecting the value of the apparent dissociation
constant.

    In order to test whether the low-affinity site for cytochrome c
on cytochrome c oxidase is the sole site for entering electrons
from cytochrome c, we have also studied the electron transfer between cytochrome c and cytochrome c oxidase complexed with porphyrin cytochrome c. The value of the rate and dissociation constants, calculated from Fig. 1C, is the same as that found for the
reaction with uncomplexed cytochrome c oxidase and with its complex with cytochrome c. Since the half-reduction potential of porphyrin cytochrome c is more than 1 V lower than that of cytochrome
c, it is very unlikely that porphyrin cytochrome c mediates in
rapid electron transfer. Furthermore, the explanation that the
porphyrin cytochrome c-cytochrome c oxidase complex dissociates
rapidly after which cytochrome c reacts with its oxidase, can
be rejected since the reduction of heme a is rapid and the dissociation rate constant must be small as otherwise the complex can
not be isolated. Therefore, it is concluded that the low-affinity
site for cytochrome c on cytochrome c oxidase is the primary
electron-accepting site.

    The kinetics of the reduction of cytochrome c oxidase by cyto-
chrome c have also been studied by pulse radiolysis. By this
technique hydrated electrons, being a potent reducing agent

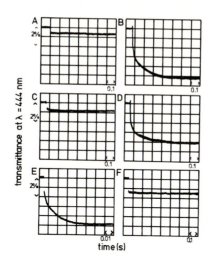

Fig. 2. Transmittance changes at 444 nm after generation of hydrated electrons in the presence of 1:1 complex of (porphyrin) cytochrome *c* - cytochrome *c* oxidase. Conditions: 5 mM potassium phosphate, pH 7.0, 1% Tween-20, 100 mM t-butanol, temp., 21-22°C. Generated hydrated electrons, 2.5 - 3.8 μM. A, Cytochrome *c* - cytochrome *c* oxidase complex, 3.0 μM; B, cytochrome *c*-cytochrome *c* oxidase complex, 3.0 μM, and cytochrome *c*, 6 μM; C, porphyrin cytochrome *c* - cytochrome *c* oxidase complex, 3.8 μM; D, porphyrin cytochrome *c* - cytochrome *c* oxidase complex, 3.8 μM, and cytochrome *c*, 6 μM; E, porphyrin cytochrome *c* - cytochrome *c* oxidase complex, 3.8 μM, and porphyrin cytochrome *c*, 4 μM; F, cytochrome *c* - cytochrome *c* oxidase complex, 2.2 μM, and porphyrin cytochrome *c*, 7.6 μM.

$(E_0 = -2.9 \text{ V})$, can be generated. The advantage of this perturbation method is that it allows a more careful study of the initial phase of the reaction between (porphyrin) cytochrome *c* and cytochrome *c* oxidase.

It has been shown by Van Buuren et al.[7] that the reduction of cytochrome *c* oxidase occurred with a low yield (3%) and that, with excess of cytochrome *c*, cytochrome *c* oxidase was reduced rapidly via cytochrome *c* with a very high yield. Trace A in Fig. 2, where the reduction of the heme of cytochrome *c* oxidase was monitored at 444 nm, shows that the complex of cytochrome *c* and its oxidase is hardly reduced by hydrated electrons. Since no reduction of cytochrome *c* was observed, as monitored at 550 nm (not shown), it is concluded that the tightly bound cytochrome *c* is also inaccessible to hydrated electrons. This inaccessibility of cytochrome *c* can be caused by masking its reductive side when tightly bound to cytochrome *c* oxidase, or by repulsion of the hydrated electron by the complex of cytochromes, which is probably strongly negatively charged.

Trace B in Fig. 2 shows the reduction of heme a of complexed cytochrome c oxidase in the presence of an additional amount of cytochrome c. Apparently, the free cytochrome c can be reduced fast by the hydrated electron that subsequently reduces the complex of cytochrome c and its oxidase.

At a higher resolution of time and at various concentrations of the complex of both cytochromes, the apparent first-order rate constant for the reduction of cytochrome c oxidase was calculated. The rate constants were plotted in a Guggenheim plot (Fig. 3, left)

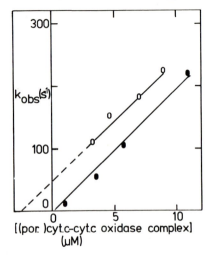

Fig. 3. Determination of rate constants of the reduction of (porphyrin) cytochrome c - cytochrome c complex by (porphyrin) cytochrome c, using pulse radiolysis. The data are obtained from experiments such as shown in Fig. 2. Conditions as in the legend of Fig. 2. O——O, cytochrome c + 1:1 complex of cytochrome c - cytochrome c oxidase; ●——●, porphyrin cytochrome c + 1:1 complex of porphyrin cytochrome c - cytochrome c oxidase.

and from this a k_{on} of $2 \cdot 10^7 M^{-1} \cdot s^{-1}$, a k_{off} of 50 s^{-1} and an apparent K_D of 2.5 μM at 5 mM phosphate, pH 7.0 and 21°C, were calculated. The value of these constants corresponds to that found by stopped flow[15] and are the same as reported by Van Buuren et al.[7], who generated hydrated electrons in a system of cytochrome c oxidase with excess cytochrome c at low-ionic strength.

Since hydrated electrons can reduce rapidly porphyrin cyto-

chrome c to its anion radical[8] with a relatively high yield, it is
of interest for the mechanism of electron transmittance between
cytochrome c and its oxidase whether the anion radical of porphyrin cytochrome c was able to reduce cytochrome c oxidase. Trace C
of Fig. 2 shows that the 1:1 complex of cytochrome c oxidase and
porphyrin cytochrome c, isolated by column chromatography at low-
ionic strengh[4], is hardly reduced by hydrated electrons, whereas
the same complex with excess porphyrin cytochrome c was reduced
rapidly upon generation of hydrated electrons (trace E of Fig. 2).
These results indicate that the anion radical of porphyrin cytochrome c reacts also with the low-affinity site for cytochrome c on
cytochrome c oxidase.

The rate constant for the reduction of the heme of cytochrome
c oxidase by the porphyrin anion radical has been calculated from
Fig. 3. It was found that the heme of cytochrome c oxidase was
reduced with a rate ($2 \cdot 10^7 M^{-1} \cdot s^{-1}$ at 5 mM potassium phosphate, pH
7.0 and 21°C), corresponding to that found when cytochrome c was
used as electron donor. The same rate of reduction of the heme of
cytochrome c oxidase by cytochrome c and by its porphyrin derivative
indicates that the electron transfer occurs via a mechanism, involving a similar type of 'redox orbitals', which might be the π^*
orbitals of the porphyrin ring. The high value of the second-order
rate constant together with the independence of its value of the
midpoint potential of the reducing agent suggest that the reaction
between cytochrome c and cytochrome c oxidase is diffusion-con-
trolled.

The 'off' rate constant for the reaction between porphyrin cytochrome c and cytochrome c oxidase is too small to be measured
from Fig. 3. This very small 'off' rate constant in comparison to
that when native cytochrome c is used, is probably due to the very
large difference in midpoint potential between porphyrin cytochrome c and the heme of cytochrome c oxidase, whereas the difference in midpoint potential is small between the heme of native
cytochrome c and its oxidase[16].

Trace D of Fig. 2, shows the reduction of heme a in a 1:1 complex of porphyrin cytochrome c - cytochrome c oxidase with supple-
mentary cytochrome c. It is obvious that the cause of reduction
of heme a is similar to that of experiment B of Fig. 2, where cy-

tochrome c instead of porphyrin cytochrome c was complexed to cytochrome c oxidase. Although the result of experiment B is consistent with the conclusion that the electron-accepting site on cytochrome c oxidase is the low-affinity site for cytochrome c, we are reluctant to draw further conclusions from this experiment. By means of chromatography it was found that cytochrome c and its porphyrin derivative exchange slowly in the complex with cytochrome c oxidase under the conditions used for pulse radiolysis. An exchange between cytochrome c species in the complex with cytochrome c oxidase has also been observed by Ferguson et al.[17]. Therefore, we should assume that in experiment D we had a mixture of cytochrome c oxidase complexed with either cytochrome c or its porphyrin derivative and that free cytochrome c as well as free porphyrin cytochrome c were present. The same argument holds for experiment F of Fig. 2, where the reduction of the heme of cytochrome c oxidase was monitored in a 1:1 complex of cytochrome c and its oxidase with additional porphyrin cytochrome c.

Our results have shown clearly that reduced cytochrome c, interacting with its low-affinity site on cytochrome c oxidase, transmits electrons directly to cytochrome c oxidase without the mediation of the tightly bound cytochrome c. Furthermore, there is no evidence that the tightly bound cytochrome c functions as a modulator in this electron transfer. It was found, however, that the tightly bound cytochrome c greatly affects the electron transmittance from the reduced cytochrome c_1 to the heme of cytochrome c oxidase, since the rate constant of reduction of the 1:1 complex is 500-fold faster (k $10^6 M^{-1} \cdot s^{-1}$ at 5 mM potassium phosphate, 1% Tween-20, pH 7.0 and $10^\circ C$) than in the absence of cytochrome c (k $2 \cdot 10^3 M^{-1} \cdot s^{-1}$). Whether cytochrome c serves as a redox mediator or a modulator in electron transfer between cytochrome c_1 and cytochrome c oxidase remains to be investigated.

REFERENCES

1. Henderson, R., Capaldi, R.A. and Leigh, J.S. (1977) J. Mol. Biol. 112, 631-648.
2. Ferguson-Miller, S., Brautigan, D.L. and Margoliash, E. (1976) J. Biol. Chem. 251, 1104-1115.
3. Errede, B., Haight, Jr., G.P. and Kamen, M.D. (1976) Proc. Natl. Acad, Sci. USA 73, 113-117.

4. Orii, Y., Sekuzu, I. and Okunuki, K. (1962) J. Biochem. Tokyo 51, 204-215.

5. Vanderkooi, J.M. and Erecińska, M. (1975) Eur. J. Biochem. 60, 199-207.

6. Wilting, J., Van Buuren, K.J.H., Braams, R. and Van Gelder, B.F. (1975) Biochim. Biophys. Acta 376, 285-297.

7. Van Buuren, K.J.H., Van Gelder, B.F., Wilting, J. and Braams, R. (1974) Biochim. Biophys. Acta 333, 421-429.

8. De Kok, J., Butler, J., Braams, R. and Van Gelder, B.F. (1977) Biochim. Biophys. Acta 460, 290-298.

9. Fowler, L.R., Richardson, S.H. and Hatefi, Y. (1962) Biochim. Biophys. Acta 64, 170-173.

10. MacLennan, D.H. and Tzagoloff, A. (1965) Biochim. Biophys. Acta 96, 166-168.

11. Van Buuren, K.J.H. (1972) Binding of cyanide to cytochrome aa_3. Ph.D. thesis, Amsterdam, Gerja, Waarland.

12. Margoliash, E. and Walasek, O.F. (1967) in Methods in Enzymology (Estabrook, R.W. and Pullman, M.E., eds), Vol. 10, pp. 339-348, Academic Press, New York.

13. Van Gelder, B.F. (1966) Biochim. Biophys. Acta 118, 36-46.

14. Van Gelder, B.F. and Slater, E.C. (1962) Biochim. Biophys. Acta 58, 593-595.

15. Van Gelder, B.F., Van Buuren, K.J.H., Wilms, J. and Verboom, C.N. (1975) in Electron-Transfer Chains and Oxidative Phosphorylation (Quagliariello, E., Papa, S., Palmieri, F., Slater, E.C. and Siliprandi, N., eds), North-Holland Publ. Comp,, Amsterdam, pp. 63-68.

16. Tiesjema, R.H., Muijsers, A.O. and Van Gelder, B.F. (1973) Biochim. Biophys. Acta 305, 19-28.

17. Ferguson-Miller, S., Brautigan, D.L. and Margoliash, E. (1978) J. Biol. Chem. 253, 149-159.

Cytochrome Oxidase, T.E. King et al. eds.
© *1979 Elsevier/North-Holland Biomedical Press*

CYTOCHROME C OXIDASE, THE PRIMARY REGULATORY SITE OF MITOCHONDRIAL
OXIDATIVE PHOSPHORYLATION

DAVID F. WILSON AND MARIA ERECIŃSKA
Department of Biochemistry and Biophysics, University of Pennsylvania,
Philadelphia, Pennsylvania, 19104 USA

Cytochrome c oxidase is one of the most important enzymes in eukaryotic
organisms. It plays an essential role in oxidative phosphorylation by being
the site for the reduction of molecular oxygen to water as well as the source
of adenosine triphosphate (ATP). Evidence has recently been presented that
cytochrome oxidase is responsible for the regulation of mitochondrial respiration[1,2] and efforts have been directed towards elucidating the mechanism of the
reaction and its kinetic properties. Advances have occurred in our knowledge
of both these areas.

The mechanism of cytochrome c oxidase and its reaction with oxygen. One of
the important questions concerning cytochrome oxidase is the nature of the
active site for oxygen reduction. An important observation concerning the
active site was that of Lindsay and coworkers[3,4]; when potentiometric titrations of the oxidation and reduction of cytochromes a and a_3 were carried out
in the presence of CO, two reducing equivalents were required for the formation
of the reduced cytochrome a_3-CO compound (n=2.0). This provided evidence that
CO binds to cytochrome a_3 with high affinity only when two oxidation-reduction
components are reduced, cytochrome a_3 and the "invisible" copper. Titration of
both isolated cytochrome oxidase and submitochondrial particles with reductant
(NADH) and oxidant (O_2) confirmed the requirement for two equivalents per cytochrome a for the formation of the cytochrome a_3-CO compound [5,6] see however 7.
Additional evidence for a role of the "invisible" copper in the active site
was obtained from potentiometric titrations in which samples were anaerobically
transferred into EPR sample tubes and frozen in order to measure the behavior
of the EPR active components of cytochrome oxidase[8]. It was found that the
"invisible" copper with a half reduction potential of 0.35 V[4] must be reduced
in order to observe the characteristic g=2.94 signal of low spin ferric heme-
azide complex which is formed on partial reduction of cytochrome oxidase in the
presence of azide[8].

The presence of two oxidation-reduction components in the active site, in particular a heme and copper, led to a proposed mechanism[2,4] in which the oxygen molecule formed a bridged compound between the two metal atoms. This mechanism allows for two electron reduction of oxygen to a bound peroxide and further reduction to water with facilitated breaking of the oxygen-oxygen bond. Reduction of oxygen to water must occur at a potential of near 0.6 V[1,2]. One electron reduction of O_2 to O_2^- has a half-reduction potential of approximately -0.32 V at pH 7[9,10] and is thermodynamically very unfavorable. Two electron reduction of O_2 to O_2^{2-}, on the other hand, has a half-reduction potential near 0.8 V[9,10] and is thermodynamically favorable.

The kinetics of oxidation of reduced cytochrome by molecular oxygen as catalysed by cytochrome oxidase. Cytochrome $\underline{c}$ oxidase catalyses the reaction

$$ADP + Pi + 2c^{2+} + 1/2\ O_2 + 2H^+ \longrightarrow 2c^{3+} + H_2O + ATP \quad (1)$$

In intact mitochondria at high (>100 μM) concentrations of oxygen the reaction was shown[2] to follow the rate expression:

$$\upsilon = \frac{k_2 a_{3T}[O_2]}{1 + \alpha[O_2]} \quad (2)$$

where

$$\alpha = \frac{k_2}{k_1[c^{2+}]} \left\{ 1 + \frac{k_1(1+K_3)}{k_{4b} + K_3 k_{4a}} + K_5^{-1} \frac{[ATP]}{[ADP][Pi]} \frac{[c^{3+}]^2}{[c^{2+}]^2} \right\}$$

The constants are defined as given by Wilson et al[2] and have two important properties. 1. At low values of $[ATP]/[ADP][Pi]$ the rate expression becomes

$$\upsilon = [c^{2+}][a_{3T}]\ k_1 k_{4a}/(k_1 + k_{4a}) \quad (3)$$

The steady state respiratory rate is proportional to $[c^{2+}]$ and the measured oxidation of reduced cytochrome $\underline{c}$ is a 1st order reaction such as is observed for suspensions of uncoupled mitochondria[2] and for suspensions of submitochondrial particles and isolated cytochrome oxidase[11]. 2. In suspensions of well coupled mitochondria at high [ATP]/[ADP][Pi], the steady state respiratory rate is a very non linear function of cytochrome $\underline{c}$ reduction in agreement with the predictions of equation 2. This non linear relationship of the mitochondrial respiration rate on the reduction of cytochrome $\underline{c}$ can be quantitatively fit by equation 2^2.

Extension of the basic rate expression to include the first two sites of oxidative phosphorylation

$$\text{NADH}_m + 2c^{3+} + 2\text{ADP}_c + 2\text{Pi}_c \rightleftharpoons \text{NAD}_m^+ + 2c^{2+} + 2\text{ATP}_c \qquad (4)$$

was accomplished by assuming that this reaction is near equilibrium when the NAD^+ and NADH are intramitochondrial and ATP, ADP and Pi are cytosolic (for data and discussion see[12-14]). The resulting rate expression gives an excellent fit to the observed respiratory behavior not only in suspensions of isolated mitochondria but also in intact cells and tissues [14,15], including perfused rat heart, isolated rat liver cells, cultured kidney and neuroblastoma cells, Sarcoma 180 ascites tumor cells, Tetrahymena pyriformis and Paracoccus denitrificans.

SUMMARY

A model has been developed for the active site of cytochrome oxidase in which oxygen forms a bridged compound between the iron of cytochrome a_3 and the "invisible" copper. Steady state rate expressions based on this model are able to fit accurately the observed behavior of isolated cytochrome c oxidase and of suspensions of uncoupled mitochondria. In addition a fit is obtained for the kinetics of cytochrome c oxidase in suspensions of mitochondria under phosphorylating conditions and in mitochondria of intact cells.

ACKNOWLEDGEMENTS

This work was supported by USPHS grants GM12202 and GM21524. M.E. is an Established Investigator for the American Heart Association.

REFERENCES

1. Owen, C.S. and Wilson, D.F. (1974) Arch. Biochem. Biophys. 161, 581-591.

2. Wilson, D.F., Owen, C.S. and Holian, A. (1977) Arch. Biochem. Biophys. 169, 199-208.

3. Lindsay, J.G., and Wilson, D.F. (1974) FEBS Lett. 48, 45-49.

4. Lindsay, J.G., Owen, C.S. and Wilson, D.F. (1975) Arch. Biochem. Biophys. 169, 492-505.

5. Wilson, D.F. and Miyata, Y. (1977) Biochim. Biophys. Acta 461, 218-230.

6. Wever, R., van Drooge, J.H., Muijsers, A.O., Bakker, E.P. and van Gelder, B.F. (1977) Eur. J. Biochem. 73, 149-154.

318

7. Anderson, J.L., Kuwana, T. and Hartzell, C.R. (1976) Biochemistry 15, 3847-3855.

8. Wilson, D.F., Erecińska, M. and Owen, C.S. (1975) Arch. Biochem. Biophys. 175, 160-172.

9. George, P. (1965) in Oxidases and Related Redox Systems, King, T.E., Mason, H.S. and Morrison, M. eds., vol. 1 Wiley, New York, pp. 3-33.

10. Wood, P.M. (1974) FEBS Lett. 44, 22-24.

11. Smith, L. and Conrad, H. (1956) Arch. Biochem. Biophys. 63, 403-413.

12. Wilson, D.F., Stubbs, M., Veech, R.L., Erecińska, M. and Krebs, H.A. (1974) Biochem. J. 140, 57-64.

13. Wilson, D.F., Stubbs, M., Oshino, N. and Erecińska, M. (1974) Biochemistry 13, 5305-5311.

14. Erecińska, M. and Wilson, D.F. (1978) TIBS 3, 219-223.

15. Erecińska, M., Wilson, D.F. and Nishiki, K. (1978) Amer. J. Physiol: Cell Physio. 3(2), C82-C89.

Cytochrome Oxidase, T.E. King et al. eds.
© 1979 Elsevier/North-Holland Biomedical Press

BEHAVIOR OF CYTOCHROME OXIDASE IN LIVING LIVER TISSUE: DIRECT ANALYSIS OF TURN-
OVER OF CYTOCHROME aa_3 IN LIVER IN SITU BY REFLECTANCE SPECTROPHOTOMETRY

NOBUHIRO SATO, MOTOAKI SHICHIRI, NORIO HAYASHI, TAKENOBU KAMADA, HIROSHI ABE
AND BUNJI HAGIHARA*
The First Department of Medicine and Department of Biochemistry*, Osaka Uni-
versity Medical School, Osaka 553 (Japan)

ABSTRACT

Reflectance spectra of living liver tissue were measured in anesthetized, air-
breathing rats using optical fibers which coupled the liver to a spectrophotometer equipped with a memory circuit for reference and with an image sensor as a
detector. Pressure exerted on the liver surface in situ caused spectral changes
which were explained as being due to the complete blocking of blood flow, resulting in total deoxygenation of hemoglobin and subsequent almost complete reduction
of the respiratory enzymes. The reduced cytochromes were again oxidized upon
release of pressure.

By analyzing these spectra of the liver in situ during pressurization the rate
of O_2 consumption and the concentrations of the respiratory cytochromes were determined. The measured turnover of cytochrome aa_3 was 20-26 e^-/sec per mole of
cytochrome aa_3 in the livers of normal, anesthetized, air-breathing rats and
15-22 e^-/sec per mole of cytochrome aa_3 in 3'-methyl-4-dimethylaminoazobenzene
(3'-DAB)-induced hepatomas of rats.

INTRODUCTION

The hemoproteins such as hemoglobin (Hb) and respiratory chain cytochromes
show spectroscopically detectable signals which may serve as indicators of tissue
O_2 supply and utilization. The extensive studies of Warburg[1], Keilin[2] and Chance
e.g.[3] on these hemoproteins have led the foundation of our knowledge of cellular
energetics. The metabolic state of mitochondria of living tissue can be adequate-
ly defined by the turnover number of the cytochromes (rate of O_2 consumption/cyto-
chrome), the cytoplasmic phosphorylation state and the oxidation-reduction potential of intramitochondrial NAD couple[4]. Therefore the measurement of the rate
of O_2 consumption and the concentration of respiratory chain enzymes in vivo is
of great value for understanding of patho-physiology of aerobic tissues. In this
communication we describe a non-destructive method of measuring the rate of O_2
consumption and the concentrations of respiratory chain cytochromes in liver in
vivo by analyzing these hemoproteins with reflectance spectrophotometry. More-

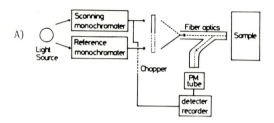

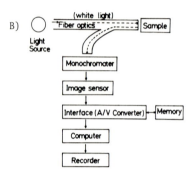

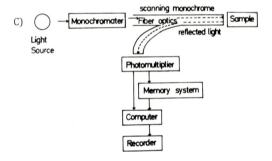

Fig. 1. Block diagram of the spectrophotometer system. Type A, double-beam system; B, scanning monochromatic light-photomultiplier-memory system; C, white light-image sensor-memory system.

over experimental measurements of the turnover number of cytochrome aa₃ in normal livers and 3'-DAB-induced hepatomas of air-breathing, anesthetized rats are presented and discussed.

MATERIALS AND METHODS

Spectrophotometer. Three types of spectrophotometers could be used to obtain reflectance spectra of intact tissues in situ. The block diagrams of these spectrophotometers are shown in Fig. 1. In all types, branched optic fiber bundles

(diameter 8 mm) were used for coupling the liver surface to a spectrophotometer.
Halon white board (reflectance lying between 0.97 and 0.98 in the spectral region
400-700 nm) was used as the reference standard material. The type A spectrophotometer which is a modification of the commercially available double-beam spectrophotometer[5] records the difference in absorption between two wavelengths where
the first is the scanning sample beam and the second, a fixed reference beam.
Types B and C are equipped with a memory circuit which stores the spectrum of the
reference material, and then, subtracts it from the sample spectrum. In types A
and B a photomultiplier is used as detector, while in type C an image sensor (Si
photodiode) is used. The latter system is very quick as the whole spectrum (100-
200 nm) is scanned in 0.01-1.0 sec. Moreover the advantage of type C is that the
spectra of sample tissue are obtained simultaneously at each wavelength, and thus
this type is suitable for measuring kinetics of oxidation and reduction of the
respiratory components.

 Animals. Non-fasting, male Wistar rats weighing 150-250 g were anesthetized
with pentobarbital (Nembutal, Abbott Laboratory) administered intraperitoneally
at a concentration of 35-40 mg/Kg body weight. The abdomen was opened through a
midline incision, and flexible, light-conducting fiber optic bundles were guided
to the liver surface. Tracheotomy was carried out to maintain normal respiration.

RESULTS AND DISCUSSION

Redox cycling of cytochrome aa$_3$ in liver in situ. Fig. 2 shows reflectance
spectra of the liver of an air-breathing, anesthetized rat taken in the type B
spectrophotometer. The group of spectra labeled A was taken at one min intervals
under the condition when the optic guide was gently touching the surface of the
liver at an angle of 90°. The spectra were reproducible, and showed two peaks
at 543 and 577 nm in a region 500-700 nm with a strong absorption (and partly
light scattering) at the wavelengths shorter than 470 nm, reflecting mainly oxy-
Hb absorbance in living liver tissue. The absorbances from cytochromes and other
intracellular pigments should also contribute to these spectra, although under
these conditions they apparently revealed no appreciable bands presumably because
they were in oxidized forms under normoxic condition.

 The spectrum B was obtained during 12 sec pressurization of the liver in situ
by the tip of the optic guide at a pressuring power of 400-800g/cm^2. The spectral pattern differs entirely from that of the non-pressed liver. It shows absorption bands at around 550 to 560 nm with shoulders at 520 to 530 nm, and 606
nm, which suggests that the spectrum of the pressed liver is composed of the deoxy-Hb as well as of the reduced respiratory enzymes. It was found that this
spectrum was quite similar to that of the excised liver of the same rat (spectrum

322

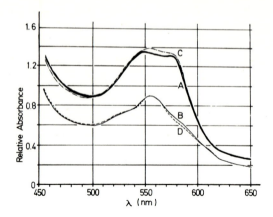

Fig. 2. Reflectance spectra of the liver of an air-breathing, anesthetized rat taken in the type B spectrophotometer. A series of spectra lebeled A was taken when the light guide was gently placed on the liver surface. The spectrum B was taken 2 sec after pressurization of the liver by the tip of the light guide. The spectrum C was taken 2 sec after release of pressure. The dotted curve D is the reflectance spectrum of the excised liver of the same rat when the tissue was placed under an atmosphere of nitrogen. Scanning speed, 20 nm/sec.

D), where the liver was made anoxic under an atmosphere of nitrogen. Thus, 12 sec pressurization of the liver in situ resulted in complete blocking of in- and outflow of blood which lead to tissue anoxia. The lack of blood flow in the pressed liver was confirmed by observing no essential change in the spectrum when all blood vessels to the liver were ligated during pressurization.

When pressure was released, the spectrum returned to that of the control liver in about 30 sec, although the spectrum taken immediately after release of pressure revealed a shift of one of the peaks toward the red (from 543 to 552 nm)(C). This was probably due to a greater percentage of deoxy-Hb and reduced cytochromes c and c_1. Thus the cytochromes in liver in situ revealed redox cycling following external application of pressure followed by release of pressure.

Measurement of the concentrations of cytochromes in liver in situ. Attempts were made to separate the absorbance of Hb from the absorbances of other pigments in the reflectance spectrum of the pressed liver. Fig. 3 shows a variety of the pressed liver spectra as well as those of anoxic, blood-free, perfused livers of rats treated with various drugs and hepatotoxins such as phenobarbital, thyroxin, carbon tetrachloride, acetaminophen and 3'-DAB. The in vitro spectra of the blood (Hb)-free, perfused livers show almost equal absorbances at around 545 and 526 nm. Therefore, the absorbance difference between 545 and 526 nm which is seen in the pressed liver must be due to deoxy-Hb. Using this difference in ab-

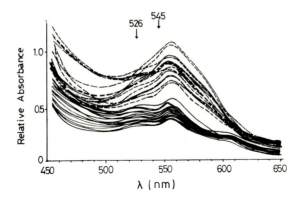

Fig. 3. Reflectance spectra of the pressed livers in vivo (dotted) and anoxic, blood-free, perfused livers (solid) of rats which were treated with various drugs and hepatotoxins. The type B spectrophotometer; temperature, 37°.

sorption ($\Delta E_{545-526}$=4.7 mM^{-1})[6], the deoxy-Hb concentration in the in vivo spectrum of the pressed liver was determined. The absorption spectrum of deoxy-Hb in the region 450-650 nm was then computed using the extinction coefficients given by Assendelft[6] for the various wavelengths is shown in Fig. 4-B. Then, subtracting this computed deoxy-Hb spectrum from the in vivo spectrum, a spectrum of the remaining absorbing materials could be determined (Fig. 4-C). This spectrum (C) appeared quite similar to that of an anoxic, blood-free, perfused liver of the same rat (Fig. 4-D) in a region between 450 and 650 nm. Absorption peaks were seen at 606, 565-550 and 530-520 nm and were identified as reduced alpha bands of cytochromes aa₃, b (b_K+b_T), c and c₁ and beta bands of cytochromes b, c and c₁.

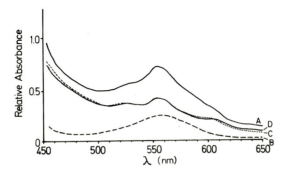

Fig. 4. The absorption spectra of hemoglobin (B) and of liver pigments (C) calculated from the reflectance spectra of the pressed liver in vivo (A). Curve D is the anoxic, blood-free, perfused liver of the same rat.

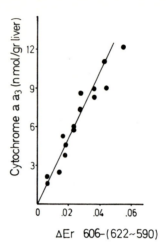

Fig. 5. A relationship of the relative absorbance at 606 nm from a straight line drawn between 622 and 590 nm in the calculated absorbance of liver pigments and the concentration of cytochrome aa_3 in livers in situ.

Thus, it was concluded that the in vivo spectrum of the pressed liver was composed of the absorbances due mainly to deoxy-Hb and to the reduced respiratory enzymes.

From the calculated spectrum, the concentration of the respiratory enzymes could be determined. The relative absorbances at each peak of the respiratory enzymes calculated from straight lines drawn between the two reference points were proportional to the concentration of each enzyme. Fig. 5. shows a case of cytochrome aa_3 in the liver, where the concentration of cytochrome aa_3 in Hb-containing livers was measured by transmission spectrophotometry by the method of Sato et al[7,8]. The cytochrome aa_3 concentration in livers of normal rats is approx. 6-9 n moles/g liver, which is in agreement with the results of Wilson et al[9] and is a little bit higher than the results of Schollmeyer and Klingenberg[10]. In 3'-DAB-induced hepatomas in situ, it is between 2-4 n moles/g fresh weight of hepatoma, which is also consistent with our previous findings in various tumors[11-13]. The concentrations of cytochromes b and $c+c_1$ were also determined from the calculated absorbance of liver pigments, as shown elsewhere[14].

Measurement of the rate of O_2 consumption in liver in situ. It was found that when the liver was subjected to pressure from the tip of a light guide, the tissue became anoxic due to a complete blockage of the blood flow. Hence, the spectral analysis of O_2-saturation of Hb may allow us to determine the rate of O_2 consumption in the liver in situ during pressurization. Fig. 6 shows the sequen-

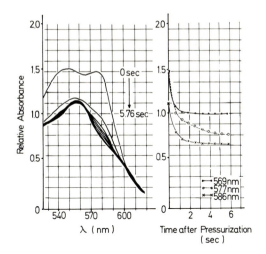

Fig. 6. (Left) The reflectance spectra of the liver of an anesthetized rat taken sequentially during pressurization of the liver in situ. Each spectrum was obtained in 0.64 sec with 1.28 sec intervals between the scans using an image sensor type spectrophotometer. The top curve shows the absorbance prior to pressurization. (Right) The plot of absorbance changes at 569, 577 and 586 nm with time during pressurization.

tial changes in the spectra with time during pressurization. Each spectrum was taken in 0.64 sec with a 1.28 sec interval in type C spectrophotometer. The initial change in the spectrum following pressurization showed a general decrease of absorbance presumably due to the expulsion of blood in situ. This was followed

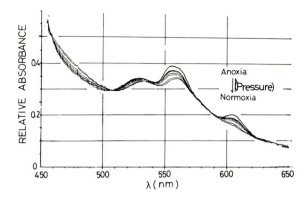

Fig. 7. Reflectance spectra of blood-free, perfused livers of a rat during normoxic and anoxic cycling. The perfusate was Krebs-Ringer-Bicarbonate (pH 7.4) equilibrated with a gas mixture of 95% O_2 and 5% CO_2. Temperature, 30°.

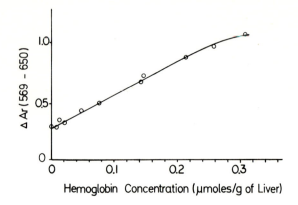

Fig. 8. The differences in absorption between 569 and 650 nm in the reflectance spectra of perfused rat livers having various concentrations of Hb as a function of Hb concentration in livers in situ.

by a change in the spectral pattern which could be attributed to the change from the normoxic to the anoxic state of the liver. After 1-2 sec of pressurization, the decrease in absorbances at 569 and 586 nm became negligible, whereas the decrease of absorbance at 577 nm continued for the next 4-5 sec. Since the 577 nm absorbance is that of the alpha peak of oxy-Hb, this decrease in absorbance at 577 nm, without a change at 569 and 586 nm (isobestic points for oxy- and deoxy-Hb), could be attributed to a decrease in O_2 saturation of Hb with time.

Hence, the O_2 consumed by the liver in situ could be determined provided that the concentration of Hb as well as that of the various intracellular pigments which contribute to the absorbances at these wavelengths were known. The absorbance of anoxic-liver pigments was calculated from the spectrum of the pressed liver as described above. A study of perfused livers (Fig.7) revealed that the absorbance of intracellular pigments showed essentially the same during normoxic and anoxic cycling in a region between 569 and 586 nm. Therefore, the contribution of absorbance of intracellular pigments could be calculated after subtraction of the spectrum obtained on the anoxic liver.

The Hb concentration in livers in situ can be determined as follows. The reflectance spectra of anoxic livers having various concentrations of Hb were taken, and the difference in absorption between 569 and 650 nm were plotted as a function of the Hb concentration as determined by transmission spectrophotometry[15] (Fig. 8). When the Hb concentration is relatively low, one can obtain a linear correlation between them.

The rate of O_2 consumption of the liver in situ was determined by the following equations:

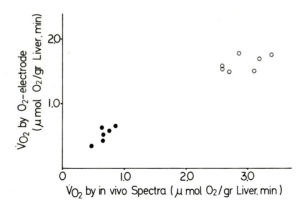

Fig. 9. The rate of O_2 consumption of normal livers (○) and 3'-DAB-induced hepatomas (•) of rats, as measured by in vivo spectra (abscissa) and by polaro-graphy (ordinate).

$$VO_2 \text{ (moles } O_2/g \text{ liver x min)} = K \times 4C \times (b_1-b_2)/(a-a')/\Delta t$$

where a and a' are the absorbance differences between 569 and 586 nm in the in vivo reflectance spectrum as in Fig. 6 and in the calculated spectrum of liver pigments as in Fig. 4-C, respectively; b_1 and b_2 are the absorbance at 577 nm in the in vivo two spectra; C is the Hb concentration (moles/g liver); Δt is a time interval between two spectra (min); K is a constant (0.673)[6]. $K \times (b_1-b_2)/(a-a')$ is the difference in O_2-saturation of Hb between the two spectra concerned.

The abscissa of Fig. 9 shows the rate of O_2 consumption, calculated by the present method, in livers in situ of normal rats and in 3'-DAB-induced hepatomas of rats. The ordinate of the Fig. 9 shows values of the sliced livers of the identical rats which were obtained in the O_2-electrode chamber equipped with an electrode covered with a cellulose acetate membrane[16]. The calculated rate in normal, air-breathing anesthetized rats showed the values of 2.5-3.5 μ moles/g liver x min, which seemed quite reasonable as compared to those obtained in liver slices and in perfused livers[9,17]. As expected, the rate of O_2 consumption in 3'-DAB-induced hepatomas was much smaller than the normal (0.5-1.0 μ moles/g liver x min).

Turnover number of cytochrome aa_3 in liver in situ. The present spectrophoto-metric technique can measure steady state O_2 consumption rate in livers in situ. If the concentration of cytochromes is also measured simultaneously, the turn-over numbers of cytochromes in liver in situ can be precisely estimated. The measured turnover number of cytochrome aa_3 was 20-26 e⁻/sec per mole of cyto-chrome aa_3 in the "pressed" livers of normal rats, and 15-22 e⁻/sec per mole of cytochrome aa_3 in the "pressed" 3'-DAB-induced hepatomas. Although the hepatoma showed considerable decreases of the rate of O_2 consumption and of the concentrations of respiratory components, the turnover number of the hepatoma does not differ significantly from that of normal livers, suggesting that the cytochrome oxidase in the hepatoma is working at almost full capacity as in normal livers if the measured O_2 consumption is all via the cytochrome chain.

The present turnover number in normal livers of rats is higher than that measured at 25° in suspensions of isolated mitochondria at state $3^{e.g.18}$ and in isolated liver cells of rats[19], and is close to that of Wilson et al[9] in suspensions of isolated liver cells measured at 37° in the absence of added substrates, where the calculated degree of reduction of cytochromes c and aa_3 in the cells was between 16-26 %. The redox levels of cytochromes prior to pressurization, the $(ATP)/(ADP)(P_i)$ and the oxidation-reduction potential of intramitochondrial NAD couple are to be further clarified in living livers under the present condition.

SUMMARY

The rate of O_2 consumption and the concentrations of the respiratory cytochromes in liver in situ were determined in anesthetized rats, by analyzing the reflectance spectra of livers obtained by a spectrophotometer which was equipped with a memory circuit for reference and with an image sensor as a detector.

ACKNOWLEDGEMENTS

The authors are grateful to Dr. M. Erecinska for her many interesting discussions and her aid in preparation of the manuscript.

REFERENCES

1. Warburg, O. (1949) Heavy Metal Prosthetic Groups and Enzyme Action (Trans. A. Lawson), Clarendon Press, Oxford

2. Keilin, D. (1966) The History of Cell Respiration and Cytochromes, Cambridge University Press, Cambridge

3. Chance, B. (1973) in Oxygen Transport to Tissues. Bicher, H.I. and Bruley, D.F. eds., Plenum, New York, pp. 277-292

4. Wilson, D.F., Stubbs, M., Veech, R.L., Erecinska, M. and Krebs, H.A. (1974) Biochem. J., 140, 57

5. Chance, B. (1957) in Methods in Enzymol., Colowick, S.P. and Kaplan, N.O. eds. Vol. 4, Academic Press, New York, pp.273-329

6. Assendelft, D.W. (1970) Spectrophotometry of Hemoglobin Derivatives, Assen, Royal Vangoreum

7. Sato, N., Hagihara, B., Kamada, T. and Abe, H. (1976) Anal. Biochem. 74, 105

8. Sato, N., Kamada, T. Abe, H., Suematsu, T., Kawano, S., Hayashi, N., Matsumura, T. and Hagihara, B. (1977) Clin. Chem. Acta, 80, 243

9. Wilson, D.F., Stubbs, M., Oshino, N. and Erecinska, M. (1974) Biochemistry, 13,. 5305

10. Scollmeyer, P. and Klingenberg, M. (1962) Biochem. Z. 335, 426

11. Sato, N. and Hagihara, B. (1970) Cancer Res. 30, 2061

12. Oyanagui, Y., Sato, N. and Hagihara, B. (1974) Cancer Res. 34, 458

13. Sato, N., Hagihara, B., Kamada, T., Abe, H., Senoh, H. and Kitagawa, M. (1976) Biochim. Biophys. Acta, 423, 557

14. Sato, N., Shichiri, M., Hayashi, N., Matsumura, T., Kamada, T., Abe, H. and Hagihara, B. (1978) in Frontier in Bioenergetics. From Electron to Tissues. Scarpa, A., Dutton, P.L. and Leigh, J. eds. Academic Press, New York, in press

15. Sato, N., Kamada, T., Shichiri, M., Kawano, S., Abe, H. and Hagihara, B. (1978) Gastroenterology, in press

16. Hagihara, B., Ishibashi, F., Sasaki, K. and Kamigawara, Y. (1978) Anal. Biochem. 86, 417

17. Sestoft, L. (1974) Biochim. Biophys. Acta, 343, 1

18. Chance, B. and Hess, B. (1959) J. Biol. Chem. 234, 2404

19. Erecinska, M., Wilson, D.F. and Nishiki, K. (1978) Am. J. Physiol. 234, C82-C89

Cytochrome Oxidase, T.E. King et al. eds.
© *1979 Elsevier/North-Holland Biomedical Press*

ALTERNATIVE PATHWAYS FOR THE REDUCTION OF MOLECULAR OXYGEN BY CYTOCHROME OXIDASE

YUTAKA ORII

Department of Biology, Faculty of Science, Osaka University, Toyonaka, Osaka 560
(Japan)

ABSTRACT

On mixing with an air-saturated medium in a stopped-flow rapid scan apparatus
cytochrome oxidase which had been reduced with either dithionite or NADH in the
presence of a catalytic amount of phenazine methosulfate (PMS) yielded a product
having absorption maxima at 425 and 601-2 nm within 5 msec. The product was
identified as Comp. III[1], a conformational variant of the oxidized oxidase.
When an amount of dithionite was still in a large excess after mixing the
reduced enzyme was regenerated, and the spectral changes showed a single isobestic point at 432 nm in the Soret region. This process contrasted with the
reduction of the resting oxidized oxidase which showed multiple isobestic
points. When NADH remained in excess after mixing, on the contrary, the absorption of Comp. III at 603 nm increased to a certain level within 300 msec.
When ferrocytochrome c was an electron donor to Comp. III the maximal extent of
the absorbance increase amounted to 40-50% of the total change due to complete
reduction and it occurred within a few milliseconds. This was followed by a
decrease in absorbance. Based on spectral properties as well as kinetic
behavior the transiently formed species with an enhanced α-peak was concluded
to be Comp. I.

The formation of Comp. III by reaction of fully reduced cytochrome oxidase
with oxygen is a one-shot process, because as soon as a half-reduced species is
formed as a consequence of electron supply to Comp. III it reacts with oxygen
forming Comp. I instead of receiving two more electrons to become fully reduced.
Comp. I further receives electrons and reduces oxygen to water thus itself
going back to Comp. III. This cyclic turnover of cytochrome oxidase must be
of physiological significance and it should be noticed that in the catalytic
turnover neither resting oxidized nor fully reduced cytochrome oxidase is
involved.

INTRODUCTION

The classical way of preparing oxygenated cytochrome oxidase was to aerate
a solution of the dithionite-reduced oxidase[2,3,4]. When the spectrum was taken

332

on a conventional spectrophotometer the Soret peak appeared at 426-8 nm and the α-peak at 603 nm, which was slightly higher than that of the oxidized oxidase as prepared. In 1965 Orii and Okunuki obtained the oxygenated form which showed an enhanced α-peak and pointed out a possibility to obtain a mixture of the oxidized and oxygenated forms depending on the extent of decay of the latter[5]. Beinert *et al.* collected spectra which were diverse in shape and were claimed to represent the oxygenated form, and suggested that "this form is not a single species but a mixture of species"[6].

A rapid reaction of the fully reduced form of purified cytochrome oxidase with molecular oxygen was first analyzed by a stopped-flow technique by Gibson and Greenwood, who identified the product as the oxidized oxidase[7]. It was confirmed later, however, that the product showed a spectrum quite similar to but distinctly different from that of the oxidized oxidase as prepared[1,8]. The product was also different from the oxygenated form that showed an enhanced α-peak.

Based on the analyses of spectral changes of a reaction system consisting of electron donors, cytochrome oxidase, cytochrome c and oxygen, Orii and Okunuki speculated that another type of ferricytochrome a* but not the resting oxidized oxidase was involved in a catalytic turnover of the enzyme, and proposed a reaction scheme of cyclic turnover[4]. More than ten years later from its original proposal, "another type of ferricytochrome a" has been revived under the name of oxygen pulsed oxidase[8,9]. Whatever the name of this species may be, this can be regarded as a conformational variant of the resting oxidized oxidase[10,11,12,13].

In the present investigation it is intended to make clear experimental conditions under which two types of oxygenated compounds (low and high α-peak) are formed. Also it is suggested that reaction of the fully reduced oxidase with molecular oxygen is of less physiological significance.

MATERIALS AND METHODS

Bovine heart cytochrome oxidase and cytochrome c were purified as described previously[14]. The oxidase was dissolved in 0.05 M sodium phosphate buffer, pH 7.4, containing 0.25% (v/v) Emasol 1130 and 0.1% (w/v) sodium cholate. The concentration was determined by using a millimolar extinction coefficient difference of 16.5 ($\Delta\varepsilon_{605-630nm}$, reduced). The concentration of cytochrome c was determined spectrophotometrically by using $\Delta\varepsilon_{550nm}$(red-oxd) = 18.5 $mM^{-1}cm^{-1}$.

---

*Cytochrome a of Okunuki's unitarian theory represents cytochrome oxidase.

Rapid reactions were followed either on Union stopped-flow rapid scan analyzer
model RA-1300 or RA-601.

RESULTS AND DISCUSSION

Dithionite-reduced cytochrome oxidase was mixed with an air-saturated medium
in a stopped-flow apparatus. An initial product, the spectrum of which was
recorded at 2 msec after mixing, showed absorption peaks at 425 and 600-1 nm,
confirming the previous observation of Orii and King[1]. The spectrum was quite

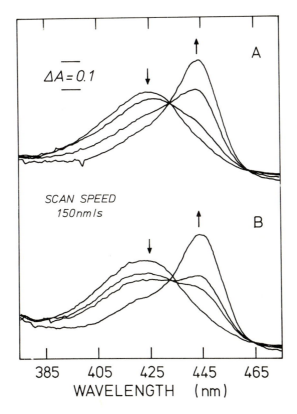

Fig. 1. Spectral changes during reaction of dithionite-reduced cytochrome oxidase with oxygen (A) and of the resting oxidized oxidase with dithionite (B).
(A) Cytochrome oxidase (9.2 μM) which had been reduced with sodium dithionite
(10 mg/5 ml) was mixed with an equal volume of air-saturated medium in the
stopped-flow apparatus, and the spectra were scanned repeatedly from 350 to
500 nm at a scan speed of 150 nm/sec. The times of initiation of recording at
350 nm were 2, 40, 80 and 160 sec. (B) Cytochrome oxidase (9.2 μM) in the
resting oxidized state was mixed with an equal volume of a solution containing
10 mg/5 ml sodium dithionite. Other conditions were the same as in (A).

334

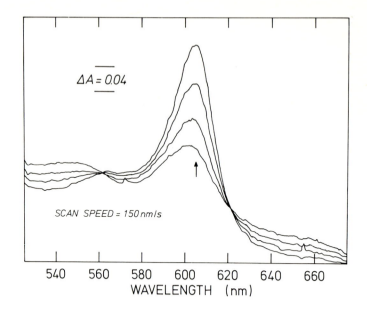

Fig. 2. Spectral changes during reaction of dithionite-reduced cytochrome oxidase with oxygen. Experimental details were the same as described in Fig. 1A except that 37 μM cytochrome oxidase was used. The spectra were recorded from 525 to 675 nm.

similar in shape to that of the resting oxidized oxidase, although the Soret peak was at longer wavelengths by 3-5 nm. The product was reduced further with remaining dithionite and the spectra recorded had isobestic points at 432 nm (Fig. 1A), 562 and 621 nm (Fig. 2), indicating the occurrence of two spectrally distinguishable species. On the other hand, the spectra recorded in an early phase of reduction of the resting oxidized oxidase had an isobestic point at 435 nm and this shifted to 432 nm with time, possibly including intermediate ones. If we neglect such monor differences and compare the spectral changes in Fig. 1A and 1B, the initial product in Fig. 1A can be concluded to be in the oxidized state. This would be a conformational variant of the resting oxidized oxidase and corresponds to "another type of ferricytochrome a" or to Comp. III as designated previously[1].

When cytochrome oxidase was reduced anaerobically with a slight excess of NADH in the presence of a catalytic amount of PMS and mixed with an air-saturated medium it was oxidized rapidly within 2 msec. The initial product was also Comp. III since the spectrum showed peaks at 425 and 602 nm in good agreement with the result of the dithionite-reduced oxidase. After the initial

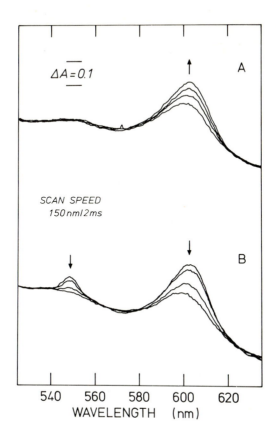

Fig. 3. Spectral changes during reaction of NADH-PMS-reduced cytochrome oxi-dase with oxygen in the absence (A) and presence (B) of cytochrome *c*. A mix-ture of 72.5 μM cytochrome oxidase and 15.7 μM PMS in a medium was bubbled with nitrogen gas for 3 min and a deaerated solution of NADH was added to a final concentration of 1.2 mM. The mixture was allowed to stand in a reservoir at least for 15 min until the cytochrome was reduced completely, and mixed with an equal volume of air-saturated medium with (B) or without (A) 15.7 μM ferricyto-chrome *c*. The spectra were recorded from 525 to 675 nm at 150 nm/2 msec. The times of initiation of recording at 525 nm were (A) 2, 80, 160 and 320 msec, and (B) 40, 160, 320 and 1280 msec.

rapid oxidation the α-peak increased in intensity with an apparent first order rate constant of 14 sec^{-1} and reached the maximal level at 300 msec after mix-ing (Fig. 3A). The amplitude of absorbance increase at 603 nm was 2.5 $mM^{-1}cm^{-1}$. Then the peak again decreased with a rate constant of 0.7 sec^{-1}. Dithionite in the presence of oxygen was effective in inducing absorbance increase in Comp. III at 603 nm but it took several seconds to attain the maximal level. Even so this process as well as the initial oxidation is too rapid to be recorded on a

conventional spectrophotometer. Thus it is apparent that depending on the
nature of reducing agent, its amount and the type of recording equipment different kinds of spectra are obtained although each of them may have been regarded as representing the initial reaction product. This would be one of
causes for confusion in defining spectral characteristics of an immediate reaction product between reduced cytochrome oxidase and molecular oxygen.

The rate of absorbance increase at 603 nm after the initial oxidation became
much faster when a small amount of cytochrome c had been added to an air-
saturated medium (Fig. 3B). In the presence of ferricytochrome c at 1/5 the

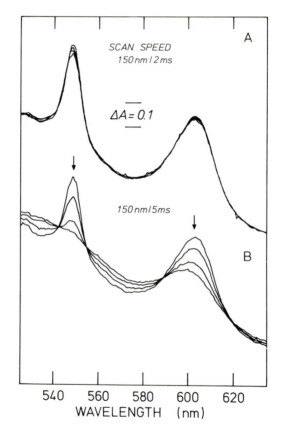

Fig. 4. Spectral changes during reaction of reduced cytochrome oxidase with
oxygen in the presence of ferrocytochrome c. Reduced cytochrome oxidase prepared as described in the legend to Fig. 3 was mixed with an equal volume of
air-saturated medium containing 81 µM ferrocytochrome c. The spectra were
recorded from 525 to 675 nm at (A) 150 nm/2 msec and (B) 150 nm/5 msec, and the
times at which recording was initiated at 525 nm were (A) 2, 10, 20 and 80 msec
and (B) 100, 200, 400 and 800 msec.

concentration of cytochrome oxidase the absorbance reached the maximal level within 40 msec after mixing, and the amplitude of the increase was 3.6 $mM^{-1}cm^{-1}$. Meanwhile about 40% of added ferricytochrome c had been reduced.

When the anaerobically reduced oxidase was mixed with an air-saturated medium containing equimolar ferrocytochrome c in the stopped-flow apparatus, the immediate spectrum already showed an increased absorbance at 603 nm (Fig. 4A). This was also the confirmation of the previous observation[1], and the same result was obtained when an anaerobic solution containing reduced oxidase and ferrocytochrome c was mixed with an air-saturated medium. The spectrum changed later on, as illustrated in Fig. 4B, sharing isobestic points at 543, 556, 590 and 617 nm. The latter two points were distinctly different from those at 562 and 621 nm that were observed during the conversion of Comp. III to the reduced form (Fig. 2). Although it remaines to be determined whether the shift of iso-bestic point is due to spectral overlap of cytochrome c to the redox change of cytochrome oxidase, it is probable that a new species different from the par-tially reduced oxidase is involved in the spectral change. The spectral pro-files and behavior of the immediate product in the presence of ferrocytochrome c coincide with those assigned to Comp. I previously[1]. Thus the immediate product is called Comp. I hereafter. The decay of Comp. I gave an apparent rate constant of 6 sec^{-1} and the maximal absorbance decrease at 603 nm was 4.5-5 $mM^{-1}cm^{-1}$. The present result may be taken as indicating that the immediate reaction product of reduced cytochrome oxidase and oxygen is Comp. I, as was interpreted previously[1]. However, on a separate stopped-flow measurement a tail of absorbance increase at 603 nm was detected immediately after the dead time of the instrument and an apparent rate constant was estimated roughly to be 700 sec^{-1}. Also it is to be noticed that during the dead time more than half of added ferrocytochrome c was oxidized. Thus it is reasonable to conclu-de that ferrocytochrome c gives electron to a certain species of cytochrome oxidase that cannot be the reduced form. Prior to this electron transfer the reaction of reduced oxidase with oxygen must have occurred yielding Comp. III.

Orii and King[15] observed that when reduced cytochrome c_1 was added to oxidiz-ed cytochrome oxidase under the air the α-peak of the oxidase increased in intensity, reached 40-50% of the total absorbance change due to the complete reduction and then decreased, while the α-peak of reduced cytochrome c_1 kept decreasing. Since the rate of electron transfer from ferrocytochrome c_1 to the oxidized oxidase was so slow that the absorbance increase of the α-peak of the enzyme was followed on a conventional spectrophotometer. During the subsequent decrease of the α-peak, a new isobestic point appeared at 585-595 nm, coinciding

338

with the 590-nm point in Fig. 4B.

It is to be noticed that under a continuous supply of electron from either ferrocytochrome c or ferrocytochrome c_1 to Comp. III in the presence of oxygen the absorbance at 603 nm increased at first approaching a 40-50% level of the total change due to the complete reduction and decreased further. Cytochrome oxidase exists as dimer in solution if we define monomer as a unit having one heme a and one copper[16], and the four redox centers are classified into high and low potential pairs; that is, high potential heme a and copper have a common midpoint potential of 340 mV and low potential heme a and copper have potentials of 220 and 240 mV, respectively[17]. For formation of a carbon monoxide complex of cytochrome oxidase two electrons must be taken up for reduction of two high potential components[18,19]. By analogy with this it is inferred that possibly when the two high potential centers are reduced this species reacts rapidly with oxygen forming an oxygenated compound, Comp. I. In fact, Greenwood et al. indicated that partially reduced cytochrome oxidase reacted with oxygen as rapidly as the fully reduced enzyme yielding a so-called oxygenated compound, although these authors regarded that only one electron was retained in the partially reduced oxidase[20]. Until all of the high potential centers are reduced and occupied with oxygen, Comp. I will be accumulated. This mechanism will explain the steady increase of the absorbance at 603 nm. After all of the high potential centers are occupied with electrons further electrons will be supplied to Comp. I, which in turn decays to Comp. III releasing water.

The above consideration leads to a suggestion that electrons compete with oxygen for half-reduced cytochrome oxidase. Usually complete reduction of cytochrome oxidase occurs after consumption of dissolved oxygen, and this situation does not occur under physiological conditions. Therefore, it is unlikely that the reduced oxidase is involved in the catalytic turnover even though its reaction with oxygen is so rapid. This reaction is a one-shot process, since if the reaction is initiated by mixing the fully reduced enzyme with oxygen the reduced form is not regenerated as long as a supply of electron and oxygen does continue.

Figure 5 summarizes the reaction cycle of cytochrome oxidase. An intermediate which appears after reaction of half-reduced cytochrome oxidase with molecular oxygen is Comp. I, although it is not indicated by a true electronic state. Its elucidation will be the problem of further studies. In this scheme it is indicated that both resting oxidized and reduced species do not participate in the catalytic turnover of physiological significance. Instead, another

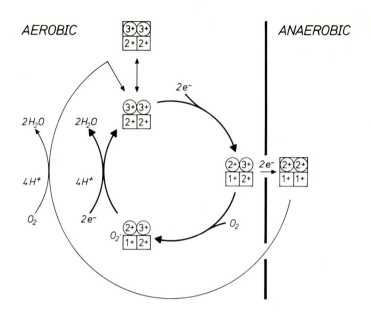

Fig. 5. Alternative pathways for cytochrome oxidase-catalyzed reduction of molecular oxygen. Small circle and square represent heme a and copper, respectively. A pathway indicated by a fine line is a one-shot process of less physiological importance. Another pathway indicated by a heavy line represents the reaction cycle of the enzyme in the physiologically functioning state. An oxygenated compound in the scheme is not indicated by the true electronic state. Enzyme species framed by a large squre are not involved in the physiological turnover.

type of oxidized form or Comp. III and a half-reduced species are involved.

Chance and coworkers identified spectrally an oxygen-carrying species of cytochrome oxidase as a reaction intermediate at low temperatures[21,22,23]. This species was obtained starting from the fully reduced oxidase. Therefore, although they claim that this species is an active reaction intermediate, this cannot be a compound of physiological significance. An argument against their claim along the similar line as the present one has been put forward by Erecińska and Wilson[24].

ACKNOWLEDGEMENT

This work was supported in part by a grant from the Ministry of Education, Science and Culture of Japan, No. 248132.

REFERENCES

1. Orii, Y. and King, T. E. (1972) FEBS Lett., 21, 199-202; (1976) J. Biol. Chem., 251, 7487-7493.

2. Okunuki, K., Hagihara, B., Sekuzu, I. and Horio, T. (1958) in Proc. Intern. Symposium on Enzyme Chemistry, Ichihara, K. ed., Maruzen, Tokyo, pp. 264-270.

3. Sekuzu, I., Takemori, S., Yonetani, T. and Okunuki, K. (1959) J. Biochem., 46, 43-49.

4. Orii, Y. and Okunuki, K. (1963) J. Biochem., 53, 489-499.

5. Orii, Y. and Okunuki, K. (1965) J. Biochem., 57, 45-54.

6. Beinert, H., Hartzell, C. R. and Orme-Johnson, W. H. (1971) in Probes of Structure and Function of Macromolecules and Membranes, Vol. 2, Chance, B., Yonetani, T. and Mildvan, A. S. eds., Academic Press, New York, N. Y., pp. 575-592.

7. Gibson, Q. H. and Greenwood, C. (1963) Biochem. J., 86, 541-554.

8. Rosén, S., Brändén, R., Vänngård, T. and Malmström, B. G. (1977) FEBS Lett., 74, 25-30.

9. Antonini, E., Brunori, M., Colosimo, A., Greenwood, C. and Wilson, M. T. (1977) Proc. Natl. Acad. Sci. USA, 74, 3128-3132.

10. Williams, G. R., Lemberg, R. and Cutler, M. E. (1968) Can. J. Biochem., 46, 1371-1379.

11. Tiesjema, R. H., Muijsers, A. O. and van Gelder, B. F. (1972) Biochim. Biophys. Acta, 256, 32-42.

12. Kornblatt, J. A., Kells, D. I. C. and Williams, G. R. (1975) Can. J. Biochem., 53, 461-466.

13. Okunuki, K., Yamamoto, T., Tsudzuki, T. and Orii, Y. (1972) in Structure and Function of Oxidation Reduction Enzymes, Akeson, A. and Ehrenberg, A. eds., Pergamon Press, Oxford, pp. 171-177.

14. Orii, Y., Manabe, M. and Yoneda, M. (1977) J. Biochem., 81, 505-517.

15. Orii, Y. and King, T. E. (1978) in Frontiers of Biological Energetics: From Electrons to Tissues, Dutton, P. L., Leight, J. S. and Scarpa, A. eds., Academic Press, New York, N. Y., in press.

16. Orii, Y., Matsumura, Y. and Okunuki, K. (1973) in Oxidases and Related Redox Systems, King, T. E., Mason, H. S. and Morrison, M. eds., University Park Press, Baltimore, pp. 666-672.

17. Schroedl, N. A. and Hartzell, C. R. (1977) Biochemistry, 16, 1327-1333.

18. Wilson, D. F. and Miyata, Y. (1977) Biochim. Biophys. Acta, 461, 218-230.

19. Wever, R., van Drooge, J. H., Muijsers, A. O., Bakker, E. P. and van Gelder, B. F. (1977) Eur. J. Biochem., 73, 149-154.

20. Greenwood, C., Wilson, M. T. and Brunori, M. (1974) Biochem. J., 137, 205-215.

21. Chance, B., Saronio, C., Leigh, J. S. (1975) J. Biol. Chem., 250, 9226-9237.

22. Chance, B. and Leigh, J. S. (1977) Proc. Natl. Acad. Sci. USA, 74, 4777-4780.

23. Chance, B., Saronio, C., Leigh, J. S., Ingledew, W. J. and King, T. E. (1978) Biochem. J., 171, 787-798.

24. Erecińska, M. and Wilson, D. F. (1978) Arch. Biochem. Biophys., 188, 1-14.

Cytochrome Oxidase, T.E. King et al. eds.
© *1979 Elsevier/North-Holland Biomedical Press*

THE MECHANISM OF THE FULLY REDUCED AND MIXED VALENCE STATE MEMBRANE BOUND
CYTOCHROME OXIDASE-OXYGEN REACTIONS IN THE 173-176 K TEMPERATURE RANGE

G. MARIUS CLORE

Department of Biochemistry, University College London, Gower Street, London
WC1E 6BT, U.K.

ABSTRACT

The kinetics of the fully reduced and mixed valence state membrane bound
cytochrome oxidase-O_2 reactions obtained in the Soret, α and near infrared
regions at 176 and 173 K respectively are presented and analysed by non-linear
numerical integration and optimization techniques. In the case of both
reactions, the only mechanism which is found to satisfy the triple requirement
of a standard deviation within the standard error of the data, good determination of the optimized parameters and a random distribution of residuals, is a
three intermediate sequential mechanism.

INTRODUCTION

Cytochrome oxidase (E.C.1.9.3.1) catalyses the terminal reaction (i.e. the
reduction of molecular O_2 to water) in the respiratory chain of all higher organisms. The minimum functioning unit of mammalian cytochrome oxidase is thought
to contain four metal centres consisting of two haems, a and a_3, and two copper
atoms, Cu_A and Cu_B[1].

The development of multi-channel dual wavelength optical spectroscopy[2] and a
low temperature kinetic method known as triple trapping[3] led to the identification of two spectroscopically distinct species in both the fully reduced (compounds A_1 and B) and mixed valence state (compounds A_2 and C) cytochrome oxidase
-O_2 reactions[4]. Difference spectra of compounds A_1 and B minus fully reduced
cytochrome oxidase ($a_3^{2+}Cu_B^+ \cdot a^{2+}Cu_A^+$), and compounds A_2 and C minus mixed
valence state cytochrome oxidase ($a_3^{2+}Cu_B^+ \cdot a^{3+}Cu_A^{2+}$)[4,7] (M. Denis, personal
communication). Compound A_1 has a 444 nm trough in the Soret region, a 591 nm
peak and a 611 nm trough in the α region, and no absorption bands in the near IR
region (700-1000 nm). Compound B has a 444 nm trough in the Soret region of
approximately twice the intensity of that of compound A_1, a 606 nm trough and no
peak in the α region, and a 790 nm peak in the near IR region. Compound A_2 has
a 444 nm trough in the Soret region of approximately the same intensity as that
of compound A_1, a 590 nm peak and a 612 nm trough in the α region, and no absorption bands in the near IR region. Compound C has identical features to

compounds A_1 and A_2 in the Soret region, an intense absorption band in the α region at 606-609 nm, and a 740 nm peak in the near IR region.

LOW TEMPERATURE KINETICS IN THE 173-176 K RANGE

On the basis of the optical spectra of the trapped compounds, detailed low temperature kinetic studies have been carried out at appropriate wavelength pairs by means of dual wavelength multi-channel spectroscopy[8-10]. Typical examples of the kinetics of the fully reduced and mixed valence state membrane bound cytochrome oxidase-O_2 reactions in the 173-176 K range are shown in Fig. 1. The 173-176 K temperature range was chosen for the following reason. Owing to the turbidity of the mitochondrial suspension the signal-to-noise ratio is too low at times less than 0.1 sec to obtain meaningful data. We therefore chose a temperature range at which the reactions proceeded at an optimum rate for monitoring at the time resolution available. Further, at higher temperatures cytochrome c oxidation occurs[6], further complicating the system under consideration. The mixed valence state cytochrome oxidase-O_2 reaction was monitored at five wavelength pairs simultaneously (444-463, 590-630, 604-630, 608-630 and 740-940 nm). In the case of the fully reduced cytochrome oxidase-O_2 reaction, the data in Fig. 1 represent two sets of experiments; in both experiments the 444-463, 590-630, 604-630, 608-630 and 740-940 nm wavelength pairs were monitored and used to determine the reproducibility of the results, the differences between experiments being less than 2%; in addition to these five wavelength pairs, the kinetics were monitored at 444-463 nm in the first experiment, and at 790-940 and 830-940 nm in the second. All the traces are multiphasic.

Analysis of low temperature kinetic data

The data presented in Fig. 1 are highly complex in that more than one species contributes to each progress curve, and both the number of species and the relative contributions of the species to the absorbance at the measured wavelength pairs are unknown. In order to fit such data it is essential to have a set of strict quantitative criteria upon which to base one's choice of model. Such criteria have been developed[8] and consist of the following triple requirement:-

(1) The standard deviation (SD) of the fit should be within the standard error of the data. We minimize χ^2 given by

$$\chi^2 = \sum_{i=1}^{n} \sum_{j=1}^{m} R_{ij}^2$$

$$= \sum_{i=1}^{n} \sum_{j=1}^{m} \{(v_{ij} - u_{ij})/\sigma_i\}^2 \qquad (1)$$

where j identifies the time point and i the data curve, R_{ij} are the residuals, v_{ij} the observed values, u_{ij} the corresponding calculated values and σ_i the standard error of curve i. The SD which, unlike χ^2, is independent of the number of data points, is calculated from χ^2 using the equation

$$SD = \phi\sqrt{\chi^2(d-p)} \qquad (2)$$

where d is the total number of experimental points, p the number of optimized parameters and ϕ the overall standard error of the data given by the weighted mean of the standard errors of the individual curves

$$\phi = \Sigma\sigma_i r_i/\Sigma r_i \qquad (3)$$

(where r_i is the range of curve i).

(2) The optimized parameters should be well determined. A quantitative measure of how accurately an unknown parameter has been determined by optimization is given by the standard deviation of the natural logarithm (SD_{ln}) of the unknown parameter. Since rate constants and other parameters need to be varied over a large range of values, the natural logarithm of the unknown parameter is varied, and, consequently, the SD_{ln} is calculated from the non-linear covariance. Because of the linearity of logarithms less than 0.2, a parameter x whose SD_{ln} lies below this value has a relative error, $\Delta x/x$, of $\pm SD_{ln}$ and is considered to have a well determined minimum in multi-dimensional parameter space. For larger values of SD_{ln}, up to 1 in magnitude, the parameter values is determined to within a factor $e \approx 2.72$, and so its order of magnitude is known. Significantly larger values of SD_{ln} show that the observations are inadequate to determine the parameter.

(3) The distribution of residuals should be random. A measure of the distribution of residuals for the overall fit is given by the mean absolute correlation index ($\bar{C}$):

$$\bar{C} = \frac{1}{n} \sum_{i=1}^{n} | \sum_{j=1}^{m} R_{ij}/(\sum_{j=1}^{m} R_{ij}^2)^{\frac{1}{2}}| \qquad (4)$$

where m is the number of time points for each curve and n the number of curves. A value of $\bar{C}$ significantly greater than 1.0 (the expected root mean square value of $\bar{C}$ if the residuals for each curve were all independent random variables of zero mean and the same variance) indicates that the departures between calculated and observed values are systematic[8].

This triple requirement excludes a large number of alternative models. In fact, in stiff non-linear problems, it is usually the case that only a single model will satisfy this triple requirement. Thus, for a given set of data, although there may be many models with a SD within the standard error of the data, models with too many degrees of freedom will fail such an analysis because of under-determination, whereas models with too few degrees of freedom will fail such an analysis because of the introduction of systematic errors in the distribution of residuals.

Preliminary attempts at non-linear optimization of the coefficients of the differential equations representing a reaction system with two intermediates was found not to fit the data for both the fully reduced and mixed valence state cytochrome oxidase-O_2 reactions on the basis of criteria (1) and (3): the SD of the fits were greater than 10% compared to a 2% standard error of the data and there were systematic errors in the distribution of residuals[8,9]. To exclude the possibility that the failure to fit the data to a two intermediate sequential mechanism was due to superimposed spectrophotometer drift, we carried out optimizations in which a zero order process was added to simulate drift. This resulted in no improvement in either the SD of the fits or the distribution of residuals. This clearly indicated the presence of other kinetically competent intermediates which had not been identified by low temperature trapping and wavelength scanning optical spectroscopy.

For both the fully reduced and mixed valence state cytochrome oxidase-O_2 reactions in the 173-176 K range, the only model which is found to satisfy the triple requirement set out above is a three intermediate sequential mechanism which is stated as follows[8,9]:

$$E_{(M)} + O_2 \underset{k_{-1}}{\overset{k_{+1}}{\rightleftharpoons}} I_{(M)} \underset{k_{-2}}{\overset{k_{+2}}{\rightleftharpoons}} II_{(M)} \underset{k_{-3}}{\overset{k_{+3}}{\rightleftharpoons}} III_{(M)} \qquad (5)$$

where E and E_M are free fully reduced and mixed valence state cytochrome oxidase respectively produced by flash photolysis of the fully reduced and mixed valence state cytochrome oxidase-CO complexes; I, II and III the intermediates in the fully reduced cytochrome oxidase-O_2 reaction; I_M, II_M and III_M the intermediates in the mixed valence state cytochrome oxidase-O_2 reaction. Although more complicated models involving more intermediates and/or branching pathways are possible, the corresponding increase in the number of parameters which have to be optimized results in an under-determined system. Further, the only alternative models involving the same number of parameters are stated as:

$$E_{(M)} + O_2 \rightleftharpoons I_{(M)} \underset{\rightleftharpoons III_{(M)}}{\overset{\rightleftharpoons II_{(M)}}{}} \qquad (6)$$

$$E_{(M)} + O_2 \overset{\rightleftharpoons I_{(M)} \rightleftharpoons II_{(M)}}{\underset{\rightleftharpoons III_{(M)}}{}} \qquad (7)$$

$$E_{(M)} \cdot CO \underset{h\nu}{\overset{CO}{\rightleftharpoons}} E_{(M)} \overset{O_2}{\rightleftharpoons} I_{(M)} \rightleftharpoons II_{(M)} \qquad (8)$$

and, like the two intermediate reaction system, fail to fit the data on the basis of SD ($\gg 10\%$) and systematic errors in the distribution of residuals[8,9]. (It should be noted that the model given by eqn. 8 represents a system where photolysis by the xenon flash is incomplete and slow photolysis of the remaining CO complex, $E_{(M)} \cdot CO$, produced by the measuring light beam, takes place during the entire course of the reaction.)

The triphasic nature of the fully reduced cytochrome oxidase-O_2 reaction is clearly indicated by the characteristic plateau phase of the 608-630 nm curve which is emphasized when plotted on a logarithmic time base (Fig. 1A). The triphasic nature of the mixed valence state cytochrome oxidase-O_2 reaction, however, is not immediately obvious from the progress curves plotted on a logarithmic time base in Fig. 1B. In Fig. 2, we therefore show the progress curves at 590-630, 604-630 and 608-630 nm of the mixed valence state cytochrome oxidase-O_2 reaction at 173 K on a slow linear time base. Looking at the 608-630 nm trace, we see that following the increase in absorbance produced by photolysis of the mixed valence state CO complex, the first phase is characterized by a rapid decrease in absorbance; this is followed by a relatively rapid second phase and a slow third phase during which the absorbance increases. The same changes are seen in the 590-630 nm trace except that the absorbance changes are reversed. In the case of the 604-630 nm trace, however, no clearly distinct third phase is seen. Further, the times at which the 604-630 and 608-630 nm traces reach their minimum and the 590-630 nm trace its maximum are all different. This too is indicative of the triphasic nature of the mixed valence state cytochrome oxidase-O_2 reaction and is best seen with the data plotted on a logarithmic time base in Fig. 1B.

The contribution of each intermediate to the absorbance at each wavelength

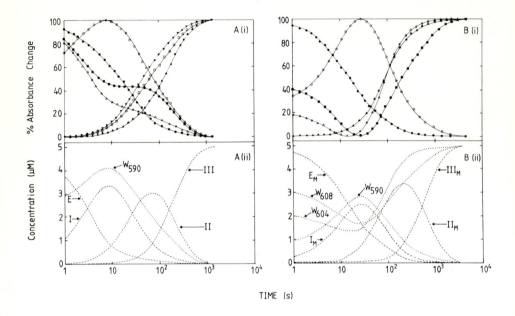

Fig. 1. (See legend on the opposite page).

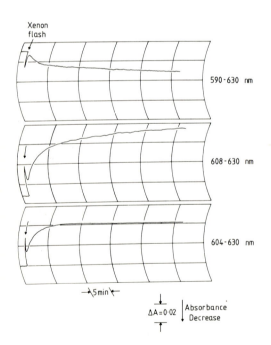

Fig. 2. Observed kinetics of the mixed valence state membrane bound cytochrome oxidase-O_2 reaction at 173 K recorded on a slow linear time scale.
Experimental conditions: as in Fig. 1B except that the O_2 concentration was 500 μM.

Fig. 1. Observed and computed kinetics of the fully reduced (A) and mixed valence state (B) membrane bound cytochrome oxidase-O_2 reactions at 176 and 173 K respectively.

The normalized experimental data, digitized by the method of Clore & Chance[8], are shown as: ●, 444-463 nm; ○, 590-630 nm; ▲, 604-630 nm; ■, 608-630 nm; ▲, 740-940 nm; □, 790-940 nm; ▼, 830-940 nm. The computed normalized absorbance changes are shown as solid lines (——); the computed kinetics of the intermediates as interrupted lines (– – –); and the crude computed absorbance changes, in units of concentration, at 590-630, 604-630 and 608-630 nm as dotted lines (······). The computed curves were obtained by numerical integration of the differential equations representing a three intermediate sequential mechanism (eqn. 12) using the optimized values of the rate constants and relative extinction coefficients given in tables 1 and 2 respectively. The SD of the fits are less than 2%, the overall standard error of the data for both reactions. The mean absolute correlation index (eqn. 4) has a value less than 1.0 in both cases indicating a random distribution of residuals[8].

Experimental procedure. Beef heart mitochondria[17] were suspended at 298 K in a medium containing 0.1 M mannitol, 50 mM sodium phosphate buffer pH 7.2 (which was stabilized down to temperatures as low as 143 K by the high concentration of protein present in the mitochondrial suspension[8]), 10 mM succinate, and left for 10 min until all the O_2 in the preparation was exhausted. The system was then cooled to 273 K and saturated with 100% CO at 1 atm. for 10 min. Ethylene glycol was added (final concn. 30% v/v) and the preparation resaturated with 100% CO at 1 atm. for a further 20 min to ensure full anaerobiosis and CO saturation. Samples were then transferred into 2 mm optical path cuvettes previously deaerated with CO for optical studies. Oxygenation was carried out by the addition of O_2 saturated 30% v/v ethylene glycol in 50 mM sodium phosphate buffer pH 7.2 (containing 1.2 and 2 mM O_2 when saturated at 296 and 250 K respectively). The cuvette was then transferred to an ethanol/solid CO_2 bath at 195 K and the suspension stirred in the dark until the viscosity increased and freezing occured. This procedure prevents ligand exchange between O_2 and the CO inhibited system[4]. The mixed valence state oxidase was prepared by adding $K_3Fe(CN)_6$ (final concn. 5 mM) 30 sec before oxygenation[9]. The cuvette was then placed in the Dewar flask of the spectrophotometer through which was flowing thermoregulated N_2 of the desired temperature. A copper constantan thermocouple was used to monitor the temperature of the measuring chamber which varied less than ± 0.2 °C during the course of each experiment.

The data were recorded using a Johnson Foundation dual wavelength multi-channel spectrophotometer[2]. The wavelengths of light were isolated by filters of appropriate spectral intervals and interlaced, one with another, by synchronized 60 Hz rotating discs. The measuring beam was provided by a Tungsten iodide lamp; the intensity was not sufficient to perturb the measured kinetics. The transmitted light was monitored using a multi-alkali photomultiplier for the 400-700 nm range (EMI 9592b) and a silicon diode detector for the 700-1000 nm range (United Detector Technology PIN-10). The reaction was activated by photolysis of the CO complex (t=0 sec) using a 200 J Xenon flash with a pulse width of 1 msec. The flash was approx. 99% saturating and CO did not recombine to a detectable extent in the presence of the relatively high O_2 concentration employed as shown by control experiments where repeated flashes over the course of the experiment only produced approx. 1% further photolysis of the CO complex, the O_2 intermediates not being susceptible to photolysis at the flash entensity used.

Final concentrations in the samples were: A, 15 mg/ml beef heart mitochondria containing 5 μM cytochrome oxidase (calculated from $\varepsilon_{red.ox}^{605}$=24 mM^{-1}cm^{-1}, ref. 18), 0.05 M mannitol, 50 mM sodium phosphate buffer pH 7.2, 5 mM succinate, 0.6 mM CO and 750 μM O_2; B, as in A plus 5 mM $K_3Fe(CN)_6$.

pair is represented by a relative extinction coefficient so that the crude computed absorbance at the ith wavelength, $W_i(t)$, in units of concentration, is given by

$$W_i(t) = \sum_\ell F_\ell(t)\varepsilon_i'(\ell) \tag{9}$$

where $F_\ell(t)$ is the concentration of the ℓth intermediate at time t and $\varepsilon_i'(\ell)$ the relative extinction coefficient of the ℓth intermediate at the ith wavelength[8]. The $\varepsilon_i'(\ell)$ are varied relative to a normalized difference extinction coefficient, $\Delta\varepsilon_i'(x-z)$, given by

$$\Delta\varepsilon_i'(x-z) = \Delta\varepsilon_i(x-z)/\Delta\varepsilon_i(x-z)$$

$$= \varepsilon_i'(x) - \varepsilon_i'(z)$$

$$= 1.0 \tag{10}$$

where $\Delta\varepsilon_i(x-z)$ is the molar difference extinction coefficient between species x and z. Thus, the relative extinction coefficients $\varepsilon_i'(x)$ and $\varepsilon_i'(z)$ of the two reference species x and z are given values of 1.0 and 0 respectively. Therefore, if $\Delta\varepsilon_i^T(x-z)$ at a temperature T is known, $\Delta\varepsilon_i^T(\ell-z)$ is given by

$$\Delta\varepsilon_i^T(\ell-z) = \varepsilon_i'(\ell)\Delta\varepsilon_i^T(x-z) \tag{11}$$

The $F_\ell(t)$ are obtained by numerical integration (using a modified version of Gear's method[11,12]) of the coupled simultaneous ordinary differential equations representing the model stated in eqn. 5:

$$dE_{(M)}/dt = k_{-1}I_{(M)} - k_{+1}E_{(M)}O_2$$

$$dO_2/dt = dE_{(M)}/dt$$

$$dI_{(M)}/dt = k_{+1}E_{(M)}O_2 + k_{-2}II_{(M)} - (k_{-1}+k_{+2})I_{(M)} \tag{12}$$

$$dII_{(M)}/dt = k_{+2}I_{(M)} + k_{-3}III_{(M)} - (k_{-2}+k_{+3})II_{(M)}$$

$$dIII_{(M)}/dt = k_{+3}II_{(M)} - k_{-3}III_{(M)}$$

with initial conditions:

$$E_{(M)} = E_{(M)}^i \qquad I_{(M)} = II_{(M)} = III_{(M)} = 0$$

$$O_2 = O_2^i \tag{13}$$

where $E_{(M)}^i$ is the concentration of $E_{(M)}$ generated by flash photolysis of the CO complex (note: the flash is approximately 99% saturating) and O_2^i the initial concentration of O_2 in the 'pocket' of cytochrome oxidase containing the active

TABLE 1

A Optimized values of the rate constants for the reaction of fully reduced membrane bound cytochrome oxidase with O_2 at 176 K[a].

$$O_2 + E^b \underset{0.043\ s^{-1}}{\overset{\overset{k_{+1}}{360\ M^{-1}\ s^{-1}}}{\underset{k_{-1}}{\rightleftharpoons}}} I \underset{0.0044\ s^{-1}}{\overset{\overset{k_{+2}}{0.034\ s^{-1}}}{\underset{k_{-2}}{\rightleftharpoons}}} II \underset{1\times10^{-6}\ s^{-1}}{\overset{\overset{k_{+3}}{0.055\ s^{-1}}}{\underset{k_{-3}}{\rightleftharpoons}}} III$$

B Optimized values of the rate constants for the reaction of mixed valence state membrane bound cytochrome oxidase with O_2 at 173 K[a].

$$O_2 + E_M^c \underset{0.023\ s^{-1}}{\overset{\overset{k_{+1}}{76\ M^{-1}\ s^{-1}}}{\underset{k_{-1}}{\rightleftharpoons}}} I_M \underset{0.00068\ s^{-1}}{\overset{\overset{k_{+2}}{0.016\ s^{-1}}}{\underset{k_{-2}}{\rightleftharpoons}}} II_M \underset{1\times10^{-6}\ s^{-1}}{\overset{\overset{k_{+3}}{0.0017\ s^{-1}}}{\underset{k_{-3}}{\rightleftharpoons}}} III_M$$

[a] The relative error of the rate constants ($\Delta k_i/k_i$) are within ± 0.10 with the exception of k_{-3} which was constrained at a value of 1×10^{-6} s^{-1} on the basis of initial optimizations in which its value was found to be small and very poorly determined ($SD_{1n} \gg 10$).

[b] E is fully reduced cytochrome oxidase: $a_3^{2+}Cu_B^+ \cdot a^{2+}Cu_A^+$.

[c] E_M is mixed valence state cytochrome oxidase: $a_3^{2+}Cu_B^+ \cdot a^{3+}Cu_A^{2+}$.

site. The observation of pseudo-first order kinetics of the reaction of O_2[4,7] and CO[13-15] with cytochrome oxidase at temperatures as low as 168 K, together with the demonstration of gas exchanges in the active site at low temperatures[16], indicates that there is trapped within the active site a population of ligand molecules that is proportional to that of the solvent at the time of freezing. The linearity of Arrhenius plots of k_{+1} from 163 K to the melting point of the solvent in the fully reduced and mixed valence state cytochrome oxidase-O_2 reactions[4,7] suggests that the constant of proportionality is 1. We have therefore assumed, on the basis of the above evidence, that the concentration of O_2 in the 'pocket' of cytochrome oxidase (i.e. the initial O_2 concentration) is equal to that in the solvent.

As the digitized data are scaled between 0 and 1, $W_i(t)$ is converted into a normalized absorbance change, $N_i(t)$, for comparison with the normalized data by

TABLE 2

A Optimized values of the relative extinction coefficients of the intermediates in the reaction of fully reduced membrane bound cytochrome oxidase with O_2[a].

Relative extinction coeffcients at:	E	I	II	III
444-463 nm $\{\varepsilon'_{444}(\ell)\}$	1.0^b	0.66	0^c	0^b
590-630 nm $\{\varepsilon'_{590}(\ell)\}$	0.46	1.0^b	0.36	0^b
604-630 nm $\{\varepsilon'_{604}(\ell)\}$	1.0^b	0.17	0.20	0^b
608-630 nm $\{\varepsilon'_{608}(\ell)\}$	1.0^b	0.28	0.52	0^b
740-940 nm $\{\varepsilon'_{740}(\ell)\}$	0^b	0^c	0.25	1.0^b
790-940 nm $\{\varepsilon'_{790}(\ell)\}$	0^b	0^c	0.57	1.0^b
830-940 nm $\{\varepsilon'_{830}(\ell)\}$	0^b	0^c	0.72	1.0^b

B Optimized values of the relative extinction coefficients of the intermediates in the reaction of mixed valence state membrane bound cytochrome oxidase with O_2[a].

Relative extinction coefficients at:	E_M	I_M	II_M	III_M
444-463 nm $\{\varepsilon'_{444}(\ell)\}$	1.0^b	0^c	0^c	0^b
590-630 nm $\{\varepsilon'_{590}(\ell)\}$	0.15	1.0^b	0.13	0^b
604-630 nm $\{\varepsilon'_{604}(\ell)\}$	0.42	0^b	1.0	1.0^b
608-630 nm $\{\varepsilon'_{608}(\ell)\}$	0.64	0^b	0.70	1.0^b
740-940 nm $\{\varepsilon'_{740}(\ell)\}$	0^b	0^c	0.93	1.0^b

[a] The relative errors of the optimized relative extinction coefficients are within ± 0.10.

[b] The relative extinction coefficients are varied relative to a normalized difference extinction coefficient $\Delta\varepsilon'_i(x-z)$ defined by eqn. (10) so that $\varepsilon'_i(x)$ and $\varepsilon'_i(z)$ are given values of 1.0 and 0 respectively. In the case of both the fully reduced and mixed valence state cytochrome oxidase-O_2 reactions, the $\varepsilon'_{444}(\ell)$, $\varepsilon'_{590}(\ell)$, $\varepsilon'_{740}(\ell)$, $\varepsilon'_{790}(\ell)$ and $\varepsilon'_{830}(\ell)$ are varied relative to $\Delta\varepsilon'_{444}(E_{(M)}-III_{(M)})$, $\Delta\varepsilon'_{590}(I_{(M)}-III_{(M)})$, $\Delta\varepsilon'_{740}(III_{(M)}-E_{(M)})$, $\Delta\varepsilon'_{790}(III_{(M)}-E_{(M)})$ and $\Delta\varepsilon'_{830}(III_{(M)}-E_{(M)})$ respectively. The $\varepsilon'_{604}(\ell)$ and $\varepsilon'_{608}(\ell)$ are varied relative to $\Delta\varepsilon'_{604}(E-III)$ and $\Delta\varepsilon'_{608}(E-III)$ in the case of the fully reduced cytochrome oxidase-O_2 reaction, and relative to $\Delta\varepsilon'_{604}(III_M-I_M)$ and $\Delta\varepsilon'_{608}(III_M-I_M)$ in the case of the mixed valence state cytochrome oxidase-O_2 reaction.

[c] These relative extinction coefficients were constrained at a value of zero on the basis of initial optimizations in which their values were found to be small and very poorly determined ($SD_{1n} \gg 10$).

means of a scale factor (S_i) and offset (D_i):

$$N_i(t) = \{W_i(t) - D_i\}/S_i \qquad (14)$$

Examples of fits to typical data, obtained by optimization of the relative extinction coefficients and rate constants given in eqns. 9 and 12, are shown in Fig. 1 together with the computed kinetics of the individual intermediates. The SD of the fits for both the fully reduced and mixed valence state cytochrome oxidase-O_2 reactions are less than 2%, the overall standard error of the data. The mean absolute correlation index (eqn. 4) has a value less than 1.0 in both cases indicating a random distribution of residuals[8]. The optimized parameters are well determined with a relative error $\Delta x/x$ of approximately ± 0.1. The optimized values of the rate constants for the fully reduced and mixed valence state cytochrome oxidase-O_2 reactions at 176 and 173 K respectively are shown in table 1, and the optimized values of the relative extinction coefficients of the intermediates in table 2. The effect of O_2 concentration under pseudo-first order conditions on the calculated maximum concentrations of intermediates I and II, and I_M and II_M is shown in Fig. 3.

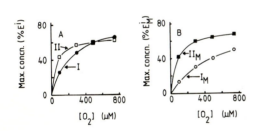

Fig. 3. Effect of O_2 concentration on the calculated maximum concentrations of intermediates I ($\bullet$) and II ($\square$), and I_M (o) and II_M ($\blacksquare$) in the fully reduced (A) and mixed valence state (B) cytochrome oxidase-O_2 reactions at 176 and 173 K respectively as a % of the initial concentrations of fully reduced (E^i) and mixed valence state (E_M^i) cytochrome oxidase.

The chemical nature of the intermediates in the fully reduced and mixed valence state membrane bound cytochrome oxidase-O_2 reactions and the assignment of valence states to the metal centres in these intermediates is discussed in detail elsewhere[8-10].

ACKNOWLEDGEMENTS

I wish to thank Prof. B. Chance and Drs. E.M. Chance and M.R. Hollaway for several useful discussions, Profs. S.P. Datta and B.R. Rabin for support, the Johnson Research Foundation for experimental facilities and the University College London Computer Centre for computing facilities. Part of this work was carried out during the tenure of a Boehringer Mannheim Travelling Fellowship.

REFERENCES

1. Caughey, W.S., Wallace, W.J., Volpe, J.A. and Yoshikawa, S. (1976) in The Enzymes, 3rd edn. (Boyer, P.D., ed.) vol. 13, pp. 299-344, Academic Press, New York.

2. Chance, B., Legallais, V., Sorge, J. and Graham, N. (1975) Analytical Biochem. 66, 498-514.

3. Chance, B., Graham, N. and Legallais, V. (1975) Analytical Biochem. 67, 552-579.

4. Chance, B., Saronio, C. and Leigh, J.S. (1975) J. Biol. Chem. 250, 9226-9237.

5. Chance, B. and Leigh, J.S. (1977) Proc. Nat. Acad. Sci. U.S.A. 74, 4777-4780.

6. Chance, B., Leigh, J.S. and Waring, A. (1977) in Structure and Function of Energy Transducing Membranes (Van Dam, K. and Van Gelder, B.F., eds.) pp. 1-10, Elsevier North-Holland, Amsterdam.

7. Chance, B., Saronio, C., Leigh, J.S., Ingledew, W.J. and King, T.E. (1978) Biochem. J. 171, 787-798.

8. Clore, G.M. and Chance, E.M. (1978a) Biochem. J. 173, 799-810.

9. Clore, G.M. and Chance, E.M. (1978b) Biochem. J. 173, 811-820.

10. Clore, G.M. and Chance, E.M. (1978c) Biochem. J. in the press.

11. Gear, C.W. (1971) Commun. Assoc. Comp. Mach. 14, 176-179.

12. Chance, E.M., Curtis, A.R., Jones, I.P. and Kirby, C.R. (1977) Atomic Energy Research Establishment (Harwell) Report No. R.8775.

13. Sharrock, M. and Yonetani, T. (1976) Biochim. Biophys. Acta 434, 333-344.

14. Sharrock, M. and Yonetani, T. (1977) Biochim. Biophys. Acta 462, 718-730.

15. Clore, G.M. and Chance, E.M. (1978d) Biochem. J. in the press.

16. Hartzell, C.H. and Beinert, H. (1974) Biochim. Biophys. Acata 368, 318-338.

17. Low, H. and Vallin, I. (1963) Biochim. Biophys. Acta 69, 371-374.

18. Van Gelder, B.F. (1963) Biochim. Biophys. Acta 73, 663-665.

Cytochrome Oxidase, T.E. King et al. eds.
© 1979 Elsevier/North-Holland Biomedical Press

THE ROLE OF PEROXIDASE-LIKE INTERMEDIATES IN THE ENZYMATIC FUNCTION OF CYTO-
CHROME OXIDASE OVER A VARIETY OF TEMPERATURES

BRITTON CHANCE AND ALAN WARING

The Johnson Research Foundation, University of Pennsylvania, Phila., PA 19104

LINDA POWERS

Bell Laboratories, 600 Mountain Ave., Murray Hill, NJ 07974

ABSTRACT

This contribution proposes a mechanism for the reactions with oxygen of the
normal complement of the membrane-bound cytochromes of intact beef heart mitochondria. At low temperatures the complications involved in the function of
the total respiratory chain are minimized and activation of electron transfer
appears to be limited to cytochrome oxidase and its "nearest neighbor" electron
donors cytochromes a, c, c_1, and Fe-S protein. At the lowest temperatures
(-80°), cytochrome c and, at higher temperatures, cytochromes c and a and its
associated copper aCu_a serve as electron donors to the intermediate peroxide
Compound B_2 of cytochrome a_3 and its associated copper Cu_{a_3}. Oxygen binding
and intramolecular electron transfer to oxygen by cytochrome $a_3Cu_{a_3}$ are tempo-
rally distinguished from intermolecular electron transfer from cytochrome c
and cytochrome aCu_a to Compound B_2. The function of the oxygen reduction products as pero- and μ-oxo-ligands is described as are the higher valence states
of iron in intermediates similar to those of peroxidase peroxide Compounds I
and II. The provision of pools of reducing power minimizes the concentration
of oxidized and intermediate states of cytochrome oxidase and maximizes the
concentration of the reduced state - and thereby maximizes the rate of its
reaction with oxygen. These ideas afford a basis for a first approximation to
the cytochrome oxidase reaction mechanism in situ in the biological membrane
at low temperatures and illustrate the possible application of this reaction
mechanism in affording rapid recovery from anoxic-ischemic episodes under
physiological conditions.

INTRODUCTION

Low temperature experiments serve to identify intermediates in oxygen reduction by cytochrome oxidase and in addition reveal new characteristics of the
electron transfer to cytochrome oxidase not hitherto revealed[1,2]. The specific
findings are the identification of intermediate compounds of cytochrome oxidase

in which only heme $\underline{a}_3$ and its copper $(\underline{a} \cdot Cu_{a_3})$* appear to be involved (Compounds B_2 and C_2)[3]; the identification of copper associated with heme $\underline{a}_3$ (Cu_{a_3}) as a reactive and functional component similar to Type I "blue" copper[4,5]; and the unrecognized ability of $\underline{a}_3Cu_{a_3}$ to accept electrons from cytochrome $\underline{c}$ — transferring electrons at very low temperatures under conditions where no change in heme $\underline{a} \cdot Cu_a$ can be detected[3]. This result identifies $\underline{a}_3 \cdot Cu_{a_3}$ in the form of Compound B_2 as the much more important electron transfer component of cytochrome oxidase function. With these new data in hand, an attempt has been made to formulate a set of equations which describe a possible mechanism for the sequential reduction of oxygen by cytochrome oxidase at low temperatures, and the relation of the low temperature mechanism to the function of cytochrome oxidase at high temperatures.

THE LOW TEMPERATURE REACTIONS

In the temperature range $-100°$ to $-40°$, $\underline{a}_3 \cdot Cu_{a_3}$ and the Compound B_2 derived therefrom are the principal participants in the electron transfer reactions[1] and $\underline{a}Cu_a$ is omitted from the equations:

$$\underline{a}_3^{2+} \; Cu_{a_3}^{1+} + O_2 \leftrightarrow \underset{A_1}{\underline{a}_3^{2+} - O_2 \; Cu_{a_3}^{1+}} \tag{1}$$

Eqn. 1 represents the initial step. Unliganded $\underline{a}_3^{2+} \cdot Cu^{1+}$ obtained by flash photolysis of the CO compound reacts with oxygen at temperatures of $-125°$ and upwards. The reaction product is oxy-cytochrome oxidase (Compound A_1) in which oxygen binds the heme iron and forms a compound similar to that with carbon monoxide. Under the conditions of these experiments, a considerable stoichiometric excess of oxygen is required to form Compound A_1, presumably due to the need to populate the active site (oxygen pocket) with an adequate number of oxygen molecules at the low temperatures[3]. Oxygen interaction with the copper cannot be determined since the near infrared spectrum is not responsive to the state of reduced copper[4].

---

*The designation of iron and copper moieties in cytochrome oxidase differs from that of other authors and is based upon reinterpretations of the nature and function of the components. The heme of cytochrome $\underline{a}_3$ is designated $\underline{a}_3$ and the closely coupled copper has previously been denoted Cu^u (u for undetectable) is now revised (Cu_{a_3}) because it is now readily detectable by optical and x-ray spectroscopy. This designation $\underline{a}_3Cu_{a_3}$ indicates their close geometric association. Cytochrome $\underline{a}$ is designated $\underline{a}$ and the copper, otherwise denoted Cu_D (for epr detectable) is denoted Cu_a on account of its close kinetic and thermodynamic association with $\underline{a}$.

INTRAMOLECULAR ELECTRON TRANSFER

$$a_2^{2+} \xrightarrow{\;\;O_2\;\;} Cu_{a_3}^{1+} \longrightarrow a_3^{3+} - OO^= \underset{B_2}{Cu_{a_3}^{2+}} \qquad (2)$$

At -105° (Eqn.2) absorption decreases are observed at wavelengths characteristic of the γ and α bands of reduced cytochrome oxidase and increases are
observed at 655 nm appropriate to an intramolecular electron transfer and involving the formation of a_3^{3+} (in the absence of detectable g=6 esr signals[3,6])
and $Cu_{a_3}^{2+}$ as monitored at 750-800 nm where absorbance increases indicate the
formation of $Cu_{a_3}^{2+}$ [1,4]. Since the absorbance change observed is too small to
be due to the formation of $a^{3+} \cdot Cu_a^{2+}$, it is concluded that only two electrons
are transferred (one from the iron and one from the copper of $a_3^{2+} \cdot Cu_{a_3}^{1+}$) to give
a ferric-cupric peroxy compound termed Compound B_2 which can be a bridged or
unbridged peroxide complex[1,2,4].

INTERMOLECULAR ELECTRON TRANSFER

The first intermolecular electron transfer event that can be observed at low
temperatures (-80°) in isolated mitochondria is the oxidation of cytochrome c
as monitored at the usual wavelength pair 550-540 nm with a half-time of ~30
min[3,7] (Eqn.3) and, surprisingly, without detectable oxidation of aCu_a. At
these low temperatures only a small fraction of the total complement of cytochrome c beef heart mitochondria (two molecules per heme a) is oxidized.

Compound B_2 accepts electrons from cytochrome c and serves as an intermediate in this reaction as verified by spectroscopic observations which show
Compound B_2 to be maintained nearly in a steady state from -80° to -50° (as
in Eqn.3). Presumably both c^{3+} and a_3^{3+} (as observed at 655 nm) are products
of the reaction of two molecules of cytochrome c^{2+} and Compound B_2 (3).

Two mechanisms are considered, first electron donation from cytochrome c
directly to the peroxide:

$$c^{2+} + a_3^{3+} - O\,O^= \, Cu_{a_3}^{2+} + 2H^+ \longrightarrow c^{3+} + H_2O + a_3^{3+} - O^= - Cu_{a_3}^{2+} \qquad (3)$$

The reaction product may initially be a μ-oxo bridged compound. Alternatively, there may be a prior intramolecular oxidation-reduction reaction:

$$a_3^{3+} - OO^= \, Cu_{a_3}^{2+} + 2H^+ \longrightarrow a_3^{4+} - O^- \, Cu_{a_3}^{2+} + H_2O \qquad (4)$$

leading to the formation of a ferryl ion, Compound II or to the Compound I configuration[8]:

$$a_3^{4+} - O^- \; \underset{a_3}{Cu}^{2+} \longrightarrow a_3^{5+} - O^= - \underset{a_3}{Cu}^{2+} \qquad (5)$$

These two intermediates may react with cytochrome $\underline{c}$ in a stepwise fashion as in the case of horse-radish or yeast cytochrome $\underline{c}$ peroxidase to give the μ-oxo or resting forms of the oxidase[8,9]:

$$2\underline{c}^{2+} + a_3^{5+} - O^= \underset{a_3}{Cu}^{2+} \longrightarrow 2\underline{c}^{3+} + a_3^{3+} - O^= - \underset{a_3}{Cu}^{2+} \qquad (6)$$

THE MIDDLE TEMPERATURE RANGE

The participation of $\underline{a}Cu_a$ in the electron transfer reaction between $-50°$ and $-20°$ depends upon the presence of cytochrome $\underline{c}$. One explanation is that this occurs by a reversal of the usual reaction sequence of the cytochrome $\underline{a}$ - $\underline{c}$ electron transport reaction.

$$2\underline{c}^{3+} + \underline{a}^{2+} \; \underset{a}{Cu}^{1+} \longleftrightarrow 2\underline{c}^{2+} + \underline{a}^{3+} \; \underset{a}{Cu}^{2+} \qquad (7)$$

This could be due to a low temperature perturbation of the mid-potential to a more positive value or to an alteration of tunneling parameters such as barrier height and barrier width so that the reaction of cytochrome $\underline{c}$ and $\underline{a} \cdot Cu_a$ occurs. Whatever these changes may be, they are less at higher temperatures at which cytochrome $\underline{c}$ seems to transfer electrons through the normal pathway to $\underline{a}Cu_a$.

THE HIGHER TEMPERATURE RANGE

The principal step of the reaction mechanism occurring above $-20°$ is a rapid reduction of $a_3^{3+} - O^= - \underset{a_3}{Cu}^{2+}$ and oxidation of $\underline{a}Cu_a$ in a reaction that keeps pace with the oxidation of cytochrome $\underline{c}$ (in contrast to the region between $-80°$ and $-20°$). This step is depicted in Eqn.8 as a simple electron exchange reaction and the elimination of water:

$$a_3^{3+} - O^= - \underset{a_3}{Cu}^{2+} + \underline{a}^{2+} \underset{a}{Cu}^{1+} + 2H^+ \longleftrightarrow a_3^{2+} \; \underset{a_3}{Cu}^{1+} + \underline{a}^{3+} \underset{a}{Cu}^{2+} + H_2O \qquad (8)$$

Eqn.7 operates in the usual direction (reverse of Eqn.7 as written) for the reaction of cytochrome $\underline{c}$ with cytochrome oxidase where two molecules of cytochrome $\underline{c}$ donate one electron each to $\underline{a}^{3+} \cdot Cu_a^{2+}$, reducing them to the $\underline{a}^{2+} \cdot Cu_a^{1+}$ state and allowing the formation of oxidized cytochrome $\underline{c}$ and the recycling of $\underline{a}Cu_a$ back through Eqn.8. The sum of these reactions gives the usual over-- all reaction for cytochrome oxidase.

THE ROLE OF PEROXIDATIC REACTIONS

In Compound B_2 (Eqn.2) two of the oxidizing equivalents have been trans- ferred to the Fe and Cu and two remain the peroxide. Thereafter, a variety of pathways by which the peroxide may be reduced to water. Stepwise reduction of the peroxide and a split of the peroxide bridge causes one oxygen atom to be reduced to water while the other bears the two remaining oxidizing equiva- lents leading to stepwise oxidation of the iron to form the ferryl ion "Com- pound II" (Eqn.4) [both familiar in peroxidase reaction mechanisms and the pentavalent iron(Eqn.5) or Compound I state]. These two intermediates may well follow Compound B_2 and may accumulate in the absence of cytochrome $\underline{c}^{2+}$, and at temperatures below those at which $\underline{a}^{2+}Cu_a^{1+}$ is an effective electron donor.

The reduction of two intermediates can occur as an intermolecular reaction with cytochrome $\underline{c}$ (eqn.3), or as a purely peroxidatic reaction with $\underline{a}^{2+}Cu_a^{1+}$ or a peroxidase donor (phenols, amines, etc.) (Eqn.9).

$$AH_2 + \underline{a}_3^{5+} - O^= \underset{\underline{a}_3}{Cu^{2+}} \leftrightarrow A + H_2O + \underline{a}_3^{3+} \underset{\underline{a}_3}{Cu^{2+}} \tag{9}$$

DISCUSSION

The nature of the reaction product is conveniently indicated in these re- actions (Eqn.3) to be a μ-oxo bridge between $\underline{a}_3^{3+}$ and $Cu_{\underline{a}_3}^{2+}$. This type of formulation is discussed in this Symposium by W. Blumberg (this volume). This species could protonate and dissociate at pH = 7.0 or more rapidly at low pH. However, the μ-oxo bridge may distinguish between the oxidized "active" and the oxidized "resting" states discussed by Antonini and by Malmström and their co-workers[10,11].

The equations for the reactions emphasize a novel possibility in the cyto- chrome oxidase mechanism that either cytochrome $\underline{c}$ or $\underline{a}Cu_a$ may act as electron donors to $\underline{a}_3 \cdot Cu_{\underline{a}_3}$. Thus, at least four electron equivalents (two molecules of cytochrome $\underline{c}$ plus $\underline{a} \cdot Cu_a$) can interact directly with Compound B_2. This emphasizes a feature of cytochrome oxidase not usually recognized, namely, that the function of the reductant is to maintain a low concentration of Compound B_2 and $\underline{a}_3^{3+} \cdot Cu_{\underline{a}_3}^{2+}$ or, what is more important, to maintain the concentration of $a_3^{2+} \cdot Cu_{a_3}^{1+}$ maximal in order that the oxygen reaction proceed as rapidly as possible and that the accumulation of intermediate species of oxygen reduction be negligible.

The data further focus our attention upon the quite different roles of the two portions of cytochrome oxidase: $a_3 \cdot Cu_{a_3}$ dealing exclusively with the oxygen reduction reaction and cytochrome c, plus $a \cdot Cu_a$ functioning to maximize $a_3^{2+} \cdot Cu_{a_3}^{1+}$ and to minimize the intermediate compounds. This dichotomy of roles of the two portions of cytochrome oxidase is consonant with the different properties of the hemes: $a_3 \cdot Cu_{a_3}$ can accept oxygen as a ligand while aCu_a cannot. The reaction mechanisms may differ, too: $a_3 Cu_{a_3}$ deals with oxygen binding and oxygen reduction, intermediate compound formation, and short-range electron transfer reactions; while cytochrome c and $a \cdot Cu_a$ deal with electron transfer reactions which, in view of the large intermolecular distances, are more likely to depend upon tunneling mechanisms.

The identification of Cu_{a_3} as a Type I "blue" copper similar to that in laccase and some other copper oxidases seems appropriate to its function to react with the oxygen molecule bound to the iron[5,12]; i.e., the specific function of Cu_{a_3} is to transfer electrons to oxygen either directly or indirectly, as is suggested by the possibility of a proximal location of the a_3 and Cu_{a_3}. This is supported by the antiferromagnetic coupling of a_3 and Cu_{a_3} leading to greatly modified epr signals[8,13] and the presence of a second sphere neighbor to oxidized iron as suggested by the recent Exafs studies[5].

The function of cytochromes c, a and Cu_a is to provide an electron donor pool containing the four electron equivalents necessary for oxygen reduction. Indeed, excess reducing equivalents are provided; i.e., a total of 6, including cytochrome c_1 and the Fe-S component, appears implicit in the overall design of maintaining a maximal concentration of $a_3^{2+} \cdot Cu_{a_3}^{1+}$ and a minimal concentration of the oxidized state and intermediate Compound B_2. In this light we may consider the rationale of the 2Fe/2Cu oxidase as contrasted to a 4Fe oxidase. Copper itself may in this case simply be an electron "reservoir" in equilibrium with the iron. This is supported by the redox titrations and kinetic responses that closely link Cu_a to heme a and Cu_{a_3} to heme a_3. A possible kinetic discrepancy based on the observation of an absorbance change at 830 nm copper preceeding the oxidation of heme a may be due to a contribution to the 830 nm band from Cu_{a_3} instead of Cu_a especially since Cu_{a_3} is linked to cytochrome a_3 in the initial rapid phase of the oxygen reaction. Thus, a complete kinetic analysis must separate the two components Cu_a and Cu_{a_3} into their separate contributions at 830 nm[14].

The mechanism presented here emphasizes the function of cytochromes c, c_1 and $a \cdot Cu_a$ in affording rapid reductants for $a_3^{3+} \cdot Cu_{a_3}^{2+}$ and for Compound B_2 at low temperatures and presumably at physiological temperatures as well. However, we find the steady-state of cytochrome a_3 to be highly oxidized in resting muscle and only slightly more reduced under conditions of rapid electron transport. This is because the electron flow into the reductant pool (cytochrome c, c_1, and Fe-S protein) is limited to a fraction of the capability of cytochrome oxidase. So what is the need for such "over-powered" oxidase? It would seem that the system is "geared" for rapid response to an O_2 pulse following an ischemic/anoxic episode; for example, a rapid restoration of depleted energy related phosphate compounds during recovery from hypoxia induced by a natural or unnatural underwater episode in mammals might have an advantage for which selective pressure could be developed. Under such conditions the full activity of the oxidase working against the fully reduced pool of donors would be available for fast ATP synthesis. These rapid reactions are readily demonstrated in O_2 pulse experiments or ligand exchange (O_2 for CO) reactions as studied here at low temperatures. In this sense, we find physiological relevance for the reaction mechanism presented here.

REFERENCES

1. Chance, B., Saronio, C., Leigh, Jr., J.S.: Functional intermediates in reaction of cytochrome oxidase with oxygen. Proc. Natl. Acad. Sci. U.S.A. 72, 1635-1640 (1975)

2. Chance, B., Saronio, C., Leigh, Jr., J.S.: Functional intermediates in the reaction of membrane-bound cytochrome oxidase with oxygen. J. Biol. Chem. 250, 9226-9237 (1975)

3. Chance, B., Leigh, Jr., J.S., Waring, A.: Structure and function of cytochrome oxidase and its intermediate compounds with oxygen reduction products. In: Structure and function of energy transducing membrances. K. van Dam, B.F. van Gelder (eds.) pp. 1-10. Amsterdam: Elsevier-North Holland 1977

4. Chance, B., Leigh, Jr., J.S.: Oxygen intermediates and mixed valence states of cytochrome oxidase: Infrared absorption difference spectra of Compounds A,B, and C of cytochrome oxidase and oxygen. Proc. Natl. Acad. Sci. U.S.A. 74, 4777-4780 (1977)

5. Powers, L., Chance, B., Leigh, Jr., J.S., Smith, J., Barlow, C., and Vik, S.: Optical and x-ray absorption edge spectroscopy of copper in solubilized cytochrome c oxidase. (This volume.)

6. Beinert, H., Shaw, R.W.: On the identity of the high spin heme components of cytochrome c oxidase. Biochim. Biophys. Acta 462, 121-130 (1977)

7. Chance, B., Saronio, C., Waring, A., and Leigh, Jr., J.S.: Cytochrome c-cytochrome oxidase interaction at subzero temperatures. BBA 503, 37-55 (1978)

8. Chance, B.: Enzyme Substrate Compounds. Advances in Enzymology 12, 153-
    190. 1951

9. Yonetani, T., Schleyer, H., Ehrenberg, A. and Chance, B.: The chemical
    nature of complex es of cytochrome c peroxidase. In Hemes and Hemoproteins,
    B. Chance, R. Estabrook, T. Yonetani (eds.) Academic Press N.Y. 293-305
    (1966)

10. Antonini, E., Brunori, M., Colosimo, A., Greenwood, C., and Wilson, M.:
    Oxygen "pulsed" cytochrome c oxidase: functional properties and catalytic
    relevance. Proc. Natl. Acad. Sci. 74, 3128-3132 (1977)

11. Rosen, S., Branden, R., Vanngard, T., Malmstrom, B.: EPR evidence for an
    active form of cytochrome c oxidase different from the resting enzyme.
    FEBS Letters, 74

12. Powers, L., Blumberg, W.E., Chance, B., Barlow, C., Leigh, Jr., J.S., Smith
    J., Yonetani, T., Vik, S., Peisach, J.: The nature of the copper atoms of
    cytochrome c oxidase as studied by optical and x-ray absorption edge
    spectroscopy. BBA (in press)

13. Beinert, H., Griffiths, D.E., Warton, D.C., Sands, R.H.: Properties of the
    copper associated with cytochrome oxidase as studied by paramagnetic
    resonance spectroscopy. J. Biol. Chem. 237, 2337-2346 (1962)

14. Gibson, Q. and Greenwood, C.: Kinetic observations on the near infrared
    band of cytochrome c oxidase. J. Biol. Chem. 240, 2694-2698. (1965)

Cytochrome Oxidase, T.E. King et al. eds.
© *1979 Elsevier/North-Holland Biomedical Press*

ELECTROMOTIVE FUNCTION OF CYTOCHROME OXIDASE

PETER MITCHELL AND JENNIFER MOYLE

Glynn Research Institute, Bodmin, Cornwall, PL30 4AU, U.K.

INTRODUCTION

Some years ago[1], Mitchell suggested that the cytochrome oxidase complex is
plugged through the coupling membrane of mitochondria and bacteria so that the
reduction and protonation of O_2 to give $2H_2O$ involves the conduction of $4e^-$ from
cytochrome c at the outer surface of the membrane to $4H^+$ ions that enter the
reaction domain from the inner aqueous phase. This view was subsequently found
to be supported both by quantitative studies of the electromotive, but non-
protonmotive, property of the cytochrome oxidase reaction[2-7], and by studies of
the functional and structural sidedness and general topological organisation of
the cytochrome oxidase complex in the membrane[8-11]. Recently, however, observations by Wikström and colleagues on acid-base changes accompanying cytochrome
oxidase activity in whole mitochondria, in submitochondrial vesicles, and in
liposomes inlaid with cytochrome oxidase[12,13], encouraged the idea that the
cytochrome oxidase complex is equipped with a proton pump that may be conformationally coupled to the electron transfer reaction[14-17]. In the present
paper, we review the earlier evidence against proton-pumping by cytochrome
oxidase, and describe recent O_2-pulse experiments[18,19] which show that the $\rightarrow e^-$/O
and $\leftarrow H^+$/O ratios characteristic of the cytochrome oxidase reaction in rat liver
mitochondria are very close to 2.0 and 0.0 respectively. From this, and other
supporting evidence[20-22], one may conclude that cytochrome oxidase is not a proton pump, but that, as originally considered by Keilin[23,24], it conducts electrons from cytochrome c to oxygen. The electromotive function of the reaction
arises simply from its vectorial organisation[1,3].

METHODS

The general experimental methods used in the work summarised in this paper
were as described in the papers cited.

Abbreviations: pH_O, pH of suspension medium; pK_O, pK = $-\log_{10}(K^+$ ion activity)
of suspension medium; $\leftarrow H^+$ or $\leftarrow K^+$, outward H^+ or K^+ translocation; $\rightarrow H^+$, $\rightarrow K^+$ or
$\rightarrow e^-$, inward H^+, K^+ or electron translocation; ΔH_O^+, quantity of H^+ entering the
suspension medium; TMPD, N,N,N'N'-tetramethyl-p-phenylenediamine; WB, oxidised
TMPD; $DADH_2$, 2,3,5,6,-tetramethyl-p-phenylenediamine; FCCP, carbonylcyanide
trifluoromethoxyphenylhydrazone; NEM, N-ethylmaleimide; dig, digitonin;
val, valinomycin; RLM, rat liver mitochondria.

362

RESULTS

$\leftarrow H^+/2e^-$ Ratios using oxygen, ferricyanide or ferricytochrome c

Figure 1 illustrates respiratory pulse experiments done by Mitchell and
Moyle[2,3] ten years ago with suspensions of rat liver mitochondria in a 150 mM
KCl medium, using (A) succinate (+ rotenone), or (B) β-hydroxybutyrate as respiratory substrate, and injecting either air-saturated saline, or anaerobic
saline containing potassium ferricyanide, as the pulsed oxidant. The data show
that the observed $\leftarrow H^+/2e^-$ ratio, measured as equivalents of acid ΔH_O^+ exported
into the suspension medium, is independent of whether the respiratory substrate
is oxidised by oxygen via cytochrome oxidase, or by ferricyanide via cytochrome
c, bypassing cytochrome oxidase. It follows that cytochrome oxidase does not
pump protons across the membrane. Similar work by Papa and colleagues (see
refs. 7,22), using beef heart mitochondria with duroquinol as reductant and
ferricytochrome c or O_2 as alternative oxidants, led to the same conclusion.

Poising of cytochrome a redox potential by $\Delta\psi$

Hinkle and Mitchell[4] showed, by spectrophotometric observations on antimycin-

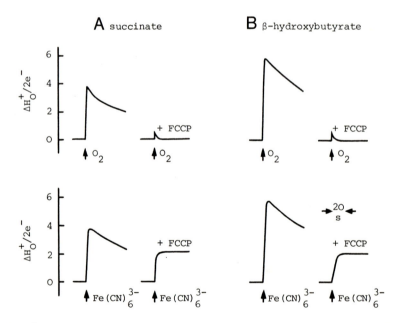

Fig. 1. $\Delta H_O^+/2e^-$ ratios in rat liver mitochondria (about 6 mg protein/ml) in a
150 mM KCl medium at pH_O 7.2-7.3 and 25°, using (A) 2 mM succinate (+ 25 μg
rotenone/g protein) or (B) 2 mM β-hydroxybutyrate as respiratory substrate, and
with either oxygen or ferricyanide as oxidant. Where indicated, FCCP (1 μM)
was present. Further details in Mitchell and Moyle[2].

treated rat liver mitochondria in the presence of CO, that the redox potential
of cytochrome a, measured as an apparent redox midpoint E''_m, was shifted relative
to cytochrome c by the application of an electric potential difference $\Delta\psi$ across
the membrane. In some experiments, $\Delta\psi$ was applied, negative inside, by treating
the K^+-filled mitochondria with valinomycin in a medium of low K^+ ion content.
In this case, the total protonmotive potential difference Δp was initially equal
to the applied $\Delta\psi$. In other experiments, the mitochondria were treated with the
proton-conducting agent FCCP, and an electric diffusion potential negative or
positive inside, was created by suddenly raising or lowering pH_o respectively.
In this case, the $\Delta\psi$ initially created across the membrane, was equal and
opposite to the applied pH potential difference $-Z\Delta pH$, and Δp was zero. The
data from the two kinds of experiment, replotted in Figure 2, fall on the same
straight line, and show that the E''_m value of cytochrome a is a linear function

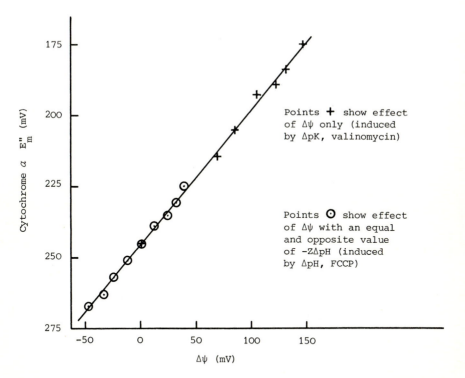

Fig. 2. Effect of $\Delta\psi$ only, and of $\Delta\psi$ with an equal and opposite $-Z\Delta pH$, on the
redox poise of cytochrome a in anaerobic suspensions of antimycin-treated rat
liver mitochondria in the presence of CO, using ferricyanide/ferrocyanide to fix
the redox potential of cytochrome c. The redox poise of cytochrome a,
represented by the apparent midpoint E''_m, was measured by the absorbance
difference at 605-630 nm. Further details in Hinkle and Mitchell[4].

of the electric potential difference $\Delta\psi$, but is independent of the total protonic potential difference Δp, across the membrane. This observation is in keeping with the electron-transferring function of cytochrome a through the cytochrome oxidase, but it excludes any energetic role of the redox changes of cytochrome a in proton pumping through cytochrome oxidase.

The observation of Wrigglesworth[21] that respiratory control in liposomes inlaid with cytochrome oxidase depends on $\Delta\psi$, and is influenced by $-Z\Delta pH$ (or Δp) only to a minor extent, if at all, is likewise more consistent with an electro-motive function than with a protonmotive function of cytochrome oxidase. The recently discovered[25] complexation of the K^+-valinomycin complex with FCCP may, incidentally, explain some apparent anomalies in the uncoupling actions of FCCP, valinomycin and other ionophores in suspensions of liposomes inlaid with cytochrome oxidase[21,26].

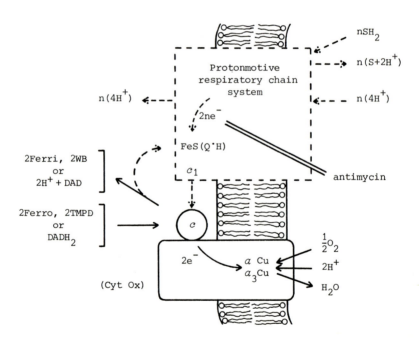

Fig. 3. Diagram of mitochondrial cytochrome oxidase with natural and artificial redox connections. Ferri and ferro stand for ferricyanide or ferri-cytochrome c and ferrocyanide or ferrocytochrome c respectively, and WB means Wurster's blue, the free-radical oxidised form of TMPD. An unidentified res-piratory reductant is represented by SH_2, and FeS stands for the Rieske iron-sulphur protein.

Ferrocyanide oxidation in aerobic antimycin-inhibited mitochondria

When aerobic suspensions of antimycin-inhibited rat liver mitochondria (about 6 mg protein/ml) in a 150 mM KCl medium near pH_O 7 at 25° without added substrate are pulsed with high concentrations of ferrocyanide (0.5 mM final concentration), the rate of reduction of cytochrome c is increased, and the electronic throughput of cytochrome oxidase is increased about fivefold. We found that this actuation of cytochrome oxidase is not accompanied by significant proton translocation[2,3,18,19]. Wikström[12] recently confirmed our observation, but he went on to show that, if the antimycin-inhibited mitochondria were preincubated with NEM before the addition of the pulse of ferrocyanide, the ferrocyanide oxidation was accompanied by outward proton translocation. We, in turn, confirmed that proton translocation does, indeed, occur during ferrocyanide oxidation by the NEM-treated mitochondria[18]. However, measurements of the relative net rates of consumption of O and of $2H^+$, which should be equal during the ferrocyanide oxidation in control experiments in the presence of FCCP, revealed that, after NEM treatment, there was some 30% excess O consumption[18]. As illustrated in Figure 3, it was thus evident that in the NEM-treated mitochondria some acid-producing endogenous hydrogenated reductant (SH_2) was oxidised in addition to the ferrocyanide, as previously observed by Papa et al[22].

We found that, in this aerobic type of experiment, the acid-producing reaction accompanying ferrocyanide oxidation could be induced, not only by preincubating the mitochondria with NEM, but also by adding respiratory substrate[18]. The type of mechanism outlined in Figure 3, where n represents the excess O consumption or shortfall in $2H^+$ consumption, was confirmed in principle by our observation that high concentrations of ferricyanide could be reduced through a protonmotive antimycin-insensitive reaction[18]. But whatever the details of the protonmotive mechanism may be, the proton translocation is not correlated with cytochrome oxidase turnover during ferrocyanide oxidation, and must be attributed to oxidation of the interfering endogenous reductant.

Wikström and Krab[27] confirmed that they could observe excess O consumption by some 12 to 18%, even though they used only a fifth of the mitochondrial suspension density employed in our experiments. We think the excess O consumption effects are much too large to be accounted for by relaxation of binding of H^+ and/or H_3O^+ by ferrocyanide anions on oxidation, as Wikström and Krab[27] have suggested. At all events, this type of explanation would not account for the fact that, in our experiments, the excess O consumption (and the associated proton translocation) do not occur significantly in the aerobic mitochondrial suspensions without respiratory substrate, and can be reproducibly induced by respiratory substrate and/or by preincubating the mitochondria with NEM[18].

Oxygen-pulse experiments with antimycin-inhibited mitochondria preincubated with reductants of cytochrome c anaerobically

In this type of experiment, the quantity of O_2 injected in the pulse gives a more accurate measurement of O_2 consumption than can generally be obtained in the aerobic type of experiment of the previous section, where O_2 consumption is measured with an O_2 electrode. However, we found that the anaerobic

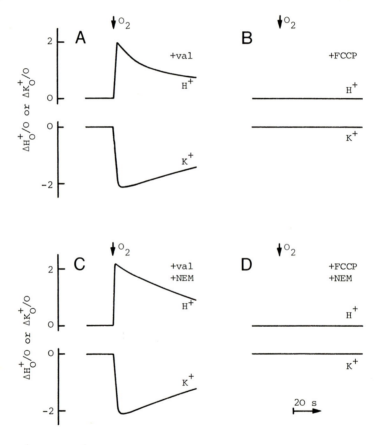

Fig. 4. $\Delta H_O^+/O$ and $\Delta K_O^+/O$ ratios characteristic of electron translocation by cytochrome oxidase, using 0.6 mM $DADH_2$ as reductant in O_2-pulse experiments (2.5 μg atom O injected/g protein) with rat liver mitochondria (about 6 mg protein/ml) in a 240 mM sucrose, 3.3 mM glycylglycine, 10 mM $MgSO_4$, 1 mM EGTA medium at pH_O 7.0-7.1 and 25°. Antimycin (36 μg/g protein) and rotenone (0.4 μM) were present in all experiments; and valinomycin (200 μg/g protein), FCCP (1 μM) and NEM (0.2 mM) were present where indicated. Further details in Moyle and Mitchell[19].

preincubation (like NEM under aerobic conditions) induced excess O consumption by the unidentified endogenous reductant, and a corresponding translocation of protons, when ferrocyanide, sulphite or TMPD was used as reductant (refs. 18,19 and unpublished work). This excess O consumption depends on the reductant and on the suspension medium. With TMPD, the factor n (Figure 3), representing the excess O or $2H^+$ shortfall, can amount to as much as 40% of the O reduced. Fortunately, however, n is virtually zero when $DADH_2$ is used as reductant in appropriate media, or when added cytochrome c is used as reductant[18,19].

Figure 4 illustrates the results of O_2-pulse experiments on antimycin-treated and rotenone-treated rat liver mitochondria using $DADH_2$ as reductant in a 240 mM sucrose, 10 mM $MgSO_4$, 1 mM EGTA medium at pH_O 7.0-7.1[19]. Both pH_O and pK_O changes were measured to enable us to estimate $\rightarrow e^-/O$ and $\leftarrow H^+/O$ values by appropriate use of valinomycin (A,C) and FCCP (B,D). The effect of NEM was also observed (C,D). After a small correction for the normally protonmotive antimycin-insensitive respiration that proceeds at about 5% of the rate of $DADH_2$ oxidation, the $\Delta H_O^+/O$ ratio was found to be 1.98 ± 0.08 (8 values) and 1.98 ± 0.05 (8 values) in the absence (A) and presence (C) of NEM respectively. The expected pH_O change for the oxidation of $DADH_2$ to DAD + $2H^+$ at the outer surface of the mitochondria is 2.0. It follows that the $\leftarrow H^+/O$ ratio of cytochrome oxidase is zero. This is confirmed by the experiments in the presence of FCCP (B,D) where the deprotonation of $DADH_2$ at the C side exactly compensates the protonation of O from the M side of the cytochrome oxidase complex during the transfer of $2e^-$ from $DADH_2$ to O.

The electrophoretic $\rightarrow K^+/O$ ratio was found to be 2.03 ± 0.10 (8 values) and 2.01 ± 0.07 (8 values) in the absence and presence of NEM respectively. We conclude that the $\rightarrow e^-/O$ ratio of cytochrome oxidase is 2.0.

Wikström suggested (unpublished proceedings of 12th FEBS Meeting, Dresden, 1978) that the above conclusions drawn from our use of $DADH_2$ as reductant of cytochrome c were invalid, because he observed that $DADH_2$ oxidation yielded a product that was only partially deprotonated at pH 7. Figure 5 shows that we could reproduce Wikström's observation, when no rat liver mitochondria were present. However, in the presence of rat liver mitochondria (3 mg protein/ml), oxidation of $DADH_2$ to DAD by ferricyanide caused the net dissociation of very close to 2.0 H^+ ions. The final pair of curves in Figure 5 show that the net dissociation of 2.0 H^+ ions on oxidation of $DADH_2$ in the presence of the mitochondria could not be due to re-reduction of DAD by an endogenous mitochondrial reductant SH_2 (giving S + $2H^+$) synchronously with $DADH_2$ oxidation. We also found that the steady-state O_2 consumption rate of the rotenone- and antimycin-inhibited mitochondria in the presence of 0.6 mM

DAD/DADH$_2$ and 1 µM FCCP was not more than 5% of the rate of O$_2$ consumption during pulsed oxidation of DADH$_2$. This is consistent with our observation that reduction of O by DADH$_2$ via cytochrome c and cytochrome oxidase in antimycin-inhibited mitochondria in the presence of FCCP causes zero change of pH$_O$, and it confirms the validity of our rationale.

Oxygen-pulse experiments with added ferrocytochrome c as reductant produced a remarkable net acidification artefact in antimycin-inhibited rat liver mitochondria suspended in a 150 mM KCl, 1 mM EDTA medium, especially when digitonin (100 mg/g mitochondrial protein) was present[18]. The digitonin was originally included to facilitate permeation of ferrocytochrome c through the outer membrane during anaerobic preincubation of the mitochondria, but it was later found to be unnecessary. The special interest of this artefact stems from its resemblance to the acidification seen by Hinkle[20] and by Wikström and Saari[13] during oxidation of cytochrome c by suspensions of liposomes inlaid with cytochrome oxidase. Figure 6A illustrates the net acidification artefact, corrected for protonmotive antimycin-insensitive respiration. Under these favourable con-

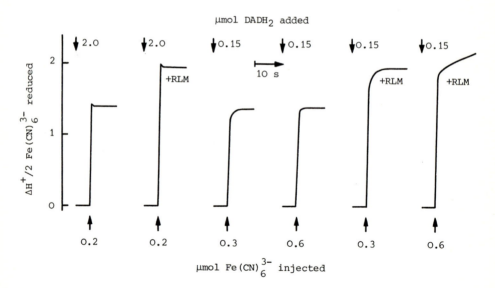

Fig. 5. Effect of the presence of rat liver mitochondria on $\Delta H^+/2e^-$ ratios for DADH$_2$ oxidation by potassium ferricyanide. The experiments were done anaerobically in 3.3 ml medium containing 150 mM KCl, 3.3 mM glycylglycine, 1 mM EDTA and 30 µg carbonic anhydrase/ml at 25° and at pH$_O$ 7.0-7.1. Where indicated (RLM), rat liver mitochondria with antimycin (36 µg/g protein), rotenone (0.4 mM) and FCCP (1 µM) were also present. Each curve represents a separate experiment in which a given quantity of DADH$_2$ was titrated with a given pulse of anaerobic K$_3$Fe(CN)$_6$ solution.

ditions for producing it, the artefact amounts to 1 µg ion H^+/g protein —— i.e. about 10 H^+ per cytochrome oxidase complex, but not more than $2H^+$ per cytochrome c molecule oxidised as it is produced. At first sight, the artefact appears to be (partially) sensitive to FCCP, but the decrease in the apparent extent of the net acidification by FCCP is accounted for by the superposition of a more rapid internal alkalinisation (due to inward H^+ conduction by FCCP) on the given rate of external acidification. This remarkably large acidification cannot be due to proton pumping by cytochrome oxidase, because it can be quantitatively reproduced (Figure 6A) by using anaerobic ferricyanide in place of O_2, even in the presence of KCN, when there could be no cytochrome oxidase activity. Moreover, it is produced during the oxidation of only part of the ferrocyto-chrome c that reduces the pulse of O_2 or ferricyanide; and it persists until this fraction of the cytochrome c is re-reduced by endogenous reductants that leak reducing equivalents through the antimycin-inhibited site[28].

Ferricytochrome c binds anions whereas ferrocytochrome c binds cations[29]. We

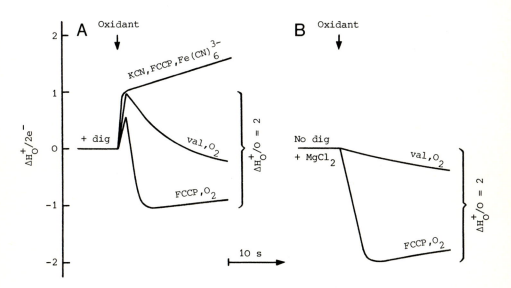

Fig. 6. Net-acidification artefact in O_2-pulse experiments with rat liver mito-chondria, using added ferrocytochrome c as reductant. In A, the conditions were favourable for producing the artefact: rat liver mitochondria (about 6 mg protein/ml) in an anaerobic 150 mM KCl, 3.3 mM glycylglycine, 1 mM EDTA medium, containing digitonin (100 mg/g protein), antimycin (36 µg/g protein) and rotenone (0.4 µM), and with 25 µM horse heart ferrocytochrome c as reductant. Either O_2 (150 mM KCl saturated with air) or anaerobic $K_3Fe(CN)_6$ were used as oxidant; and valinomycin (10 µg/g protein), FCCP (1 µM), and KCN (0.1 mM) were present where indicated. In B, the artefact was suppressed: conditions as in A, but omitting the digitonin and adding 5 mM $MgCl_2$ to the suspension medium.

therefore attribute the net acidification artefact to the effective pK_a decrease of certain anionic and/or cationic cytochrome c-binding components at the surface of the mitochondrial cristae membrane, induced by oxidation of the bound cytochrome c.

As shown in Figure 6B, the net acidification artefact is completely suppressed when the 150 mM KCl medium is supplemented with 5 mM $MgCl_2$ and contains no digitonin. It is likewise suppressed in a 240 mM sucrose medium containing 10 mM $MgSO_4$ and 1 mM EGTA. Figure 7 shows the results of O_2-pulse experiments in the latter medium, like those done with $DADH_2$ (Figure 4), but using 25 μM horse heart ferrocytochrome c as reductant[19]. In this type of experiment, protonmotive antimycin-insensitive respiration proceeds at about 10% of the rate of ferrocytochrome c oxidation; and it accounts quantitatively for the small

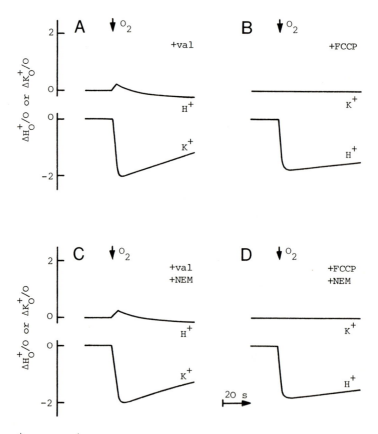

Fig. 7. $\Delta H_O^+/O$ and $\Delta K_O^+/O$ ratios characteristic of electron translocation by cytochrome oxidase, using 25 μM horse heart ferrocytochrome c as reductant in O_2-pulse experiments with rat liver mitochondria. Other details as in Fig. 4.

ΔH_O^+ pulses observed in A and C. The secondary electrophoretic $\rightarrow K^+/O$ ratio (A) and $\rightarrow H^+/O$ ratio (B) were found to be 1.98 ± 0.06 (8 values) and 1.96 ± 0.07 (10 values) respectively. NEM had no significant effect[18,19]. These results confirm our conclusion, in agreement with Hinkle's results[6,20] on liposomes inlaid with cytochrome oxidase using ferrocytochrome c as reductant, that the $\leftarrow H^+/O$ and $\rightarrow e^-/O$ ratios for cytochrome oxidase are zero and 2.0 respectively.

CONCLUSION

The observations summarised in this paper show that two major types of artefact may interfere with experiments designed to measure the electromotive and protonmotive function of the mitochondrial cytochrome oxidase. One type of artefact arises from the accidental actuation of extraneous protonmotive reactions in the respiratory chain system in concert with the cytochrome oxidase reaction that one is endeavouring to isolate and study. This type of artefact is liable to apply, not only to experiments on whole mitochondria, as described in this paper, but also to experiments, such as those of Wikström and Saari[13], and Sorgato and Ferguson[15], on submitochondrial vesicles. The other type of artefact arises from remarkably large net acid-base changes induced by oxidation of complexes of ferrocytochrome c with mitochondrial components. This type of artefact may explain the rapid pH_O changes attributed by Wikström[12,13,27] to a cytochrome oxidase proton pump in experiments with liposomes inlaid with cytochrome oxidase, using ferrocytochrome c as reductant[13].

Our experimental data show that, having identified and avoided these major experimental artefacts, one can conclude with a fair degree of certainty that the $\leftarrow H^+/O$ and $\rightarrow e^-/O$ ratios of cytochrome oxidase are very close to zero and 2.0 respectively. In other words, cytochrome oxidase conducts electrons from cytochrome c to oxygen, as originally considered by Keilin[23,24]. The electromotive function of cytochrome oxidase arises simply from its anisotropic topological configuration[1,3,8-11], as indicated in Figure 3.

ACKNOWLEDGEMENTS

We thank Mr. Robert Harper and Mrs. Stephanie Key for expert technical assistance and help in preparing the manuscript. We gratefully acknowledge the financial support of Glynn Research Ltd.

REFERENCES

1. Mitchell, P. (1966) Chemiosmotic Coupling in Oxidative and Photosynthetic Phosphorylation, Glynn Research, Bodmin, pp. 1-192.

2. Mitchell, P. and Moyle, J. (1967) in Biochemistry of Mitochondria, Slater, E.C. et al. eds., Academic Press/PWN, London/Warszawa, pp. 53-74.

3. Mitchell, P. (1969) in The Molecular Basis of Membrane Transport, Tosteson, D.C. ed., Prentice-Hall, Englewood Cliffs, New Jersey, pp. 483-518.

4. Hinkle, P. and Mitchell, P. (1970) J. Bioenergetics, 1, 45-60.

5. Mitchell, P. and Moyle, J. (1970) in Electron Transport and Energy Conservation, Tager, J.M. et al. eds., Adriatica Editrice, Bari, pp. 575-587.

6. Hinkle, P.C. (1973) Fed. Proc. Fed. Am. Soc. Exp. Biol., 32, 1988-1992.

7. Papa, S. (1976) Biochim. Biophys. Acta, 456, 39-84.

8. De Pierre, J.W. and Ernster, L. (1978) Ann. Rev. Biochem., 46, 201-262.

9. Dockter, M.E., Steinmann, A. and Schatz, G. (1978) J. Biol. Chem., 253, 311-317.

10. Eytan, G.D. and Broza, R. (1978) J. Biol. Chem., 253, 3196-3202.

11. Erecinska, M. and Wilson, D.F. (1978) Arch. Biochem. Biophys., 188, 1-14.

12. Wikström, M.K.F. (1977) Nature, 266, 271-273.

13. Wikström, M.K.F. and Saari, H.T. (1977) Biochim. Biophys. Acta, 462, 346-361.

14. Artzatbanov, V.Yu., Konstantinov, A.A. and Skulachev, V.P. (1978) FEBS Lett., 87, 180-185.

15. Sorgato, M.C. and Ferguson, S.J. (1978) FEBS Lett., 90, 178-182.

16. Alexandre, A. and Lehninger, A.L. (1978) in Frontiers of Biological Energetics: From Electrons to Tissues, Dutton, P.L. et al. eds., Academic Press, New York, in press.

17. Azzone, G.F., Pozzan, T., Bragadin, M. and Di Virgilio, F. (1978) in Frontiers of Biological Energetics: From Electrons to Tissues, Dutton, P.L. et al. eds., Academic Press, New York, in press.

18. Moyle, J. and Mitchell, P. (1978) FEBS Lett., 88, 268-272.

19. Moyle, J. and Mitchell, P. (1978) FEBS Lett., 90, 361-365.

20. Hinkle, P.C. (1978) FEBS Symp., 45, 79-83.

21. Wrigglesworth, J.M. (1978) FEBS Symp., 45, 95-103.

22. Papa, S., Lorusso, M., Guerrieri, F., Izzo, G. and Capuano, F. (1978) in The Proton and Calcium Pumps, Azzone, G.F. et al. eds., Elsevier/North Holland, Amsterdam, pp. 227-237.

23. Keilin, D. (1930) Proc. Roy. Soc. B, 106, 418-444.

24. Keilin, D. and Hartree, E.F. (1939) Proc. Roy. Soc. B, 127, 167-191.

25. O'Brien, T.A., Nieva-Gomez, D. and Gennis, R.B. (1978) J. Biol. Chem., 253, 1749-1751.

26. Hansen, F.B., Miller, M. and Nicholls, P. (1978) Biochim. Biophys. Acta, 502, 385-399.

27. Wikström, M.K.F. and Krab, K. (1978) FEBS Lett., 91, 8-14.

28. Mitchell, P. and Moyle, J. (1978) in Frontiers of Biological Energetics: From Electrons to Tissues, Dutton, P.L. et al. eds., Academic Press, New York, in press.

29. Margalit, R. and Schejter, A. (1974) Eur. J. Biochem., 46, 387-391.

BIOSYNTHESIS

Cytochrome Oxidase, T.E. King et al. eds.
© *1979 Elsevier/North-Holland Biomedical Press*

SWEET POTATO CYTOCHROME c OXIDASE: ITS PROPERTIES

AND BIOSYNTHESIS IN WOUNDED ROOT TISSUE

MASAYOSHI MAESHIMA AND TADASHI ASAHI

Laboratory of Biochemistry, Faculty of Agriculture

Nagoya University, Nagoya 464, Japan

ABSTRACT

Purified sweet potato cytochrome c oxidase was composed of at least five
polypeptides with molecular weights of 39,000, 33,500, 26,000, 20,000, and
5,700. Experiments with labeled leucine and the antibody against the purified
enzyme suggested that an increase in cytochrome c oxidase activity during aging
of sliced root tissue is achieved by assembly of cytoplasmically made subunits
preexisting in intact root tissue with subunits newly synthesized on mitochondrial ribosomes during aging of slices.

INTRODUCTION

Active biogenesis of mitochondria takes place without accompanying cell
division, when slices of bulky plant storage tissues, such as potato tubers and
sweet potato roots, are incubated at moderate temperatures[1,2]. An increase in
cytochrome c oxidase activity during the biogenesis in sweet potato root tissue
slices is inhibited by chloramphenicol but not by cycloheximide under conditions
where an increase in succinate dehydrogenase activity during aging of slices is
almost completely inhibited by cycloheximide (Figure 1). It is well known that
cytochrome c oxidase is composed of subunits synthesized on both cytoplasmic and
mitochondrial ribosomes[3]. Consequently, we have conducted a series of experiments to test if the increase in cytochrome c oxidase is achieved by assembly of
cytoplasmically synthesized subunits preexisting in intact root tissue with
subunits newly synthesized on mitochondrial ribosomes during aging of slices,
and first purified cytochrome c oxidase from sweet potato root tissue to prepare
the antibody.

In the present paper, we describe some properties of sweet potato cytochrome
c oxidase and data suggesting the presence of excess amounts of cytoplasmically
synthesized subunits in intact root tissue.

MATERIALS AND METHODS

Sweet potato roots (Ipomoea batatas, Kokei No. 14) harvested in autumn were

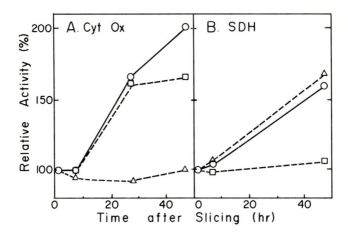

Fig. 1. Effects of antibiotics on increases in cytochrome $\underline{c}$ oxidase and succinate dehydrogenase activities during aging of sliced sweet potato root tissue. Slices were incubated for one hour in 1 mM phosphate buffer, pH 6.5, containing cycloheximide (10^{-5} M) or chloramphenicol (6×10^{-3} M) at room temperature with continuous stirring. Control was run with the buffer containing no antibiotics. Crude mitochondrial fractions were assayed for cytochrome $\underline{c}$ oxidase (A) and succinate dehydrogenase (B) activities. (O) control, (□) cycloheximide, (△) chloramphenicol.

stored at 13 to 16°C until used. Slices (3 mm in thickness and 19 mm in diameter) prepared from inner parenchymatous tissue of the roots were incubated at 29°C under moist conditions to yield aged slices. The administration of [^{3}H]-leucine to tissue slices was performed as described in a previous paper[4]. Eight µCi of [^{3}H]-leucine (112 Ci/mmol) in 40 µl of 10 mM phosphate buffer, pH 7.0, was applied on the surface of each of 40 slices which had been aged for 7 hours, and then the slices were further aged for 17 hours.

Crude mitochondrial and purified mitochondrial fractions were prepared from sweet potato root tissue as described by Nakamura and Asahi[5]. Sweet potato cytochrome $\underline{c}$ oxidase was purified by solubilization of the enzyme from submitochondrial particles by deoxycholate, twice diethylaminoethyl-cellulose column chromatographies, fractionation with ammonium sulfate, and sucrose density gradient centrifugation as described previously[6]. The enzyme preparation from ammonium sulfate fractionation ("step 4" preparation) was used for studies on properties of the enzyme. Antibody was raised in rabbit by injection of 3 mg of the cytochrome $\underline{c}$ oxidase (final "step 5" preparation, heme $\underline{a}$ 12.4 nmol/mg protein) which was homogenized in an equal volume of Freund's complete adjuvant. The second and third injections of purified enzyme in Freund's incomplete adjuvant were given three and four weeks later, respectively. The antiserum thus obtained was subjected to ammonium sulfate fractionation and diethylamino-ethyl-cellulose column chromatography.

Protein was determined by the method of Lowry et al[7] after precipitation with trichloroacetic acid, using bovine serum albumin as standard. Cytochrome c oxidase activity was determined as described previously[6]. Polyacrylamide gel electrophoresis was run in the presence of sodium dodecylsulfate (SDS) and 8 M urea in gels containing 10% total acrylamide on the basis of the general procedure of Swank and Munkres[8] with some modifications as described before[6]. Gel containing labeled polypeptides was sliced into 1 mm slices, which were then assayed for the radioactivity as described elsewhere[9].

RESULTS AND DISCUSSION

Cytochrome c oxidase has been isolated from animal cells[10-13], fungal cells[10,14] and photosynthetic bacteria[15], but not from higher plants, and was only partially purified from potato tubers[16]. Sweet potato cytochrome c oxidase, which was purified in a pure form and in a good yield by a rapid and handy procedure, contained more than 12 nmoles of heme a per mg of protein and 2.5% phospholipids, and its specific activity was almost the same as that of a pure preparation from yeast[10] and was much higher than that of a pure preparation from Rhodopseudomonas palustris[15].

The oxidized form of sweet potato cytochrome c oxidase exhibited an α-band at approximately 601 nm and a γ-band at 421 nm. In the reduced form, the α- and γ-bands were at 601 nm and 438 nm, respectively, with a small peak at 517 nm (β-band). The difference spectrum between the reduced and oxidized forms showed maxima at 443 and 601 nm. There were similarities in the wavelength of the bands between cytochrome c oxidases from sweet potato and white potato[16]. Thus the wavelength of γ-band of the enzyme in higher plants seemed different from that in other organisms[17].

As shown in Figure 2, cytochrome c oxidase activity in submitochondrial particles from sweet potato root tissue was stimulated by Triton X-100 and phospholipid. In case of the purified cytochrome c oxidase, the activity was depressed by high concentrations of Triton X-100 and stimulated about three-fold by phospholipids as shown with the enzymes isolated from other sources[10-12]. In addition, the pH optimum was altered by Triton X-100 and phospholipids. It should be noted that pH optima with phosphatidylcholine and phosphatidylethanol-amine were different from each other. The pH optimum appeared to depend upon the environment surrounding the enzyme protein. However, it remains unsolved how the environment affects the pH optimum.

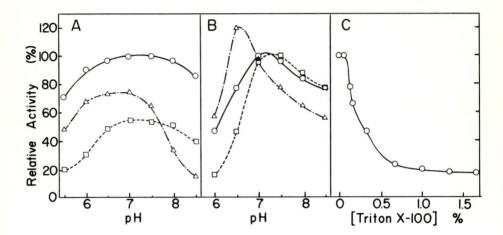

Fig. 2. Effects of phospholipids and Triton X-100 on the activity of sweet potato cytochrome c oxidase.
A. Effects of phospholipid and Triton X-100 on the activity in submitochondrial particles. The enzyme activity was assayed in 15 mM potassium phosphate buffer containing 0.08% Triton X-100 (△) or egg yolk phosphatidylcholine (O, 6.6 µg of phosphorus/assay cuvette). Control (□) was carried out in the buffer containing neither Triton X-100 nor phospholipid. The activity is expressed as the percentage of activity at a given pH relative to that at pH 7.0 with egg yolk phosphatidylcholine.
B. Effects of phospholipids on the activity of purified enzyme ("step 4" preparation). The assay mixture contained 15 mM phosphate, 0.05% Triton X-100 and sufficient amounts of phospholipids. O, egg yolk phosphatidylcholine (6.6 µg of phosphorus/assay cuvette); △, phosphatidylethanolamine (6.9 µg of phosphorus/assay cuvette); □, soybean phosphatidylcholine (9.7 µg of phosphorus/assay cuvette).
C. Inhibition of the activity of purified enzyme ("step 4" preparation) by Triton X-100. The activity was assayed in 10 mM potassium phosphate buffer, pH 7.5, containing phosphatidylcholine micelles (6.6 µg of phosphorus/assay cuvette).

Sweet potato cytochrome c oxidase contained at least five subunits with molecular weights of 39,000 (I), 33,500 (II), 26,000 (III), 20,000 (IV), and 5,700 (V) (Figure 3). Judging from the intensity of staining, subunit V with the lowest molecular weight seemed to be composed of more than one subunit. Cytochrome c oxidase from fungal cells is comprised of seven subunits with apparent molecular weights ranging from 42,500 to 4,600[10,14], while the enzyme from animal cells has six to eight subunits in a similar molecular weight range[10-13]. In sweet potato cytochrome c oxidase, subunits with molecular weights ranging from 20,000 to 10,000 were not detected, although the presence of two or three subunits with the molecular weight range has been shown with

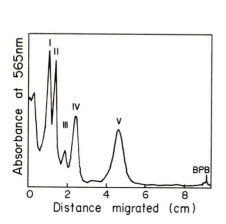

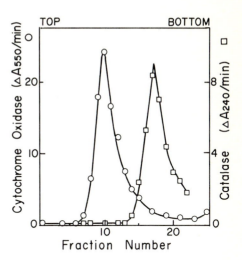

Fig. 3 (left). The polypeptide profile of sweet potato cytochrome c oxidase run on SDS-urea polyacrylamide gel. The enzyme preparation from step 5 was dissociated by incubation at 37°C for one hour in 2% SDS, 2% 2-mercaptoethanol and 2 M urea, and the mixture was applied to the gel. The apparent molecular weights of the peaks are: I,39,000; II, 33,500; III, 26,000; IV, 20,000; and V, 5,700.

Fig. 4 (right). Sucrose density gradient centrifugation of sweet potato cytochrome c oxidase[19]. A mixture of the enzyme preparation from step 4 (0.12 mg) and catalase (0.4 mg) in 10 mM Tris-acetate buffer, pH 7.5, containing 0.1% Triton X-100 was applied to a 5-ml linear sucrose density gradient, 5 to 20%, containing 0.1% Triton X-100, 0.1 M KCl, and 10 mM Tris-acetate buffer, pH 7.5, The gradient was centrifuged for 9 hours at 180,000 xg at 4°C.

cytochrome c oxidase from othe sources[13].

We suppose that the molecular weight of sweet potato cytochrome c oxidase is about 150,000. The sum of the molecular weights of five subunits is 120,000 to 140,000. The fact that heme a content of the purified enzyme was approximately 12 nmoles per mg of protein indicates that the minimum molecular weight is 81,000. Since active cytochrome c oxidase should contain at least two molecules of heme a per one enzyme molecule, the minimum molecular weight would be 162,000. The molecular weight was calculated as about 100,000 by sucrose density gradient centrifugation (Figure 4), but this low value would be due to an abnormal value of the partial specific volume as reported by Rubin and Tzagoloff[10].

The purified enzyme preparation formed a single precipitin line against its antibody in Ouchterlony double diffusion tests (Figure 5). However, an addi-

380

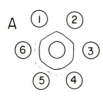

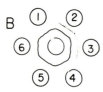

Fig. 5. Ouchterlony double diffusion tests[20]. The plates were prepared with 1% agar in the presence of 0.5% Triton X-100, 0.7% NaCl, 20 mM Tris-acetate buffer, pH 7.5, and 0.01% NaN_3.

A. Center well, immunoglobulin to cytochrome c oxidase; wells 1 and 2, purified oxidase; wells 3 and 4, Triton X-100 extracts of purified mitochondrial fractions from tissues aged for 48 and 24 hours, respectively; wells 5 and 6, Triton X-100 extracts of purified mitochondrial fractions from intact tissue.

B. Center well, anti-oxidase; well 1, Triton X-100 extract of submitochondrial particles from tissue aged for 48 hours; well 2, Triton X-100 extract of submitochondrial particles from intact tissue; wells 3, 4, 5 and 6, Triton X-100 extracts of crude mitochondrial fractions from tissues aged for 0, 10, 24 and 48 hours, respectively.

tional faint precipitin line was formed between the antibody and a Triton X-100 extract of mitochondria from intact tissue (Figure 5). The faint precipitin line was only scarcely observed with mitochondrial preparations from slices aged for 10 hours and not with those from slices aged for 48 hours. In case of slices aged for 24 hours, the faint line was detectable in some experiments (Figure 5A) but not in the others (Figure 5B). The results indicate that protein forming the faint line disappears during aging of slices. It remains obscure, however, what kind of protein forms the faint line. Only about a half of cytochrome c oxidase activity was inhibited by the antibody in amounts causing maximal amounts of antigen-antibody precipitate in both cases of purified preparations and mitochondrial preparations treated with Triton X-100.

It was tested by using the antibody whether all the subunits of cytochrome c oxidase are labeled when aged slices are administrated by [^{3}H]-leucine. Mitochondria prepared from labeled slices were extracted first with 0.06% deoxycholate in the presence of 1 M KCl to remove a large amount of protein other than the cytochrome c oxidase, and then almost all of the cytochrome c oxidase in the mitochondria was extracted by 1% Triton X-100 in the presence of 0.2 M KCl. The Triton X-100 extract was treated with immunoglobulin, and the precipitate was subjected to gel electrophoresis (Figure 6). Three of the subunits with relatively high molecular weights (I, II and III) were labeled during aging. However, no radioactivity was detected with either of subunits IV and V. There were highly labeled polypeptides which migrated much more slowly than any of the oxidase subunits. It remains unsolved whether the polypeptides are related to the oxidase or not. In any way, the results strongly suggest that

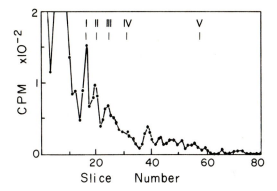

Fig. 6. Identification of cytochrome c oxidase polypeptides synthesized during aging of sliced sweet potato root tissue. Crude mitochondrial fraction from [^{3}H]-leucine labeled slices was treated with 0.06% deoxycholate containing 1 M KCl and centrifuged at 80,000 xg for one hour. The pellet was extracted with 1% Triton X-100 containing 0.2 M KCl, and the extract was treated with an optimal amount of immunoglobulin. The precipitate was dissolved in a solution containing SDS, urea and mercaptoethanol and subjected to gel electrophoresis as reported previously[6]. The Roman numerals mark positions of the subunits of purified enzyme.

subunits IV and V, which are probably of cytoplasmic origin, are present in intact root tissue to wait for aging of tissue slices to be assembled with other subunits newly synthesized on mitochondrial ribosomes. Recently, it has been proposed that in yeast cells, cytoplasmically made subunits of cytochrome c oxidase accumulate in the mitochondrial membrane fraction under some conditions and are subsequently assembled into the holoenzyme with mitochondrially made subunits[18]. We infer that an increase in cytochrome c oxidase activity during aging of sliced sweet potato root tissue is achieved by assembly of cytoplasmically made subunits preexisting in intact root tissue with subunits newly synthesized in the mitochondria during aging.

REFERENCES

1. Asahi, T., Honda, Y. and Uritani, I. (1966) Plant Physiol. 41, 1179-1184.
2. Lee, S. H. and Chasson, R. M. (1966) Physiol. Plant. 19, 199-206.
3. Schatz, G. and Mason, T. L. (1974) Annu. Rev. Biochem. 43, 51-87.
4. Sakano, K. and Asahi, T. (1969) Agr. Biol. Chem. 33, 1433-1439.
5. Nakamura, K. and Asahi, T. (1976) Arch. Biochem. Biophys. 174, 393-401.
6. Maeshima, M. and Asahi, T. (1978) Arch. Biochem. Biophys. 187, 423-430.

7. Lowry, O. H., Rosebrough, N. J., Farr, A. L. and Randall, R. J. (1951) J. Biol. Chem. 193, 265-275.

8. Swank and Munkres, K. D. (1971) Anal. Biochem. 39, 462-477.

9. Tanaka, Y. and Uritani, I. (1977) Eur. J. Biochem. 73, 255-260.

10. Rubin, M. S. and Tzagoloff, A. (1973) J. Biol. Chem. 248, 4269-4274.

11. George, K. (1976) J. Biol. Chem. 251, 6097-6107.

12. Bucher, J. R. and Penniall, T. (1975) FEBS Lett. 60, 180-184.

13. Downer, N. W., Robinson, N. C. and Capaldi, R. A. (1976) Biochemistry, 15, 2930-2936.

14. Mason, T. L., Poyton, R. O., Wharton, D. C. and Schatz, G. (1973) J. Biol. Chem. 248, 1346-1354.

15. King, M. T. and Drews, G. (1976) Eur. J. Biochem. 68, 5-12.

16. Ducet, G., Diano, M. and Denis, M. (1970) C. R. Acad. Sc. Paris, 270, 2288-2291.

17. Caughey, W. S., Wallace, W. J., Volpe, J. A. and Yoshikawa, S. (1976) in The Enzymes, Boyer, P. D., ed., Academic Press, New York, Vol. 13, pp. 299-344.

18. Poyton, R. O. and Groot, B. S. P. (1975) Proc. Nat. Acad. Sci. USA, 72, 172-176.

19. Martin, T. G. and Ames, B. N. (1961) J. Biol. Chem. 236, 1372-1379.

20. Ouchterlony, O. (1953) Acta Pathol. Microbiol. Scand. 32, 231-247.

EVOLUTION

Cytochrome Oxidase, T.E. King et al. eds.
© *1979 Elsevier/North-Holland Biomedical Press*

THE EVOLUTIONARY CONTROL OF CYTOCHROME *c* FUNCTION

NEIL OSHEROFF, D. BORDEN, W. H. KOPPENOL AND E. MARGOLIASH

Department of Biochemistry and Molecular Biology, Northwestern University, Evanston, Illinois 60201, U.S.A.

INTRODUCTION

For many years, it was considered that the cytochromes *c* from widely divergent eukaryotic species all acted identically with mammalian cytochrome *c* oxidases[1] and with cytochrome *c*-depleted mitochondria from mammalian sources.[2] Recently, it was found that variations in the primary structure of cytochrome *c* can dramatically influence the kinetics of its reaction with cytochrome oxidase.[3-12] Such differences are observed under conditions which maximize the binding between the two proteins and at cytochrome *c* to cytochrome oxidase ratios approaching those found in mitochondria.[4,13,14] Many of these differences in activity have been correlated with amino acid residue changes that affect the properties of the cytochrome oxidase binding domain on cytochrome *c*.[4,8,12] This domain is located on the front surface of the protein. It includes the exposed edge of the heme prosthetic group and is centered near the point at which the positive end of the dipole axis of the molecule crosses the surface of the protein in the region of Residue 82.[8,13-19] The magnitude and direction of the dipole moment of the protein[18,20] and the distribution of charged and hydrophobic side chains within and around the binding domain all affect the reaction.[8,14] These parameters are likely to determine the binding affinity and the alignment of cytochrome *c* with its physiological redox partners and to be responsible for the major differences in activity observed in some cases between the cytochromes *c* of widely divergent eukaryotic taxa. This appears to constitute a primary level of evolutionary control of the cytochrome *c*-cytochrome *c* oxidase reaction. However, remarkable changes in the kinetics of the reaction were also observed with primate cytochromes *c* in which the binding of the protein to the enzyme is largely unchanged and the corresponding evolutionary variations in amino acid sequence are at locations far from the interaction domain with the oxidase.[6-11] This indicates that binding alone is insufficient to provide the structure within which electron transfer can occur and that binding must be followed by a structural change with or without solvent reorganization into a productive complex. The factors which control the rate of formation of this complex constitute a second level of evolutionary control which can act as a fine adjustment of the reaction kinetics.

To examine whether the primate cytochromes c differ from the non-primate proteins in the ease with which their tertiary structures can be changed, the stabilities of their native conformations were assessed by determining the pK_a values for their 695 nm absorption bands. It was observed that cytochromes c which have similar kinetics with any particular oxidase tend to have similar pK_a values. Moreover, two hydrogen and/or ionic bonds located at the top and bottom of the heme crevice respectively, appear to have a major influence on these pK_a values and as such may represent a mechanism by which amino acid changes outside the oxidase binding domain can influence the reaction between cytochrome c and cytochrome c oxidase.[21]

These results are reviewed below.

STRUCTURAL CONTROL OF THE STABILITY OF THE HEME CREVICE

The iron atom of the heme prosthetic group of cytochrome c is hexa-coordinated, with the imidazole nitrogen of histidine 18 and the sulfur of methionine 80 providing the fifth and sixth ligands, respectively.[22] The presence of the methionine 80 sulfur-heme iron bond gives rise to a weak band in the visible absorption spectrum of ferricytochrome c which is centered at 695 nm.[23-27] When the conformation of the protein is disturbed sufficiently that the sulfur-iron bond is broken, there is a concomitant loss of the 695 nm absorbance.[27-37] If under these conditions the ε-amino group of lysyl residue 79 is available, it appears to form the sixth heme ligand, maintaining a hemochrome spectrum after the sulfur ligand is lost.[38-42] When it was observed that the cytochromes c from primate species had low activity with beef cytochrome c oxidase,[6-11] even though they bound to the enzyme as well or better than did the horse protein,[6] it became clear that the correlation between binding and activity, which had been used to explain the lack of reaction of various native and chemically-substituted cytochromes c,[4,13] was not applicable in this case. Since it had been shown that some cytochromes c have pK_a values for their 695 nm bands which range from 8.4 to 10.4,[37] it was of interest to determine whether the primate proteins similarly differed among themselves and from non-primate cytochromes c and whether such differences correlated with their activities with cytochrome c oxidase. Indeed, it could be considered that the open-closed transition of the heme crevice may reflect the ease with which cytochrome c can attain its productive conformation. As described below, the results of these experiments yielded correlations between certain changes in the amino acid sequences of these cytochromes c and their pK_α values on the one hand, and their pK_α values and activity with cytochrome c oxidase on the other.[21]

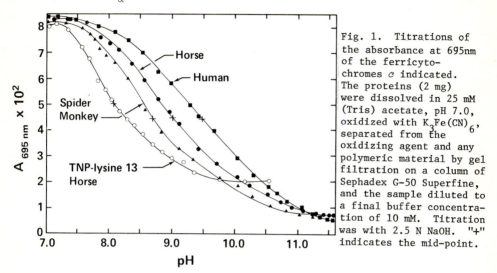

Fig. 1. Titrations of the absorbance at 695nm of the ferricyto-chromes c indicated. The proteins (2 mg) were dissolved in 25 mM (Tris) acetate, pH 7.0, oxidized with $K_3Fe(CN)_6$, separated from the oxidizing agent and any polymeric material by gel filtration on a column of Sephadex G-50 Superfine, and the sample diluted to a final buffer concentration of 10 mM. Titration was with 2.5 N NaOH. "+" indicates the mid-point.

TABLE 1

THE pK_α VALUES FOR THE 695 nm ABSORPTION BAND OF SOME FERRICYTOCHROMES c

Cytochrome c	pK_α
Human	9.5
Slow loris	9.1
Horse	9.05
Spider Monkey	8.7
TNP-lysine 13 Human[a]	9.1
TNP-lysine 13 Horse[a]	8.1
CDNP-lysine 13 Horse[a]	8.2

[a]Prepared according to References 21,43-45.

Some typical acid-base titrations of the 695 nm absorption band of various native and modified ferricytochromes c are shown in Figure 1 and pK_α values for these and other proteins are listed in Table 1. These pK_α values ranged from 8.1 to 9.5, demonstrating a considerable degree of variability for a relatively small number of amino acid sequence changes (Table 2). The most remarkable result was the lowering of the pK_α which resulted from derivatization of lysine 13 to form either a 4-carboxy-2,6-dinitrophenyl(CDNP)- or 2,4,6-trinitrophenyl(TNP)-product. Since CDNP-lysine is negatively charged

TABLE 2

PARTIAL AMINO ACID SEQUENCES OF SOME PRIMATE AND NON-PRIMATE CYTOCHROMES c[a]

Cytochrome c	\multicolumn{15}{c}{Residue Number}
	8	11	12	15	32	44	46	47	50	58	60	62	83	89	92
Human[46]	K	I	M	S	L	P	Y	S	A	I	G	D	V	E	A
Slow loris[47]	-	-	-		S	F			D	T		E	A	-	-
Horse[48]	V	Q	A			F	T	D	T	K	E	A		T	E
Spider Monkey[49]	R				I	S	F	T	E						

[a] The single letter code for amino acid residues is employed. The blank spaces denote positions where the sequence is identical to that for human cytochrome c (top line). The dashes indicate positions for which the amino acid residue has not yet been identified.

while TNP-lysine is neutral, the destabilization of the heme crevice must be due to a change in the properties of Residue 13 which is not related to the charge of the modified residue. It would appear that this effect results from the disruption of an intramolecular bond between the positively charged ε-amino group of lysine 13 and the negatively charged γ-carboxyl of glutamic acid 90.[21] This bond is situated at the top of the heme crevice with the α-carbon of lysine 13 being located slightly to the right and the α-carbon of glutamic acid 90 well to the left of the prosthetic group. Such a bond appears in the X-ray crystallographic structure of tuna cytochrome c[50,51] as shown in Figure 2. If one assumes that this structure is sufficiently similar to those of all mammalian species discussed here, this interaction must be present in the chordate cytochromes c, all of which carry the requisite lysyl and glutamyl residues.[52] It is notable that in the fungal cytochromes c, where Residue 13 has the longer side chain of arginine, Residue 90 has the shorter side chain of aspartic acid.[52] Thus, the same bond may also be important in maintaining the integrity of the heme crevice in the fungal proteins. Disruption of this band may well have been the cause of the lowering of the pK_{α} of the 695 nm band of the complex between the ferrihemopeptide 1-65 with peptide 66-104 of horse cytochrome c observed upon guanidination of the 1-65 hemopeptide.[42] Indeed, guanidination increases the basic residue by two atoms, which may push the bond off the hydrophobic surface of the protein where it normally resides, thereby weakening it. The resultant change in the pK_{α} of the 695 nm band from 8.9 to 8.1[42] is similar to the changes observed upon modification of lysine 13 of horse cytochrome c in the present study.

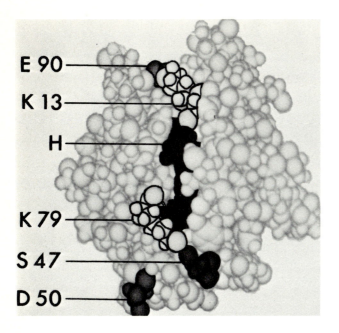

E 90

K 13

H

K 79

S 47

D 50

Fig. 2. Computer-generated space-filling model representation of tuna cytochrome c. View of front of the molecule. The following are indicated on the figure: glutamic acid 90 (E 90); lysine 13 (K13); the edge of the heme exposed on the front surface of the molecule (H); lysine 79 (K 79); serine 47 (S 47); Aspartic acid 50 (D 50). Courtesy of R. J. Feldmann and T.K. Porter of the Macromolecular Surface Display Project, National Institutes of Health.

A similar influence on the stability of the crevice appears to be exerted by a bond located near the bottom of the heme between the ε-amino of lysyl residue 79 and the backbone carboxyl of Residue 47, either a serine or a threonine in all chordate cytochromes c.[52] This bond was observed in the X-ray crystallographic structure of tuna cytochrome c[50,51] and could have presumed to occur in all those cytochromes c that have the requisite residues. However, the presence of an aspartic acid or glutamic acid, at position 50, as compared to an alanine, causes a lowering of the pK_α of the 695 nm absorption band which may result from a weakening of the bond by competition for the charged lysyl side chain by the negatively charged residue at position 50.

The human[46] and spider monkey[9,49] proteins differ by only six amino acid residues (positions 8, 32, 44, 46, 47 and 50 ; Table 2), but the pK_α values for their 695 nm absorption bands differ by 0.8 pH units, the largest variation observed among the native cytochromes c examined in this study. Clearly, one or more of these residues must have a profound influence on the pK_α. However, there appears to be no clear correlation between the pK_α of the 695 nm band and the residues at positions 8, 32, 44, 46, or 47. Thus, human and horse cytochromes c which carry lysine at position 8, leucine at position 32, and proline at position 44, have different pK_a values, and horse and spider monkey cytochromes c which carry phenylalanine at position 46 and threonine at position 47 also have different pK_a values. In contrast, a simple correlation does exist between the pK_a of the 695 nm absorption band and the amino acid residue at position 50. When position 50 is occupied by an alanine, as in human cytochrome c, the pK_a is higher than when it is occupied by an aspartic acid, as in the horse and slow loris proteins. In turn, the pK_a values of latter two are higher than that of the spider monkey, in which position 50 is a glutamic acid. Indeed, the relative spatial locations of Residues 79 and 50[50,51] are compatible with the formation of a bond between their side chains. When the side chains of these residues are rotated about their β-carbon atoms, their charges can be closely juxtaposed, so that the distance between the ε-amino nitrogen and the β- or γ-carboxylic acid oxygen is 2.8 Å or 1.5 Å, respectively, both values being within bonding distance. Thus, a glutamic acid at position 50 can weaken the lysyl 79:backbone carboxyl of Residue 47 bond more efficiently than an aspartic acid, because it can form a more stable bond with the lysyl side chain. The presence of an uncharged alanyl residue at position 50, precludes competition for the ε-amino of lysine 79, resulting in a tighter crevice and a higher pK_a value. This probably accounts for the fact that trinitrophenylation of Residue 13 in human cytochrome c results in a 0.4 pH unit decrease in pK_a as opposed to a nearly 1.0 pH unit drop in the horse protein (Table 1). Both the top and bottom heme crevice bonds are probably strengthened by their locations on the protein surface in patches of invariant hydrophobic residues.[50,51] Since the dielectric constants of these micro-environments are considerably lower than that of the solvent, the interactions between the charged residues are stronger. The top patch is made up by isoleucines 9 and 85, phenylalanine 82, leucine 94, and the top of the heme and the bottom patch contains phenylalanine or tyrosine 46, tyrosine 48, the carbon atoms of serine or threonine 47 and 49, as well as the bottom of the heme. These bonds are additionally strengthened by the direction and magnitude of the electric field generated by the changed amino acid residues of cytochrome c. This distribution of charges leads to a dipole moment on the order of 300 debye with the positive end of the dipole axis crossing the surface of the molecule near the β-carbon of phenylalanine 82 and the negative end, near the ε-amino group of lysine 99.[17,18] The top and bottom bonds are both oriented so that the electrostatic forces exerted by the electric field enhance the ionic interactions between the two amino acid pairs. It is notable that the electric field around the bottom heme crevice bond of human cytochrome c is about 15% stronger than that of the horse protein.[20] Therefore, the presence of an acidic residue at position 50 not only weakens the inter- action between Residues 79 and 47 by competiton for the ε-amino group of lysine 79, but also by diminishing the electrostatic potential around the pair.

RELATIONSHIP BETWEEN THE STABILITY OF THE HEME CREVICE AND CYTOCHROME c FUNCTION

The pK_a value for the 695 nm absorption band of ferricytochrome c also appears to be correlated with the activity of the protein in its reaction with cytochrome c oxidase.[21] For example, the cytochromes c of apes and Old World monkeys react normally with the cytochrome c oxidase of rhesus monkey. With beef oxidase, they exhibit less than 15% the V_{max} given by horse cytochrome c while binding as well or better than the horse protein.[6-11] The cytochromes c from prosimians, such as the slow loris and the tree shrew, behave similarly to the non-primate cytochromes c, while the proteins from New World monkeys, such as the spider monkey and the capuchin monkey, are intermediate in activity.[6-11] Typical examples are given in Figure 3. The corresponding pK_a values (Fig. 1 and Table 1) are high for the ape and Old World monkey

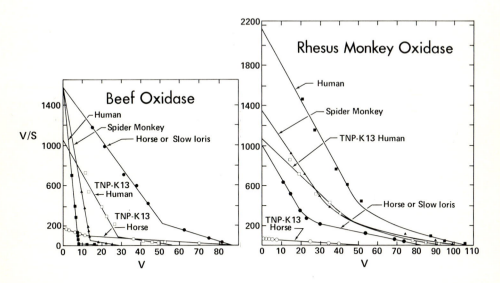

Fig. 3. Eadie-Hofstee representation of the kinetics of reaction of the indicated cytochromes c with beef and rhesus monkey cytochrome oxidases. The polarographic assay system employed TMPD/ascorbate as reductant and Keilin- Hartree particle preparations as sources of the enzymes.[4,13] The velocity of the reaction, V, is in nmoles O_2/min and the substrate, S, is the cytochrome c concentration in μmoles/1.

proteins (9.5) as compared to horse cytochrome c (9.05), low for the New World monkey proteins (8.7), and similar for the prosimian proteins (9.1). Thus, it seems that for any particular cytochrome oxidase there is an optimal value for the pK_α of the reacting cytochrome c. This value is around 9 to 9.1 for beef oxidase and clearly higher for Old World monkey oxidase. If these pK_α values are indeed related to the stability of the heme crevice, either too tight or too loose a heme crevice would be unfavorable for reaction with oxidase. Whether this phenomenon results from the cumulative effect of separate thermodynamic parameters is yet to be determined.

A further confirmation of this concept has come from the examination of the relative activities of TNP-lysine 13 human and TNP-lysine 13 horse cytochromes c (Fig. 3). As expected, upon modification of lysyl residue 13 which carries a positively charged side chain near the center of the oxidase binding domain on cytochrome c, the horse protein becomes much less active with beef as well as with rhesus monkey oxidase and the pK_α of the 695 nm absorption band shifts downward, away from the optimal value for either enzyme. The TNP-lysine 13 derivative of human cytochrome c also has a lowered pK_α (9.1), but it is now similar to that of the native horse protein (9.05). Although its activity with rhesus monkey oxidase is lowered, that with the beef enzyme is four times greater than observed with the unmodified human protein and is now over 50% that of native horse cytochrome c. Thus, the pK_α of the 695 nm absorption band reflects structural parameters that play a crucial role in the reaction with cytochrome c oxidase, as do the basic and hydrophobic side chains in and around the binding domain for the enzyme on cytochrome c.

The control of the cytochrome c-cytochrome oxidase reaction effected via changes in the stability of the heme crevice of cytochrome c necessitates a further elaboration of the kinetic scheme previously proposed for the reaction [8,13] (Fig. 4). Since the higher primate cytochromes c appear to bind to beef oxidase as well as do the non-primate proteins,[6] it is unlikely that the drastic decrease in V_{max} observed with the polarographic assay[6-11] is due to either the rate of association or dissociation of the substrate (k_1, k_{-1}) or product (k_{-4}, k_4) with the enzyme. Furthermore, in low-temperature single turnover experiments human ferrocytochrome c transfers its electron to beef oxidase as fast as does the horse protein,[53] demonstrating that once a productive complex is formed there is no hindrance to the passage of the electron. This is the reaction whose rate is governed by k_3 in the scheme below. It is therefore necessary to assume that the decreased reaction rate of the higher primate cytochromes c with beef oxidase is due to a decreased rate of attain-

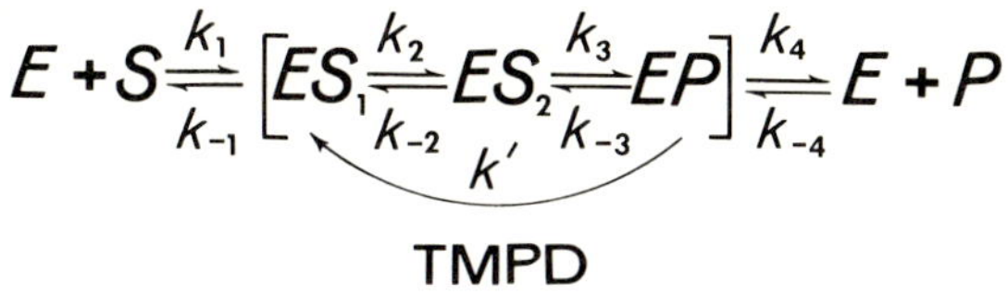

$$E + S \underset{k_{-1}}{\overset{k_1}{\rightleftharpoons}} \left[ES_1 \underset{k_{-2}}{\overset{k_2}{\rightleftharpoons}} ES_2 \underset{k_{-3}}{\overset{k_3}{\rightleftharpoons}} EP \right] \underset{k_{-4}}{\overset{k_4}{\rightleftharpoons}} E + P$$
$$\underbrace{}_{k'}$$

TMPD

Fig. 4. Simplified scheme for the reaction of cytochrome c with cytochrome c oxidase (E). The substrate (S) is ferrocytochrome c, the product (P) is ferri-cytochrome c, ES_1 represents the initial complex of enzyme and substrate, ES_2 the complex in which electron transfer occurs, and k' is the second order rate constant for the reduction of enzyme bound cytochrome c by TMPD.

ment of the productive complex (ES_2), even though the initial binding of the substrate to the enzyme is normal. This leads to the kinetic scheme proposed in Figure 4. In this scheme, ES_1 represents the initial high affinity complex between ferrocytochrome c and the oxidase and ES_2 the productive complex in which electron transfer can occur. It would appear likely that one of the differences between the ES_1 and ES_2 complexes is a change in the conformation of the ferrocytochrome c and that the rate of that change (k_2) is correlated to the pK_a of the 695 nm absorption band of ferricytochrome c. It should also be noted that the reduction of the EP complex by N,N,N',N'-tetramethyl-phenylenediamine (TMPD) leads to ES_1 and not ES_2, as in the latter case horse and human cytochromes c should react the same in the polarographic assay as they do in the low-temperature single turnover experiments.

EVOLUTIONARY CONNOTATIONS

The identification of the reaction domain for cytochrome oxidase on cytochrome c demonstrated that evolutionary changes in amino acid sequence within and around that domain, or changes that cause significant perturbations in the electric field of cytochrome c, would influence its activity.[8,13-20] Thus, a considerable proportion, but not all amino acid sequence changes that have occurred in the course of the evolution of cytochrome c were shown to have functional concomitants, making it unjustified to consider these changes to be neutral evolutionary events.[3,6,8] The present observations indicate that amino acid changes in areas remote from the reaction domain can also affect function by controlling the stability of the heme crevice.[21] Therefore, an even larger proportion of the evolutionary changes in cytochrome c are now included in the category of those that are functionally significant. Whether every one of these changes will eventually be found to be related to physiological activity is yet to be determined. The changes in activity resulting from these two sets of amino acid sequence variations can be considered to represent two separate mechanisms for the evolutionary control of cytochrome c function. Clearly, other remote mechanisms of control are conceivable and their possible existance should be investigated.

The major features of primate evolution[54-56] include a divergence from insectivore stock in the Cretaceous, about 100 million years ago, so that most prosimian groups were recognizably established by the middle of the Eocene, about 50 to 55 million years ago. The Old and New World monkeys, which derived from tarsier-like ancestors, separated from each other about 40 to 50 million years ago and are thought to represent the consequence of parallel evolution despite their obviously similar physical characteristics.[57-58] The Old World monkeys (Cercopithecoidea) and the apes (Hominoidea) are thought to have separated from each other about 30 to 40 million years ago, with the lines leading to the apes and man branching about 15 million years ago.[59] The present study supports most of these traditional interpretations of primate evolution. Thus, the prosimian, slow loris, has a cytochrome c which is un-questionably of primate character, yet it is functionally very similar to the horse protein, the rhesus monkey and human proteins are nearly identical in amino acid sequence and function, and the New World monkey cytochromes c are intermediate between these two groups. The correlations between the activities of these cytochromes c with cytochrome oxidase, the pK_a values for their 695 nm absorption bands and the character of the residue at position 50 demonstrate how particular amino acid replacements can exert major modulating influences on functional parameters. The unexpectedly large number of residue differences between the Old and New World monkeys indicates that in one or both of these groups a far more rapid rate of evolution than considered typical for cyto-chrome c ensued upon their evolutionary divergence. Thus, evolutionary vari-ations in primate cytochromes c represent a clear case in which amino acid sequence changes affect physiological activity. Together with a variable rate of fixation of mutations in the structural gene for the protein, this supports the Darwinian concept of evolution at the molecular level for cytochrome c.

ACKNOWLEDGEMENTS

Supported by Grants GM-19121 and HL-11119 from the National Institutes of Health and Grant DEB-76-81694 from the National Science Foundation. For part of the work N. O. was a trainee under Grant T32-GM-07291 from the National Institutes of Health.

REFERENCES

1. Smith, L., Nava, M.E., and Margoliash, E. (1973) in Oxidases and Related Redox Systems (King, T.E., Mason, H.S. and Morrison, M., eds.), Vol. 2, pp. 629-638, University Park Press, Baltimore.

2. Byers, V., Lambeth, D., Lardy, H.A. and Margoliash, E. (1971) Federation Proc., 30, 1286.

3. Margoliash, E., Ferguson-Miller, S., Kang, C.H., and Brautigan, D.L. (1976) Federation Proc., 35, 2124-2130.

4. Ferguson-Miller, S., Brautigan, D.L., and Margoliash, E. (1976) J. Biol. Chem., 251, 1104-1115.

5. Dethmers, J. and Ferguson-Miller, S. (1977) Federation Proc., 36, 727.

6. Margoliash, E., Ferguson-Miller, S., Brautigan, D.L., and Chaviano, A.H. (1976) in Structure-Function Relationships in Proteins (Markham, R. and Horne, R.W., eds.) pp. 145-165, North-Holland Publ. Co., Amsterdam.

7. Ferguson-Miller, S., Brautigan, D.L., Chaviano, A.H., and Margoliash, E. (1976) Federation Proc., 35, 1605.

8. Margoliash, E., Ferguson-Miller, S., Brautigan, D.L., Kang, C.H., and Dethmers, J.K. (1977) in Structure and Function of Energy-Transducing Membranes (van Dam, K. and van Gelder, B.F., eds.) pp. 69-80, Elsevier/ North-Holland Biomedical Press, Amsterdam.

9. Borden, D., Ferguson-Miller, S., Tarr, G. E., and Rodriguez, D. (1978) Federation Proc., 37, 1517.

10. Ferguson-Miller, S., Brautigan, D.L., Chaviano, A.H. and Margoliash, E. (1978) unpublished results.

11. Borden, D., Ferguson-Miller, S., Tarr, G.E., Rodriguez, D. and Margoliash, E. (1978) unpublished results.

12. Errede, B. and Kamen, M.D. (1978) Biochemistry, 17, 1015-1027.

13. Ferguson-Miller, S., Brautigan, D.L., and Margoliash, E. (1978) J. Biol. Chem., 253, 149-159.

14. Ferguson-Miller, S., Brautigan, D.L. and Margoliash, E. (1978) in the Porphyrins (Dolphin, D., ed.) Academic Press, in press.

15. Staudenmayer, N., Smith, M.B., Smith, H.T., Spies, F.K. and Millet, F. (1976) Biochemistry, 15, 3198-3205.

16. Smith, H.T., Staudenmayer, N. and Millet, F. (1977) Biochemistry, 16, 4971-4974.

17. Koppenol, W.H., Vroonland, C.A.J., Braams, R., Ferguson-Miller, S. and Brautigan, D.L. (1977), 174th National Meeting of the American Chemical Society, Chicago, IL., Biol. 85.

18. Koppenol, W.H., Vroonland, C.A.J., and Braams, R. (1978) Biochim. Biophys. Acta, in press.

19. Rieder, R. and Bosshard, H.R. (1978) J. Biol. Chem., in press.

20. Koppenol, W.H. (1978) unpublished results.

21. Osheroff, N., Borden, D., Koppenol, W.H. and Margoliash, E. (1978) unpublished results.

22. Dickerson, R.E., Kopka, M.L., Weinzierl, J., Varnum, J., Eisenberg, D. and Margoliash, E. (1967) J. Biol. Chem., 242, 3015-3017.

396

23. Bigwood, E.J., Thomas, J. and Wofters, D. (1934) Compt. Rend. Soc. Biol., 117, 220.

24. Theorell, H. and Akesson, A. (1939) Science, 90, 67.

25. Shechter, E. and Saludjian, P. (1967) Biopolymers, 5, 788–790.

26. Sreenathan, B.R. and Taylor, C.P.S. (1971) Biochem. Biophys. Res. Commun. 42, 1122–1126.

27. Gupta, R.K. and Koenig, S.H. (1971) Biochem. Biophys. Res. Commun., 45, 1134–1143.

28. Theorell, H. and Akesson, A. (1941) J. Am. Chem. Soc., 63, 1812–1818.

29. Horecker, B.L. and Kornberg, A. (1946) J. Biol. Chem., 165, 11–20.

30. Horecker, B.L., and Stannard, J.N. (1948) J. Biol. Chem., 172, 589–597.

31. Schejter, A., Glauser, S.C., George, P. and Margoliash, E. (1963) Biochim. Biophys. Acta, 73, 641–643.

32. Schejter, A. and George, P. (1964) Biochemistry 3, 1045–1049.

33. Stellwagen, E. (1968) Biochemistry, 7, 2893–2898.

34. Greenwood, C. and Wilson, M.T. (1971) Eur. J. Biochem, 22, 5–10.

35. Wilson, M.T. and Greenwood, C. (1971) Eur. J. Biochem., 22, 11–18.

36. Kaminsky, L.S., Miller, V.J. and Davison, A.J. (1973) Biochemistry, 12, 2215–2221.

37. Brautigan, D.L., Feinberg, B.A., Hoffman, B.M., Margoliash, E., Peisach, J., and Blumberg, W.E. (1977) J. Biol. Chem., 252, 574–582.

38. Dickerson, R.E., Takano, T., Eisenberg, D., Kallai, O.B., Samson, L., Cooper, A. and Margoliash, E. (1971) J. Biol. Chem., 246, 1511–1535.

39. Dickerson, R.E., Takano, T., Kallai, O.B. and Swanson, L. (1972) in Structure and Function of Oxidation-Reduction Enzymes (Akesson, A. and Ehrenberg, A., eds.) pp. 69–83, Pergamon Press, Oxford.

40. Lambeth, D.O., Campbell, K.L., Zand, R. and Palmer, G. (1973) J. Biol. Chem., 248, 8130–8136.

41. Davis, L.A., Schejter, A. and Hess, G.P. (1974) J. Biol. Chem., 249, 2624–2632.

42. Wilgus, H. and Stellwagen, E. (1974) Proc. Nat. Acad. Sci. U.S.A., 71, 2892–2894.

43. Osheroff, N., Brautigan, D.L. and Margoliash, E. (1978) unpublished results.

44. Brautigan, D.L., Ferguson-Miller, S. and Margoliash, E. (1978) J. Biol. Chem., 253, 130–139.

45. Brautigan, D.L., Ferguson-Miller, S., Tarr, G.E. and Margoliash, E. (1978) J. Biol. Chem., 253, 140–148.

46. Matsubara, H. and Smith, E.L. (1963) J. Biol. Chem., 238, 2732–2753.

47. Borden, D., Tarr, G.E., Vensel, W., Kottke, K. K. and Margoliash, E. (1978) unpublished results.

48. Margoliash, E. (1962) J. Biol. Chem., 237, 2161–2173.

49. Borden, D., Tarr, G.E., Vensel, W. and Margoliash, E. (1978) unpublished results.

50. Swanson, R., Trus, B.L., Mandel, N., Mandel, G., Kallai, O.B. and Dickerson, R.E. (1977) J. Biol. Chem., 252, 759-775.

51. Takano, T., Trus, B.L., Mandel, N., Mandel, G., Kallai, O.B., Swanson, R. and Dickerson, R.E. (1977) J. Biol. Chem., 252, 776-785.

52. Borden, D. and Margoliash, E. (1976) in Handbook of Biochemistry and Molecular Biology (Fasman, G., ed.) 3rd Ed. Vol. 3, pp. 268-279, Chemical Rubber Co. Press, Cleveland.

53. Ferguson-Miller, S., Brautigan, D.L., Chance, B., Waring, A. and Margoliash, E. (1978) Biochemistry, 17, 2246-2249.

54. Romer, A.S. (1966) Vertebrate Paleontology, University of Chicago Press, Chicago.

55. Chiarelli, A.B. (1973) Evolution of the Primates, Academic Press, New York.

56. Goodman, M. (1973) Symp. Zool. Soc. Lond., 33, 339-375.

57. Simpson, G.G. (1961) Principles of Animal Taxonomy, Columbia University Press, New York.

58. Goodman, M. (1962) Ann. N.Y. Acad. Sci., 102, 219-234.

59. Goodman, M. (1974) Ann. Rev. Anthropl., 3, 203-228.

Cytochrome Oxidase, T.E. King et al. eds.
© *1979 Elsevier/North-Holland Biomedical Press*

SOME EVOLUTIONARY ASPECTS OF CYTOCHROME OXIDASE

TATEO YAMANAKA, YOSHIHIRO FUKUMORI and KEIKO FUJII
Department of Biology, Faculty of Science, Osaka University, Toyonaka, Osaka
560 (Japan)

INTRODUCTION

In the previous studies[1-3], we have revealed that there is a distinct biological specificity in the reactions of cytochrome c with certain redox enzymes
and the reactivities with these enzymes of cytochrome c reflect the evolutionary relationship of the host organism of the cytochrome. The occurrence of
cytochromes c with different reactivities with the redox enzymes will suggest
that cytochrome oxidases with different enzymatic properties may also exist, as
cytochrome c functions as the direct electron donor for the oxidase.

From such point of veiw, we have tried to find cytochrome oxidase which resembles mitochondrial cytochrome oxidase in the spectral properties but differs
from the latter enzyme in the specificity for cytochrome c. Cytochrome oxidase
of *Thiobacillus novellus*[4] has been found to satisfy the above requirement to
some extent. Further, it has been found that cytochrome oxidase of *Nitrosomonas
europaea*[5] reacts with the "bacterial-type" cytochromes c although it has haem a
as the prosthetic group. On the basis of such specificity of the oxidase for
cytochrome c, some evolutionary aspects of the enzyme will be discussed in the
present article.

1. Reactivities of Cytochromes c with Certain Redox Enzymes

We have studied extensively the reactivities of various cytochromes c[2,3]
mainly with *Pseudomonas aeruginosa* nitrite reductase (= *Pseudomonas* cytochrome
oxidase)[6], cow cytochrome oxidase[7,8] and yeast cytochrome c peroxidase[9]. As
shown in Table 1, in general, *P. aeruginosa* nitrite reductase reacts rapidly
with cytochromes c of many prokaryotes but it reacts very poorly with eukaryotic
cytochromes c, while cow cytochrome oxidase and yeast cytochrome c peroxidase
react rapidly with eukaryotic cytochromes c but react poorly or do not react
with many prokaryotic cytochromes c. In our studies, the reactivities are calculated from the initial rates of the reactions which are generally performed
in 40 mM phosphate buffer at pH 6.5, and the comparison of the reactivities between organisms are not performed so strictly as done kinetically by a few
workers[11-13]. Therefore, *e. g.* the figures shown in Table 1 for the reactivity

TABLE 1

REACTIVITIES WITH CERTAIN REDOX ENZYMES OF VARIOUS CYTOCHROMES c WHICH PARTICIPATE IN RESPIRATION

Cytochrome c		Relative reactivity[a]		
Organism	α Peak (nm)	Ps[b]	Cow[b]	Yeast[b]
Prokaryotes				
Pseudomonas aeruginosa	551	_100_	0	0
Pseudomonas aeruginosa P6009	551	83	0	0
Paracoccus halodenitrificans	554	131	0	0
Thermus thermophilus[c]	553	140	3.7	7.0
Nitrosomonas europaea	552	56	0	0
Azotobacter vinelandii	551	25	0	0
Paracoccus denitrificans	550	2.8	3.7	35
Thiobacillus novellus	550	6.0	23	106
Eukaryotes				
Tetrahymena pyriformis	553	19	0	3.3
Wheat	550	1.5	31	33
Saccharomyces oviformis	550	4.9	_100_	_100_
Candida krusei	549	5.0	88	
Tuna	550	8.7	109	120
Bonito	550	5.4	96	
Housefly (larva)	550	2.3	69	
Horse	550	2.5	124	88
Cow	550	0.53	73	
Man	550	0.44	107	

[a]The reactivities of cytochromes c were expressed as relative values; the molecular activity (moles of cytochrome c oxidized per mole of enzyme) per min of each cytochrome c calculated from the initial reaction rate was compared with that of cytochrome c used as the standard (the value underlined) for each enzyme.
[b]Ps, _Pseudomonas aeruginosa_ nitrite reductase (used as a cytochrome oxidase); Cow, cow cytochrome oxidase; Yeast, yeast cytochrome c peroxidase.
[c]From ref. 10.

with the cow oxidase of various eukaryotic cytochromes c are regarded as quite similar to one another excepting those for _T. pyriformis_ and wheat cytochromes c. In any case, we have been led to the conclusion that most eukaryotic cytochromes *c* show similar reactivities with the cow oxidase and the yeast per-oxidase.

Smith *et al.*[11] have claimed that eukaryotic cytochromes *c* are activated by a prolonged dialysis against deionized water and that they show the very same reactivity with a particulate preparation of the cow oxidase after the dialysis.

TABLE 2

EFFECT OF THE DIALYSIS ON THE REACTIVITIES OF CYTOCHROMES *c* WITH CERTAIN REDOX ENZYMES

Cytochrome *c*		Reactivity (molecular activity/min)[a]		
Organism	Intact (I) or dialysed (D)[b]	Cow[c]	Yeast[c]	Nitroso[c]
Horse	I	195 (100)[d]	2,850 (100)	2,175 (100)
	D	179 (92)	3,100 (109)	1,914 (88)
			2,700 (95)	2,260 (104)
				2,220 (102)
				1,980 (91)
				2,045 (94)
				2,350 (108)
Cow	I	115 (100)		
	D	139 (121)		
Tuna	I	171 (100)	3,880 (100)	2,785 (100)
	D	137 (80)	3,650 (94)	2,394 (86)
		154 (90)		2,500 (90)
		200 (117)		
Yeast (*Saccharo-myces oviformis*)	I		3,240 (100)	
	D		3,300 (102)	
			3,320 (103)	

[a] Reactions were performed in 40 mM phosphate buffer at pH 6.5 for cow cytochrome oxidase and yeast cytochrome *c* peroxidase, and in 0.1 M glycine-NaOH buffer at pH 9.5 for *N. europaea* hydroxylamine oxidoreductase. Concentrations of cytochromes were about 20 μM.
[b] Dialysis was performed against deionized water for a week.
[c] Cow, cow cytochrome oxidase; Yeast, yeast cytochrome *c* peroxidase; Nitroso, *Nitrosomonas europaea* hydroxylamine oxidoreductase.
[d] Figures in parentheses show relative values.

However, in our experiments, the prolonged dialysis against deionized water did not elevate significantly the reactivity of eukaryotic cytochromes *c* either with the cow oxidase or with the yeast peroxidase. Further, the differences in their reactivities with the enzymes between organisms appear to be still observable under our experimental conditions (Table 2). We do not know whether or not the differences in the reactivities with the enzymes of eukaryotic cytochromes *c* are significant. However, we agree that most eukaryotic cytochromes *c* are not in-activated even after the dialysis for a week against deionized water. The dis-crepancy between the results by Smith *et al.* and ours will be attributable to the differences in the preparations of the cytochromes and/or the oxidase.

Among eukaryotic cytochromes *c* so far tested, cytochrome *c* of *Tetrahymena pyriformis*[14] differs greatly from the other eukaryotic cytochromes in that it does not react with the cow oxidase but reacts comparatively rapidly with *P. aeruginosa* nitrite reductase. From prokaryotes, some cytochromes *c* are obtained which show different enzymatic properties; e.g., *Thiobacillus novellus* cyto-chrome *c*[15] reacts slowly with the *P. aeruginosa* reductase while it reacts com-paratively rapidly with the cow oxidase and reacts with the yeast peroxidase as rapidly as eukaryotic cytochromes *c*. *Paracoccus denitrificans* cytochrome *c*[16] reacts poorly both with the *P. aeruginosa* reductase and with the cow oxidase, while it reacts comparatively rapidly with the yeast peroxidase[3].

2. *Thiobacillus novellus* Cytochrome *c*

As mentioned above, *T. novellus* cytochrome *c* resembles eukaryotic cytochromes *c* in its reactivity with yeast cytochrome *c* peroxidase[3,15]. Further, it reacts with *T. novellus* sulphite-cytochrome *c* reductase[18] and *N. europaea* hydroxylamine oxidoreductase[19] as rapidly as eukaryotic cytochromes *c* (ref. 15). Although its amino acid sequence has not been determined, its molecular features are not much different from those of the eukaryotic proteins[17]. Especially, it has 13 lysine residues in the molecule although its isoelectric point is at pH 7.5. As cyto-chrome *c* functions as the direct electron donor for cytochrome oxidase, it is expected that the *T. novellus* oxidase may be a little different from the mito-chondrial oxidase at least in the specificity for cytochrome *c*.

3. Purification of *Thiobacillus novellus* Cytochrome Oxidase

The cells of the organism grown autotrophically were treated with a French pressure cell, and the resulting suspension was centrifuged. The particulate fractions spun down between 10,000 x g and 100,000 x g were suspended in 0.1 M sodium phosphate buffer, pH 7.5 containing 0.2 M KCl and 0.5% Triton X-100. The suspension was allowed to stand overnight and then centrifuged at 100,000 x g. To the supernatant thus obtained was added $(NH_4)_2SO_4$ and the precipitate formed between 25 and 50% saturation collected. The precipitate was dissolved in a minimum volume of 10 mM Tris-HCl buffer, pH 8.0 containing 0.5% Triton X-100, and the resulting solution dialysed overnight against the same buffer as used for the dissolution. The dialysed solution thus obtained was charged on the DEAE-cellulose column which had been equilibrated with the same buffer as used for the dialysis and the enzyme was adsorbed on the column. The column on which the enzyme was adsorbed was washed with 10 mM Tris-HCl buffer, pH 8.0 containing Triton X-100 and 100 mM NaCl, and the enzyme moved down as a green band. By this procedure, the enzyme was separated from much amount of reddish brown substance (probably cytochrome c_1). Finally, the green solution obtained above was subjected to the chromatography with a Sephadex G-100 column. The green fraction thus obtained was used as the preparation of *T. novellus* cytochrome oxidase.

The absorption spectrum of the oxidase preparation is shown in Fig. 1; the oxidized form shows the peaks around 600 and at 428 nm and the reduced form the

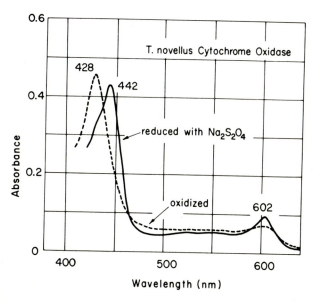

Fig. 1. Absorption spectrum of *T. novellus* cytochrome oxidase. The enzyme was dissolved in 10 mM Tris-HCl buffer, pH 8.0 containing 0.5% Triton X-100. ----, Oxidized; ——, reduced with $Na_2S_2O_4$.

404

peaks at 602 and 442 nm. The pyridine ferrohaemochrome of the enzyme has the peaks at 588 and 430 nm. When the solution of the reduced enzyme is bubbled with CO, the peaks appear at 599 and 431 nm. These spectral characteristics of the enzyme is very similar to those of the mitochondrial oxidase.

4. Reactions of Cytochromes c with *Thiobacillus novellus* Cytochrome Oxidase

The oxidation of reduced cytochrome c catalysed by the oxidase is affected greatly by the concentration of phosphate buffer; with most cytochromes c, the reaction rates decrease rapidly as the phosphate concentration increases, while with cytochromes c of *Saccharomyces oviformis* and *Candida krusei*, the rates increase first with phosphate concentration and then decrease as the salt concentration increases. In these respects, the oxidase differs from the mitochondrial enzyme which requires an optimum concentration of phosphate (or other salt) for the maximum activity when its activity is spectrophotometrically measured by the oxidation of reduced cytochrome c at the concentrations of about 20 μM (ref. 20).

When the reactions are performed in 50 mM phosphate buffer, pH 5.5, the *T. novellus* oxidase reacts rapidly with cytochromes c of yeasts (*S. oviformis* and *C. krusei*), tuna and bonito besides *T. novellus* cytochrome c. Especially, the enzyme is very reactive with *C. krusei* cytochrome c. On the contrary, it reacts very slowly with horse and cow cytochromes c. Wheat and housefly cytochromes c react with the enzyme comparatively rapidly as compared with horse and cow cytochromes c. It seems very interesting that man cytochrome c also reacts comparatively rapidly with the enzyme.

The oxidase reacts very poorly or does not react with cytochromes c derived from bacteria such as *Pseudomonas aeruginosa*, *Nitrosomonas europaea*, *Chromatium vinosum* and *Rhodospirillum rubrum*, while it reacts comparatively rapidly with *Chlorobium limicola* f. *thiosulfatophilum* cytochrome c-555. Cytochrome c-554 of *Spirulina platensis* reacts poorly with the enzyme (Table 3).

5. Comparison of the Specificity for Cytochrome c between Cytochrome Oxidases

In Fig. 2, the specificities for cytochrome c of four cytochrome oxidases are compared. In the figure, the specificity of *N. europaea* cytochrome oxidase is also included. Although the oxidase of the organism has been already highly purified by Erickson et al.[5], in the present study, the partially purified preparation was used which was obtained by the method similar to that used for the purification of the *T. novellus* enzyme. So far as tested with a few kinds of cytochromes c, the specificity of the oxidase for cytochrome c does not differ

405

TABLE 3

REACTIVITY OF VARIOUS CYTOCHROMES *C* WITH *THIOBACILLUS NOVELLUS* CYTOCHROME OXIDASE

Cytochrome *c*		Relative reactivity[a]
Organism	α Peak (nm)	
Pseudomonas aeruginosa	551	0.78
Nitrosomonas europaea	552	2.5
Chlorobium limicola **f.** *thiosulfatophilum*	555	18
Chromatium vinosum	553(550)	0
Rhodospirillum rubrum	550	1.5
Thiobacillus novellus	550	100
Spirulina platensis	554	3.8
Saccharomyces oviformis	550	77
Candida krusei	549	290
Wheat	550	26
Tuna	550	93
Bonito	550	91
Housefly (larva)	550	12
Cow	550	4.7
Horse	550	4.7
Man	550	14

[a]Reactivity of cytochromes *c* was expressed as relative value; the molecular activity per min was taken as 100% which was observed when *T. novellus* cytochrome *c* reacted with the enzyme. The reactions were performed in 50 mM phosphate buffer at pH 5.5.

between the purified and crude preparations[21].

As seen from the figure, the host organisms of cytochrome *c* which reacts rapidly with each cytochrome oxidase shift from the primitive to the higher as the host organisms of the oxidase vary from *P. aeruginosa* to cow *via N. europaea* and *T. novellus*. The *N. europaea* enzyme reacts with the "bacterial-type" cytochromes *c* such as the cytochromes of *N. europaea* and *P. aeruginosa*. Although we should not be surprised to learn that cytochrome oxidase of an organism reacts with its native cytochrome *c*, it seems to deserve attention that the oxidase occurs which has haem *a* as the prosthetic group and reacts even with the "bacterial-type" cytochromes *c*. These facts mentioned above will verify our idea that cytochrome *c* and cytochrome oxidase might have coevolved[1,2].

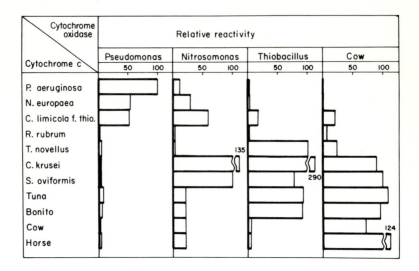

Fig. 2.   The specificities for cytochrome *c* of cytochrome oxidases derived from *P. aeruginosa, N. europaea, T. novellus* and cow.

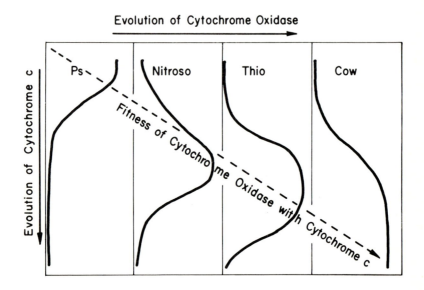

Fig. 3.   A diagrammatical presentation of coevolution of cytochrome oxidase with cytochrome *c*.

6. Prospects

    As mentioned above, the specificity for cytochrome c of cytochrome oxidase seems to vary continuously from the evolutionary point of view; the oxidase derived from a prokaryote reacts rapidly with eukaryotic cytochromes c. However, a great gap may exist in the molecular features between the prokaryotic and eukaryotic oxidases. The molecule of the mitochondrial cytochrome oxidase is composed of 5-7 subunits[22-26]. In the *S. cerevisiae* enzyme, 3 of the 7 subunits are biosynthesized on the mitochondrial ribosomes while the remaining 4 subunits are on the cytoplasmic ribosomes[27]. The endosymbiotic hypothesis in evolution assumes that mitochondria might have been derived from aerobic bacteria[28]. Therefore, it seems very interesting to examine if the bacterial cytochrome oxidase which has haem a as the prothetic group is composed of the subunits which are similar to the mitochondrial subunits of the eukaryotic oxidase. At present, the determination of the subunit structure of the *T. novellus* oxidase is in progress in our laboratory. By our preliminary study, the molecule of the bacterial oxidase seems to be composed of only two kinds of subunits with molecular weights of about 25,000 and 50,000 daltons. The study of the prokaryotic oxidase will give us also clues to elucidate the relationship between the structure and function of the cytochrome a-type oxidase.

SUMMARY

    Several kinds of cytochrome oxidases are compared on the basis of their specificity for cytochrome c. The host organisms of cytochrome c which reacts rapidly with each cytochrome oxidase shift from the primitive to the higher as the host organisms of the oxidase vary from *Pseudomonas aeruginosa* to cow *via Nitrosomonas europaea* and *Thiobacillus novellus*. This fact seems to verify our idea that cytochrome oxidase and cytochrome c have coevolved during evolution of the organisms.

ACKNOWLEDGEMENTS

    We wish to thank Professor H. Matsubara for his interest and encouragement. They are also grateful to Drs. N. Tanaka and M. Kakudo (Institute for Protein Research, Osaka University, Japan) for their generosity in supplying bonito cytochrome c. This work has been supported in part by grant-in-aid from the Kudo Scientific Foundation (T.Y.).

REFERENCES
 1. Yamanaka, T. and Okunuki, K. (1964) J. Biol. Chem., 239, 1813-1817.

408

2. Yamanaka, T. (1973) Space Life Sci., 4, 490-504.

3. Yamanaka, T. (1975) J. Biochem., 77, 493-499.

4. Yamanaka, T. and Fukumori, Y. (1977) FEBS Lett., 77, 155-158.

5. Erickson, R. H., Hooper, A. B. and Terry, K. (1972) Biochim. Biophys. Acta 283, 155-166

6. Yamanaka, T. (1964) Nature, 204, 253-255.

7. Okunuki, K. (1966) in Comprehensive Biochemistry, Florkin, M. and Stotz, E. H. eds., Elsevier, Amsterdam, vol. 14, pp. 232-308.

8. Orii, Y.and Okunuki, K. (1965) J. Biochem., 58, 561-568.

9. Yonetani, T. and Ray, G. S. (1965) J. Biol. Chem., 240, 4503-4508.

10. Hon-nami, K. and Oshima, T. (1977) J. Biochem., 82, 769-776.

11. Smith, L., Nava, M. E. and Margoliash, E. (1973) in Oxidases and Related Redox Systems, King, T. E., Mason, H. S. and Morrison, M. eds., University of Park Press, Baltimore, vol. 2, pp. 629-638.

12. Ferguson-Miller, S., Brautigan, D. L. and Margoliash, E. (1976) J. Biol. Chem., 251, 1104-1115.

13. Errede, B. and Kamen, M. D. (1978) Biochemistry, 17, 1015-1027.

14. Yamanaka, T., Nagata, Y. and Okunuki, K. (1968) J. Biochem.,63, 753-760.

15. Yamanaka, T. and Kimura, K. (1974) FEBS Lett., 48, 253-255.

16. Scholes, P. B., Mclain, G. and Smith, L. (1971) Biochemistry, 10, 2072-2076.

17. Yamanaka, T., Takenami, S., Akiyama, N. and Okunuki, K. (1971) J. Biochem., 70, 349-358.

18. Charles, A. M. and Suzuki, I. (1966) Biochim. Biophys. Acta, 128, 522-534.

19. Hooper, A. B. and Nason, A. (1965) J. Biol. Chem., 240, 4044-4057.

20. Okunuki, K., Sekuzu, I., Orii, Y., Tsuzuki, T. and Matsumura, Y. (1968) in Structure and Function of Cytochromes, Okunuki, K., Kamen, M. D. and Sekuzu, I. eds., University of Tokyo Press, pp. 35-50.

21. Yamanaka, T., Shinra, M. and Kimura, K. (1977) BioSystems, 9, 155-164.

22. Mason, T. L., Poyton, R. O., Wharton, D. C. and Schatz, G. (1973) J. Biol. Chem., 248, 1346-1354.

23. Weiss, H., Sebald, W. and Bücher, T. (1971) Eur. J. Biochem., 22, 19-26.

24. Yamanoto, T. and Orii, Y. (1974) J. Biochem., 75, 1081-1089.

25. Briggs, M., Kamp, P.-F., Robinson, N. C. and Capaldi, R. A. (1975) Biochemistry, 15, 5123-5128.

26. Maeshima, M. and Asahi, T. (1978) Arch. Biochem. Biophys., 187, 423-430.

27. Poyton, R. O. and Schatz, G. (1975) J. Biol. Chem., 250, 752-761.

28. Margulis, L. (1970) Origin of Eukaryotic Cells, Yale University Press, New Haven.

Author Index

410

Subject Index*

(Cross checking must be made, *e.g.* "High affinity sites of cytochrome *c* to
cytochrome oxidase" is listed as such rather than under "Cytochrome *c*" or
"Cytochrome oxidase.")

---

*Many subject indexes elsewhere are too brief, so it is purposely made more de-
tailed here. We are indebted to Mr. Yau-huei Wei for the painstaking task.

414